# Lecture Notes in Physics

## Bisher erschienen/Already published

Vol. 1: J. C. Erdmann. Wärmeleitung in Kristallen, theoretische Grundlagen und fortgeschrittene experimentelle Methoden. II, 283 Seiten. 1969.

Vol. 2: K. Hepp, Théorie de la renormalisation. III, 215 pages. 1969.

Vol. 3: A. Martin, Scattering Theory: Unitarity, Analyticity and Crossing. IV, 125 pages. 1969.

Vol. 4: G. Ludwig, Deutung des Begriffs „physikalische Theorie" und axiomatische Grundlegung der Hilbertraumstruktur der Quantenmechanik durch Hauptsätze des Messens. 1970. Vergriffen.

Vol. 5: Schaaf, The Reduction of the Product of Two Irreducible Unitary Representations of the Proper Orthochronous Quantummechanical Poincare Group. IV, 120 pages. 1970.

Vol. 6: Group Representations in Mathematics and Physics. Edited by V. Bargmann. V, 340 pages. 1970.

Vol. 7: R. Balescu, J. L. Lebowitz, I. Prigogine, P. Résibois, Z. W. Salsburg, Lectures in Statistical Physics. V, 181 pages. 1971.

Vol. 8: Proceedings of the Second International Conference on Numerical Methods in Fluid Dynamics. Edited by M. Holt. 1971. Out of print.

Vol. 9: D. W. Robinson, The Thermodynamic Pressure in Quantum Statistical Mechanics. V, 115 pages. 1971.

Vol. 10: J. M. Stewart, Non-Equilibrium-Relativistic Kinetic Theory. III, 113 pages. 1971.

Vol. 11: O. Steinmann, Pertubation Expansions in Axiomatic Field Theory. III, 126 pages. 1976.

Vol. 12: Statistical Models and Turbulence. Edited by C. Van Atta and M. Rosenblatt. Reprint of the First Edition. VIII, 492 pages. 1975.

Vol. 13: M. Ryan, Hamiltonian Cosmology. VII, 169 pages. 1972.

Vol. 14: Methods of Local and Global Differential Geometry in General Relativity. Edited by D. Farnsworth, J. Fink, J. Porter, and A. Thompson. V, 188 pages.

Vol. 15: M. Fierz, Vorlesungen zur Entwicklungsgeschichte der Mechanik. V, 97 Seiten. 1972.

Vol. 16: H.-O. Georgii, Phasenübergang 1. Art bei Gittergasmodellen. IX, 167 Seiten. 1972.

Vol. 17: Strong Interaction Physics. Edited by W. Rühl and A. Vancura. V, 405 pages. 1973.

Vol. 18: Proceedings of the Third International Conference on Numerical Methods in Fluid Mechanics, Vol. I. Edited by H. Cabannes and R. Temam. VII, 186 pages. 1973.

Vol. 19: Proceedings of the Third International Conference on Numerical Methods in Fluid Mechanics, Vol. II. Edited by H. Cabannes and R. Temam. VII, 275 pages. 1973.

Vol. 20: Statistical Mechanics and Mathematical Problems. Edited by A. Lenard. VIII, 247 pages. 1973.

Vol. 21: Optimization and Stability Problems in Continuum Mechanics. Edited by P. K. C. Wang. V, 94 pages. 1973.

Vol. 22: Proceedings of the Europhysics Study Conference on Intermediate Processes in Nuclear Reactions. Edited by N. Cindro, P. Kulišic and Th. Mayer-Kuckuk. XIV, 329 pages. 1973.

Vol. 23: Nuclear Structure Physics. Proceedings 1973. Edited by U. Smilansky, I. Talmi, and H. A. Weidenmüller. XII, 296 pages. 1973.

Vol. 24: R. F. Snipes, Statistical Mechanical Theory of the Electrolytic Transport of Nonelectrolytes. V, 210 pages. 1973.

Vol. 25: Constructive Quantum Field Theory. The 1973 "Ettore Majorana" International School of Mathematical Physics. Edited by G. Velo and A. Wightman. III, 331 pages. 1973.

Vol. 26: A. Hubert, Theorie der Domänenwände in geordneten Medien. XII, 377 Seiten. 1974.

Vol. 27: R. K. Zeytounian, Notes sur les Ecoulements Rotationnels de Fluides Parfaits. XIII, 407 pages. 1974.

Vol. 28: Lectures in Statistical Physics. Edited by W. C. Schieve and J. S. Turner. V, 342 pages. 1974.

Vol. 29: Foundations of Quantum Mechanics and Ordered Linear Spaces. Advanced Study Institute, Marburg 1973. Edited by A. Hartkämper and H. Neumann. VI, 355 pages. 1974.

Vol. 30: Polarization Nuclear Physics. Proceedings 1973. Edited by D. Fick. IX, 292 pages. 1974.

Vol. 31: Transport Phenomena. Sitges International Schools of Statistical Mechanics, June 1974. Edited by G. Kirczenow and J. Marro. XIV, 517 pages. 1974.

Vol. 32: Particles, Quantum Fields and Statistical Mechanics. Proceedings 1973. Edited by M. Alexanian and A. Zepeda. V, 132 pages. 1975.

Vol. 33: Classical and Quantum Mechanical Aspects of Heavy Ion Collisions. Proceedings 1974. Edited by H. L. Harney, P. Braun-Munzinger, and C. K. Gelbke. VII, 311 pages. 1975.

Vol. 34: One-Dimensional Conductors GPS Summer School Proceedings, 1974. Edited by H. G. Schuster. VII, 371 pages. 1975.

Vol. 35: Proceedings of the Fourth International Conference on Numerical Methods in Fluid Dynamics, 1974. Edited by R. D. Richtmyer. V, 457 pages. 1975.

Vol. 36: R. Gatignol, Théorie Cinétique des Gaz à Répartition Discrète de Vitesses. II, 219 pages. 1975.

Vol. 37: Trends in Elementary Particle Theory. Proceedings 1974. Edited by H. Rollnik and K. Dietz. V, 472 pages. 1975.

Vol. 38: Dynamical Systems, Theory and Applications. Proceedings 1974. Edited by J. Moser. VI, 624 pages. 1975.

Vol. 39: International Symposium on Mathematical Problems in Theoretical Physics. Proceedings 1975. Edited by H. Araki. XII, 562 pages. 1975.

Vol. 40: Effective Interactions and Operators in Nuclei. Proceedings 1975. Edited by B. R. Barrett. XII, 339 pages. 1975.

Vol. 41: Progress in Numerical Fluid Dynamics. Proceedings 1974. Edited by H. J. Wirz. V, 471 pages. 1975.

Vol. 42: H II Regions and Related Topics. Proceedings 1975. Edited by D. Downes and T. L. Wilson. XII, 488 pages. 1975.

Vol. 43: Laser Spectroscopy. Proceedings 1975. Edited by S. Haroche, J. C. Pebay-Peyroula, T. W. Hänsch, and S. E. Harris. X, 466 pages. 1975.

# Lecture Notes in Physics



84

# Stochastic Processes in Nonequilibrium Systems

Sitges International School of
Statistical Mechanics, June 1978
Sitges, Barcelona/Spain

Edited by
L. Garrido, P. Seglar and P. J. Shepherd

Springer-Verlag
Berlin Heidelberg GmbH 1978

**Editors:**
L. Garrido
P. Seglar
Departamento de Fisica Teórica
Universidad de Barcelona
Diagonal, 647
Barcelona-28 (Spain)

P. J. Shepherd
Department of Physics
University of Exeter
Exeter (England)

ISBN 978-3-540-08942-1    ISBN 978-3-540-35713-1 (eBook)
DOI 10.1007/978-3-540-35713-1


Originally published by Springer-Verlag Berlin Heidelberg New York in 1978

2153/3140-543210

# PREFACE

This book contains the lectures given at the Fifth Sitges International School of Statistical Mechanics held at Sitges (Barcelona, Spain) between June 11th and June 23rd, 1978. This year's topic - Stochastic Processes in Nonequilibrium Systems - provides concepts and mathematical tools related to many interdisciplinary problems in physics, chemistry, biology and sociology. An effort has been made to introduce students and laymen into this field.

For the first time in this School, participants were encouraged to present seminars and short communications in order to establish a better relation between the same and to promote a deeper understanding of recent developments. Most of these seminars are included in the present account of the proceedings.

I would like to take this opportunity to thank all those who have collaborated in the organization of this School. In particular, my sincere appreciation goes to all the members of the Department of Theoretical Physics of the University of Barcelona for the excellent team-work in preparing and running the School. Also I extend my warmest thanks to Doctors Seglar and Shepherd for making a heroic effort in their collaboration as editors of this volume which was completed on the day the School finished. It is pleasure for me to thank Mrs. Chester and Mrs. Baly for typing the manuscripts of the lectures with particular efficiency and care. My final thanks go to my wife for her patience and continuous cooperation.

L. Garrido

CONTENTS

SEMINARS AND SHORT COMMUNICATIONS

## PARTICIPANTS

ALDER, B., Lawrence Livermore Laboratory Livermore, Calif. (USA)

BAYER, W. University of Marburg, (Germany)

BEZ, W. University of Stuttgart, (Germany)

BIEL, J., University (Autónoma) of Barcelona (Spain)

BRACHET, M.E. University of Paris (France)

BREY, J., University of Sevilla (Spain)

v.d. BROECK, Ch., University of Brussels (Belgium)

BUHAGIAR, A. Open University, Milton Keynes (Great Britain

BUFFET, E., University of Lausanne (Switzerland)

BÜTTIKER, M., University of Basel (Switzerland)

CALDEIRA, A.O., University of Sussex (Great Britain)

CANIVELL, V., University of Barcelona (Spain)

CARMONA-RUIZ, G., Universidad Nacional Autónoma, México D.F.

CHATURVEDI, S., University of Stuttgart (Germany)

CLAVERIE, P., University of Paris (France)

COUAIRON, M., Centre d'Etudes de Limeil (France)

CUSUMANO, C., University of Palermo (Italy)

DAUDPOTA, Q.I., University of Edinburgh (Great Britain)

DENIS, A., University of Paris (France)

DOUGHERTY, J.P., University of Cambridge (Great Britain)

DUFOUR, C., University of Mons (Belgium)

EDHOLM, O., University of Stockholm (Sweden)

v. ENTER, A.C.D., University of Groningen (Holland)

ENZ, Ch., University of Geneva, (Switzerland)

ESCANDE, D., Ecole Polytechnique, Palaiseau (France)

FIORESE, G., Centre d'Etudes de Limeil (France)

GADELLA, M., University of Santander (Spain)

GARRIDO, L., University of Barcelona (Spain)

GERSCHEL, A., University Paris-Sud (France)

GRAHAM, R., University of Stuttgart (Germany)

GRAPPIN, R., Observatoire de Meudon (France)

GROSS, E.P., Brandeis University, Waltham, Mass (USA)

HAAG, G., University of Stuttgart (Germany)

HAKEN, H., University of Stuttgart (Germany)

HAKIM, R., Observatoire de Meudon (France)

JOU MIRAVENT, P., University (Autónoma) of Barcelona (Spain)

HANUSSE, P., University of Talence (France)

HAUS, J.W., Institut für Festkörperforschung, Jülich (Germany)

HESS, W., University of Konstanz (Germany)

v.d. HEYDT, N., University of Regensburg (Germany)

HONGLER, M.O., University of Geneva (Switzerland)

IRO, H., Hochschule Linz (Austria)

JOWETT, J.M., University of Cambridge (Great Britain)

JUNGLING, K., University of Heidelberg (Germany)

v. KAMPEN, University of Utrecht, (Holland)

KAUFMANN, K., Max-Planck-Institut, Göttingen (Germany)

KRAGLER, R., University of Konstanz (Germany)

KRUSZYNSKI, P., University of Warsaw (Poland)

LANGOUCHE, F., University of Leuven (Belgium)

LEAL, M., University of Santander (Spain)

LEORAT, J., Observatoire de Meudon (France)

LEVI, D., University of Rome (Italy)

MACHTA, J., MIT, Cambridge, Mass (USA)

MARRO, J., University of Barcelona,(Spain)

MARQUSEE, J., MIT, Cambridge, Mass (USA)

MAZO, R.M. University of Oregon (USA)

MENGUAL, J., University (Complutense) of Madrid, (Spain)

MISGUICH, J., CEN Fontenay aux Roses (France)

MOURITSEN, O.G., University of Aarhus (Denmark)

MURMANN, M. University of Heidelberg (Germany)

PARRA, J.M., University of Barcelona (Spain)

de PASQUALE, F., University of Rome (Italy)

PENROSE, O., Open University, Milton Keynes, (Great Britain)

PESQUERA, L., University of Paris (France)

POMEAU, Y., C.E.N. Saclay, Gif-sur-Yvette (France)

PRUZAN, P. University of Paris (France)

RODRIGUEZ, R.F., Universidad Nacional Autónoma, México, D.F.

ROEKAERTS, D., University of Leuven (Belgium)

SANCHO, J.M., University of Barcelona (Spain)

SAN MIGUEL, M., University of Barcelona (Spain)

SANTOS, E., University of Santander (Spain)

SCHRANNER, R., Max-Planck-Institut, Göttingen (Germany)

SEGLAR, P., University of Barcelona (Spain)

SHAH, S., University of Leeds (Great Britain)
SHEPHERD, P.J., University of Exeter (Great Britain)
SPOHN, H., University of Munich (Germany)
STEEB, W.H., Gesamthochschule Paderborn (Germany)
STERN, C., University of Paris (Orsay), (France)
TIRAPEGUI, E., University of Leuven (Belgium)
TOMBESI, P., University of Rome (Italy)
URUMOV, V., University of Skopje (Yugoslavia)
VARGAS, J.G., Universidad de los Andes, Bogotá Colombia (South America)
WAGENSBERG, J., University of Barcelona,(Spain)
WERNER, R., University of Marburg (Germany)
WUNDERLIN, A., University of Stuttgart (Germany)
ZAMBRINI, J.C., University of Geneva (Switzerland)
ZUMER, S., University of Ljubljana (Yugoslavia)

Picture of the Participants

1. Marro, J., Spain
2. Biel, J., Spain
3. Brey, J., Spain
4. Haag, G., Germany
5. Vargas, J., Colombia
6. Sancho, J.M., Spain
7. Denis, A., France
8. Brachet, M.E., France
9. Stern, C., France
10. Spohn, H., Germany
11. Carmona, G., Mexico
12. Mouritsen, O.G., Denmark
13. Steeb, W.H., Germany
14. Caldeira, A., Great Britain
15. Roekaerts, D., Belgium
16. Buhagiar, A., Great Britain
17. Edholm, O., Sweden
18. Schranner, R., Germany
19. Pesquera, L., France
20. Gerschel, A., France
21. Escande, D., France
22. Jüngling, K., Germany
23. Tirapegui, E., Belgium
24. Chaturvedi, S., Germany
25. Haus, J.W., Germany
26. Bez, W., Germany
27. Rodriguez, R.F., Mexico
28. San Miguel, M., Spain
29. van den Enter, A.C.D., Holland
30. Machta, J., USA
31. Haken, H., Germany
32. Kaufmann, K., Germany
33. Graham, R., Germany
34. Mürmann, M., Germany
35. Pruzan, P., France
36. Daudpota, Q.I., Great Britain
37. Gross, E.P., USA
38. Iro, H., Austria
39. Kruszynski, P., Poland
40. Dougherty, J.P., Great Britain
41. Léorat, J., France
42. Couairon, M., France
43. Grappin, R., France
44. Fiorese, G., France
45. Pomeau, Y., France
46. Joz-Miravent, D., Spain
47. Jowett, J.M., Great Britain
48. Büttiker, M., Switzerland
49. v.d. Heydt, N., Germany
50. Hakim, R., France
51. Hess, W., Germany
52. Wagensberg, J., Spain
53. Hongler, M.O., Switzerland
54. Parra, J.M., Spain
55. Garrido, L., Spain
56. Hanusse, P., France
57. Mazo, R.M., USA
58. van Kampen, N., Holland
59. Wunderlin, A., Germany
60. Kragler, R., Germany
61. Zumer, S., Yugoslavia
62. Urumov, V., Ygoslavia
63. Gadella, M., Spain
64. Misguich, J., France
65. Werner, R., Germany
66. Dufour, C., Belgium
67. Claverie, P., France
68. v.d. Broeck, Ch., Belgium
69. Mengual, J., Spain
70. Langouche, F., Belgium

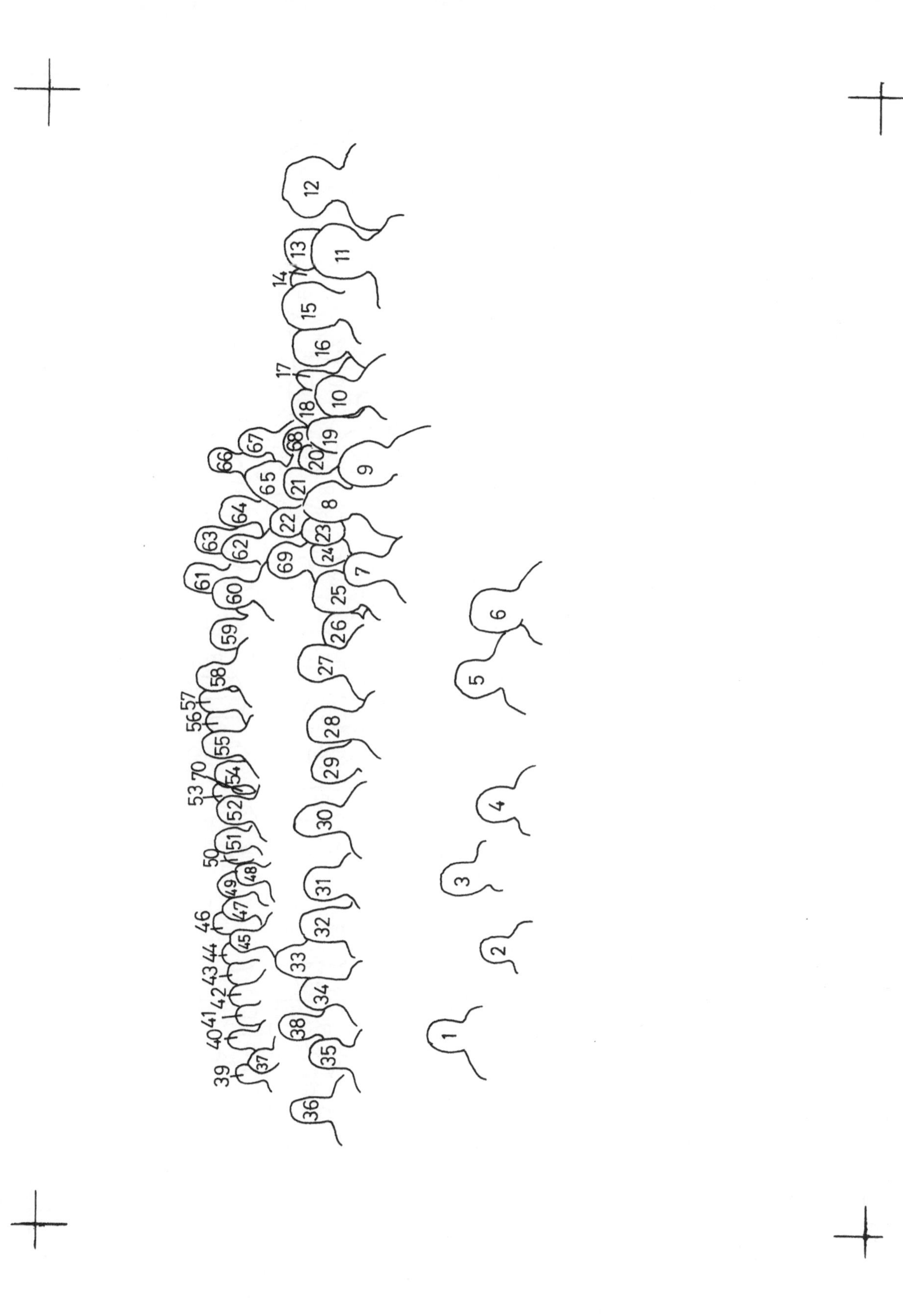

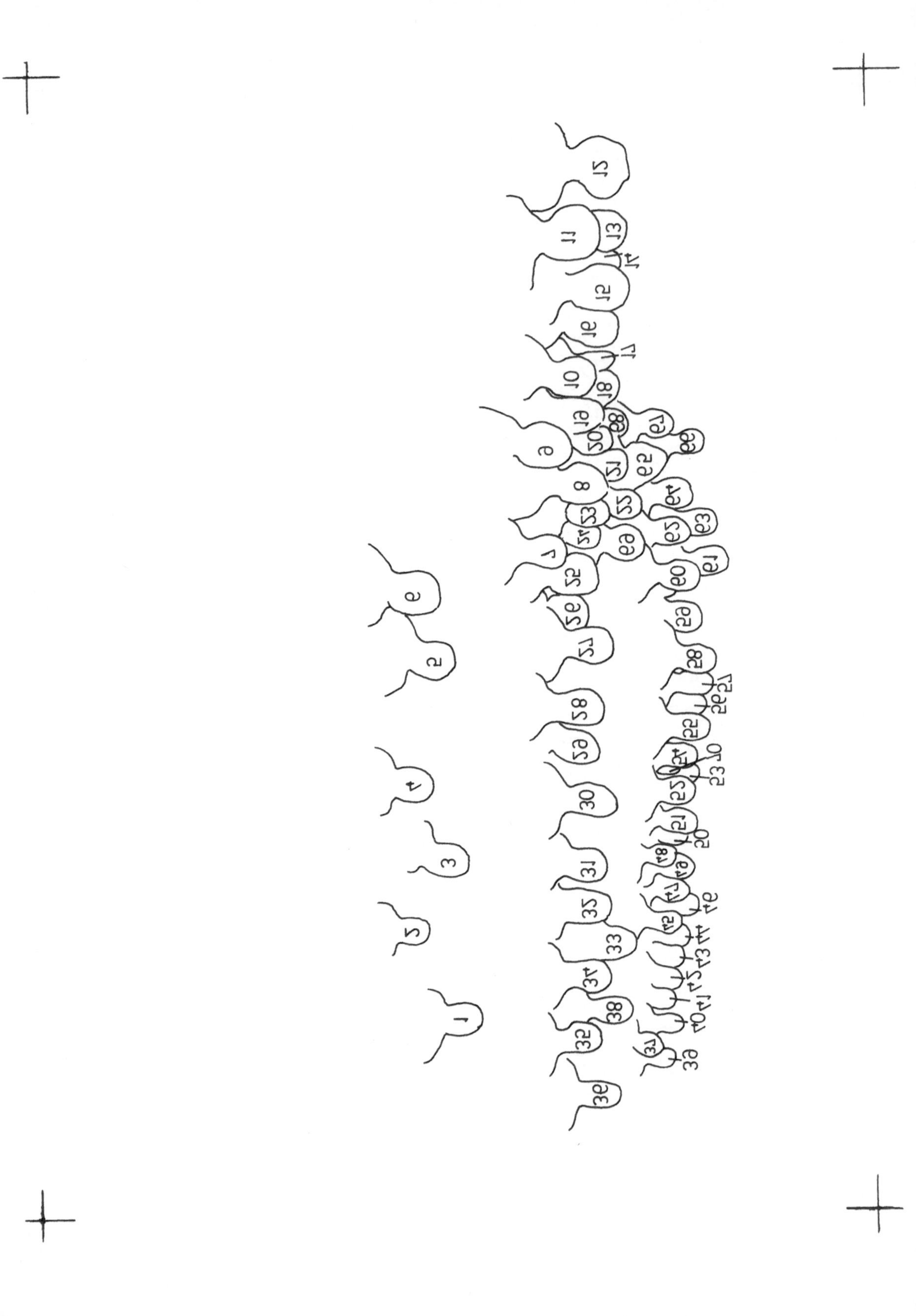

# AN INTRODUCTION TO STOCHASTIC PROCESSES FOR PHYSICISTS

N.G. van Kampen
Instituut voor Theoretische Fysica der Rijksuniversiteit
University of Utrecht, Maliesingel 23, Utrecht

## 1. PRELIMINARIES

Stochastic methods have become increasingly important in many branches of physics, technology, chemistry, biology, population dynamics, epidemiology, economics, and sociology. In spite of the diversity of the problems that come up in these fields, there are common principles and methods. They are the subject of these introductory lectures. My strategy will be to indulge as little as possible in generalities and preferably use examples from physics and chemistry to demonstrate the underlying ideas. But a few general concepts have to be established as a starting point[1].

A stochastic or random variable is an object defined by a set of possible values and a probability distribution over this set. The set of possible values (also called "range", "set of states", "sample space", or "phase space") may be discrete and finite, discrete and infinite, continuous in a certain interval in one or more variables, etc. If it is discrete and denumerable the probability distribution is given by a set of non-negative numbers $p_n$ such that $\sum_n p_n = 1$. If the range is an interval $I$ of the x-axis, the probability distribution is determined by a non-negative function $P(x)$ such that $\int_I P(x)x = 1$. This function is also called "probability density", but we do not exclude the possibility that it contains one or more delta peaks. Rather than developing a universal notation for all possible cases, we simply use the one that is most appropriate or convenient.

Let $X$ be a stochastic variable with range $(-\infty, \infty)$. Moments are defined by

$$\mu_m = \langle X^m \rangle = \int_{-\infty}^{\infty} x^m P(x) dx.$$

An indispensable tool is the characteristic function

$$G(k) = \langle e^{ikX} \rangle = \int_{-\infty}^{\infty} e^{ikx} P(x) dx \quad .$$

It serves as a generating function for the moments

$$G(k) = \sum_{m=0}^{\infty} \frac{(ik)^m}{m!} \mu_m \quad .$$

The cumulants $\kappa_m$ are combinations of the moments, defined by

$$\log G(k) = \sum_{m=1}^{\infty} \frac{(ik)^m}{m!} \kappa_m .$$

The second cumulant $\kappa_2 = \mu_2 - \mu_1^2$ is the variance of X, also denoted by $\sigma^2$. For the particular case of a Gaussian distribution

$$P(x) = \frac{1}{\sqrt{2\pi\sigma^2}} \exp\left[-\frac{(x-\mu_1)^2}{2\sigma^2}\right],$$

all cumulants beyond $\kappa_2 \equiv \sigma^2$ vanish.

For n variables $(X_1, X_2, \ldots, X_n) = \underset{\sim}{X}$, the characteristic function is

$$G(\underset{\sim}{k}) = \langle e^{i\underset{\sim}{k}\cdot\underset{\sim}{X}} \rangle = \sum_{\{m\}} \frac{(ik_1)^{m_1} \ldots (ik_n)^{m_n}}{m_1! \ldots m_n!} \langle X_1^{m_1} \ldots X_n^{m_n} \rangle$$

$$= \exp \sum_{m}{}' \frac{(ik_1)^{m_1} \ldots (ik_n)^{m_n}}{m_1! \ldots m_n!} \kappa_{m_1 \ldots m_n} .$$

The general multivariate Gaussian distribution is

$$P(\underset{\sim}{x}) = \frac{\sqrt{\operatorname{Det} A}}{(2\pi)^{n/2}} \exp\left[-(\underset{\sim}{x}-\underset{\sim}{\mu}_1)\cdot A\cdot(\underset{\sim}{x}-\underset{\sim}{\mu}_1)\right],$$

where $\underset{\sim}{\mu}_1$ is a constant vector and A a constant symmetric matrix. The "correlation matrix" is

$$\langle (x_i - \mu_i)(x_j - \mu_j) \rangle = (A^{-1})_{ij} .$$

Let X be a stochastic variable with range $n = 0, 1, 2, \ldots$ The characteristic function is now a power series in $z \equiv e^{ik}$ and one defines the probability generating function by

$$F(z) = \sum_{n=0}^{\infty} z^n p_n .$$

It now turns out that more useful than the moments and cumulants are the factorial moments $\phi_m$ defined by

$$F(1-z) = \sum_{m=0}^{\infty} \frac{(-z)^m}{m!} \phi_m$$

and the factorial cumulants $\theta_m$:

$$\log F(1-z) = \sum_{m=1}^{\infty} \frac{(-z)^m}{m!} \theta_m .$$

For the particular case of a Poisson distribution

$$p_n = \frac{\mu_1^n}{n!} e^{-\mu_1}$$

all factorial cumulants beyond $\theta_1 = \mu_1$ vanish.

## 2. STOCHASTIC FUNCTIONS

If $X$ is a well-defined stochastic variable, any quantity $Y$ related to $X$ by $Y = f(X)$ is also a well-defined stochastic variable. Its probability density $P_Y$ is related to $P_X$ by

$$P_Y(y) = \int \delta[f(x) - y] P_X(x) dx \quad .$$

In fact any mathematical object that is a function of $X$ is stochastic - in particular, any function of an auxiliary variable $t$, usually time,

$$Y(t) = f(t,X) \quad .$$

Such objects are called stochastic functions. The value of $Y$ at any one given time $t_1$ is a stochastic variable with probability density

$$P_1(y,t_1) = \int \delta[f(t_1,x)-y] P_X(x) dx \quad .$$

In classical mechanics the instantaneous state of a system of $N$ particles is a point in 6N-dimensional phase space. If at $t = 0$ the system is at a point $X$ its future is determined by the microscopic equations of motion, provided that the system is closed and isolated. Hence the value of any physical quantity $Y$ at time $t$ is a function $f(t,X)$. The transition to statistical mechanics consists in studying an appropriately chosen ensemble rather than an individual trajectory in phase space, i.e., one promotes $X$ to a stochastic variable with an appropriately chosen probability distribution. As a consequence $Y(t)$ becomes a stochastic function. The motivation for this stochastification is that the individual trajectories are incredibly complicated, but that the stochastic properties of $Y(t)$ may obey simple rules. In the venerable example of Brownian motion, $X$ is the microscopic state of the whole system, heavy particle plus all fluid molecules, $Y(t)$ is the position of the heavy particle alone, and although its precise position is an incredibly complicated function of time its average and variance are simple.

The values of $Y$ at a set of time points $t_1, t_2, \ldots, t_n$ constitute an n-dimensional stochastic variable with probability density

$$P_n(y_1,t_1;y_2,t_2;\dots;y_n,t_n)$$
$$= \int \delta[f(t_1,x)-y_1]\,\delta[f(t_2,x)-y_2]\dots\delta[f(t_n,x)-y_n]\,P_X(x)dx \quad .$$

The sequence of these functions $P_n$ has the obvious properties

(i) $P_n \geqslant 0$;

(ii) $P_n$ does not change on interchanging two pairs $(x_k,t_k)$ and $(x_\ell,t_\ell)$;

(iii) $\int P_n(y_1,t_1;\dots;y_n,t_n)dy_n = P_{n-1}(y_1,t_1;\dots;y_{n-1},t_{n-1})$;

(iv) $\int P_1(y_1,t_1)dy_1 = 1.$

Conversely, it has been proved[2)] that any such sequence of functions defines a stochastic process. Thus the $P_n$ provide an alternative tool for specifying stochastic processes, often more convenient than the original definition. For instance, the autocorrelation function is defined by

$$\langle \{Y(t_1) - \langle Y(t_1)\rangle\}\{Y(t_2) - \langle Y(t_2)\rangle\}\rangle$$
$$= \iint y_1 y_2 \{P_2(y_1,t_1;y_2,t_2) - P_1(y_1,t_1)P_1(y_2,t_2)\} dy_1 dy_2 .$$

Example I. A radioactive sample has $n_0$ active nuclei at time $t = 0$, each having a probability $\alpha$ per unit time to decay. That is, if a nucleus has not decayed at $t$, then at $t+dt$ there is a probability $\alpha dt$ that it decayed in the meantime and $1-\alpha dt$ that it is still active. The number of surviving active nuclei at $t > 0$ is a stochastic function with range $n = 0,1,2,\dots$ . Let $p_n(t)$ be its probability distribution at $t$ . Then

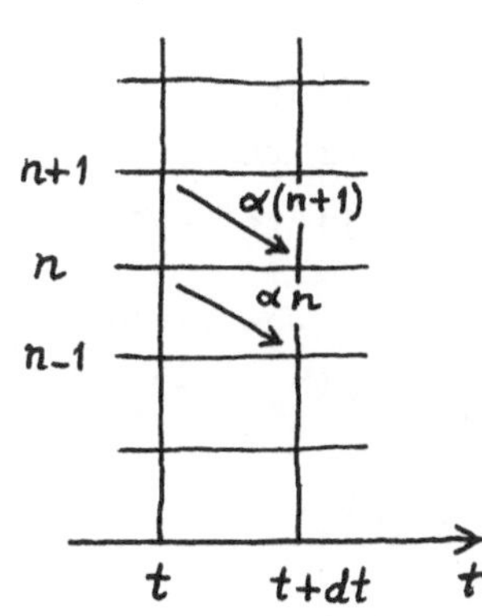


$$p_n(t+dt) = p_n(t)(1-n\alpha dt) + p_{n+1}(t)(n+1)\alpha dt \; .$$

From this follows the differential equation

$$\frac{dp_n(t)}{dt} = -n\alpha p_n(t) + (n+1)p_{n+1}(t) \; ,$$

which is called the master equation. It determines $p_n(t)$ for all $t > 0$ when $p_n(0)$ is given. In the present case $p_n(0) = \delta_{n,n_0}$ .

There are many ways of solving this equation, but the following is the most general method. Multiply by $z^n$ and sum over all $n$ ; the result is an equation for the probability generating function :

$$\frac{\partial F(z,t)}{\partial t} = -\alpha z \frac{\partial F}{\partial z} + \alpha \frac{\partial F}{\partial z} .$$

The general solution of this linear partial differential equation of first order is

$$F(z,t) = \Phi\{(1-z)e^{-\alpha t}\} ,$$

where $\Phi$ is an arbitrary function of the single argument in the curly brackets. This function is determined by the initial condition $F(z,0) = z^{n_0}$ and the final result is

$$F(z,t) = \left\{1 - (1-z)e^{-\alpha t}\right\}^{n_0} .$$

Having found $F$ it is easy to obtain $p_n(t)$ by expanding in powers of $z$. Normally, however, one is merely interested in the first few moments, which can be obtained directly from $F$, or more easily from $\log F$:

$$\langle n \rangle \equiv \Theta_1 = \left[\frac{d}{dz} \log F\right]_{z=1} = n_0\, e^{-\alpha t},$$

$$\langle n^2 \rangle - \langle n \rangle^2 - \langle n \rangle \equiv \Theta_2 = \left[\frac{d^2}{dz^2} \log F\right]_{z=1} = -n_0\, e^{-2\alpha t} .$$

The average obeys the familiar decay law and the variance is

$$\langle n^2 \rangle - \langle n \rangle^2 = n_0\, e^{-\alpha t}(1-e^{-\alpha t}) .$$

The solution $p_n(t)$ found here is more explicitly denoted by $p(n,t|n_0,0)$, i.e. the probability for having $n$ nuclei at $t$ when there were $n_0$ at 0. It is the analog of $P_1(y,t)$ for the continuous case. We shall now compute the analog of $P_2(y_1,t_1;y_2,t_2)$. Take $0 < t_1 < t_2$. Then the probability for having $n_1$ at $t_1$ is $p(n_1,t_1|n_0,0)$. If one actually has $n_1$ nuclei at $t_1$ the probability of also having $n_2$ at $t_2$ is $p(n_2,t_2|n_1,t_1) = p(n_2,t_2-t_1|n_1,0)$. Hence the joint probability of having both realized is

$$P_2(n_1,t_1;n_2,t_2) = p(n_2,t_2-t_1|n_1,0)p(n_1,t_1|n_0,0).$$

In this way the entire sequence of $P_n$ can be found as a product of factors, each of which is known once the master equation has been solved.

This argument, however, makes essential use of the physical fact that each of the $n_1$ nuclei surviving at $t_1$ still has the same decay probability $\alpha$ per unit time, regardless of the information that it has already lived a time $t_1$. If there were an aging effect we could not apply the same $p$ to the $n_1$ aged nuclei as to the original $n_0$ nuclei. Thus the same argument cannot be used for the number of surviving light bulbs out of a batch of $n_o$ light bulbs purchased at $t = 0$. The special feature that every surviving nucleus is as new, and does not remember its past, is called the Markov property. It dominates the whole subject of stochastic processes and therefore deserves a more careful definition.

Let $Y(t)$ be a stochastic process. Let $P(y_2,t_2|y_1,t_1)$ be the conditional probability for $Y(t_2)$ to have the value $y_2$, given that $Y(t_1)$ has the value $y_1$. From this definition follows the identity (Bayes' rule)

$$P_2(y_1t_1;y_2,t_2) = P_1(y_1,t_1)\ P(y_2,t_2|y_1,t_1) \quad .$$

$Y(t)$ is called a Markov process if for any set of $n$ successive times $t_1 < t_2 < \dots < t_n$ it is true that

$$P_n(y_1,t_1;y_2,t_2;y_3,t_3;\dots;y_n,t_n)$$
$$= P_1(y_1,t_1)\ P(y_2,t_2|y_1,t_1)\ P(y_3,t_3|y_2,t_2)\ \dots\ P(y_n,t_n|y_{n-1},t_{n-1}).$$

Thus a Markov process is fully determined by $P_1(y_1,t_1)$ and $P(y_2,t_2|y_1,t_1)$.

Write the defining equation with $n = 3$, integrate over $y_2$ and use some of the preceding equations; in this way one obtains the identity (for $t_1 < t_2 < t_3$)

$$P(y_3,t_3|y_1,t_1) = \int P(y_3,t_3|y_2,t_2)\ P(y_2,t_2|y_1,t_1)dy_2 \quad .$$

This is the Chapman-Kolmogorov equation for Markov processes. Any two non-negative and properly normalized functions $P_1(y,t)$ and $P(y_2,t_2|y_1,t_1)$ which obey this equation and also

$$P_1(y_2,t_2) = \int P(y_2,t_2|y_1,t_1)\ P(y_1,t_1)dy_1 \quad ,$$

define a Markov process.

## 3. SOME IMPORTANT MARKOV PROCESSES

The most famous Markov process is the Wiener-Lévy process, defined for $-\infty < y < \infty, t > 0$, by

$$P_1(y,t) = \frac{1}{\sqrt{2\pi t}} \exp\left[-\frac{y^2}{2t}\right]; \quad P(y_2,t_2|y_1,t_1) = \frac{1}{\sqrt{2\pi(t_2-t_1)}} \exp\left[-\frac{(y_2-y_1)^2}{2(t_2-t_1)}\right].$$

The reader will easily verify that the Chapman-Kolmogorov equation is satisfied and that

$$\langle y(t_1)y(t_2)\rangle = t_1 \quad \text{for} \quad 0 \leq t_1 \leq t_2 .$$

This process describes the position of a Brownian particle in one dimension. It is called a Gaussian process to indicate that all $P_n$ are (multivariate) Gaussian distributions.

The next in fame is the Ornstein-Uhlenbeck process $(-\infty < y < \infty;\ -\infty < t < \infty;\ t_2-t_1 = \tau > 0)$

$$P_1(y,t) = \frac{1}{\sqrt{2\pi}} \exp\left[-\frac{y^2}{2}\right]; \quad P(y_2,t_2|y_1,t_1) = \left[2\pi(1-e^{-2\tau})\right]^{-\frac{1}{2}} \exp\left[-\frac{(y_2-y_1e^{-\tau})^2}{2(1-e^{-2\tau})}\right].$$

It describes the velocity of a Brownian particle and is also Gaussian. Moreover, it is stationary, which means that all its $P_n$ depend on the time differences alone. Doob's theorem states that this is essentially the only stationary Gaussian Markov process, apart from changes in the scales of $y$ and $t$.[3]

The auto-correlation function of the Ornstein-Uhlenbeck process is $\langle Y(t_1)Y(t_2)\rangle = \exp[-|t_1-t_2|]$. If one sets $Y = aL$, $t = bs$ and takes the limit $b \to \infty$, $a \to \infty$, $2a^2/b = \text{const} = 1$, one has

$$\langle L(s_1)\, L(s_2)\rangle = \delta(s_1-s_2) .$$

This limiting case is called "Gaussian white noise" or "Langevin process". It is not an actual stochastic process, however, any more than the delta-function is an actual function, but it may often be treated as such. We shall now prove that its integral

$$W(t) = \int_0^t L(s)ds \qquad (t \geq 0)$$

is the Wiener process.

For this purpose consider the characteristic functional of $L(t)$, defined as a functional of an auxiliary $k(t)$ by

$$G([k]) = \left\langle \exp\left\{i\int_{-\infty}^{\infty} k(s)\, L(s)ds\right\}\right\rangle .$$

In analogy with the characteristic function of a single stochastic variable its expansion in powers of $k$ generates the moments $\langle L(s_1)L(s_2) \dots L(s_m)\rangle$, while the expansion of its logarithm generates the corresponding cumulants. But since $L(t)$ is Gaussian and has zero average, only the second cumulant survives; hence,

$$G([k]) = \exp\left[-\tfrac{1}{2}\iint_{-\infty}^{\infty} k(s_1)k(s_2)\langle L(s_1)L(s_2)\rangle ds_1 ds_2\right]$$

$$= \exp\left[-\tfrac{1}{2}\int_{-\infty}^{\infty} k(s)^2 ds\right] .$$

It is now possible to compute the characteristic functional of $W$ with an auxiliary function $K(t)$:

$$\left\langle \exp\left\{i\int_0^{\infty} K(t)W(t)dt\right\}\right\rangle = \left\langle \exp\left\{i\int_0^{\infty} L(s)ds\int_s^{\infty} K(t)dt\right\}\right\rangle$$

$$= \exp\left[-\tfrac{1}{2}\int_0^{\infty} ds\left\{\int_s^{\infty} K(t)dt\right\}^2\right]$$

$$= \exp\left[-\tfrac{1}{2}\iint_0^{\infty} K(t_1)K(t_2)dt_1dt_2.\mathrm{Min}(t_1,t_2)\right] .$$

This shows that $W$ is Gaussian with zero mean and that $\langle W(t_1)W(t_2)\rangle = \mathrm{Min}(t_1,t_2)$. As the Wiener process is also Gaussian and has the same first and second moments it is identical with $W$, q.e.d.

Viewed from a different angle this proof amounts to solving the stochastic differential equation

$$\dot{W} = L(t), \qquad W(0) = 0,$$

where $L(t)$ is given to be the Langevin process. In the same way one can solve the Langevin equation

$$\dot{V} + \beta V = \alpha L(t)$$

and the result is that $V(t)$ is the Ornstein-Uhlenbeck process. More generally one can study the equation

$$\dot{U} = f(U) + g(U)L(t) .$$

$U$ stands for one or more unknowns, $f$ and $g$ are each one or more functions of $U$. Ample use of such generalized Langevin equations has been made for the purpose of including fluctuations in the description of electronic, hydrodynamic, electromagnetic and other systems. However, not only is it hard to find solutions for nonlinear cases, it has even been shown[4)] that, unless $g$ is constant, this equation as it stands is meaningless, owing to the singular nature of $L(t)$. We therefore return to the Chapman-Kolmogorov equation.

## 4. THE MASTER EQUATION

The Chapman-Kolmogorov equation can be cast in a more manageable form by taking $t_3 = t_2 + \Delta t$ and going to the limit $\Delta t \to 0$. One has $P(y_3,t_2|y_2,t_2) = \delta(y_3-y_2)$ and may expect, therefore,

$$P(y_3,t_2+\Delta t|y_2,t_2) = \delta(y_3-y_2)(1 - A\Delta t) + \Delta t\, W(y_3|y_2) + \varepsilon ,$$

where $\varepsilon/\Delta t \to 0$. $W(y_3|y_2)$ is a transition probability per unit time from $y_2$ to $y_3$; in general it will be a function of $t_2$ as well, but not in our applications and we therefore do not indicate that dependence. Normalization tells us that

$$A(y_2) = \int W(y_3|y_2)dy_3 .$$

If $P(y_3,t_2+\Delta t|y_2,t_2)$ does indeed have this form we may substitute it in the Chapman-Kolmogorov equation and obtain after some manipulations, for $t > t_1$,

$$\frac{\partial P(y,t|y_1,t_1)}{\partial t} = \int\{W(y|y')P(y',t|y_1,t_1) - W(y'|y)P(y,t|y_1,t_1)\}dy' .$$

This is called the master equation. For the discrete case it is

$$\dot{p}_n = \sum_{n'}\{W_{nn'}\, p_{n'} - W_{n'n}\, p_n\} ,$$

which shows that it is a gain-loss equation for the probabilities $p_n$. Actually the $p_n$ are transition probabilities $p(n,t|n_1,t_1)$, but here and in the future we omit to indicate this explicitly.

The master equation for the Wiener-Lévy process is

$$\frac{\partial P(y,t)}{\partial t} = \frac{\partial^2 P(y,t)}{\partial y^2} .$$

It follows that this equation is equivalent to $\dot{y} = L(t)$. Similarly, for the Ornstein-Uhlenbeck process,

$$\frac{\partial P(y,t)}{\partial t} = \frac{\partial}{\partial y}yP + \frac{\partial^2 P}{\partial y^2} ,$$

which is therefore equivalent to $\dot{y} + y = \sqrt{2}\, L(t)$. Thus, in these two cases the master equations reduce to second-order differential equations, called the diffusion equation and the Fokker-Planck equation, respectively (although both names are also used for more general equations).

Example II. A quantized harmonic oscillator interacts with a radiation field with given density; the levels $n = 0, 1, 2, \ldots$ are the states. Since the squared matrix element between n and n-1 is proportional to n, there is a transition probability $\alpha n$ per unit time from n to n-1 through emission, and $\beta(n+1)$ from n to n+1 through absorption. Hence

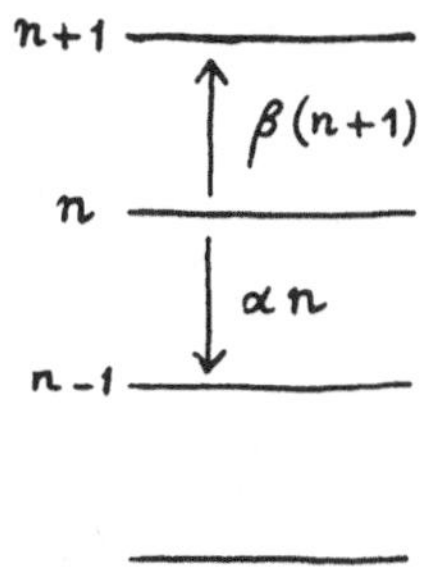

$$\dot{p}_n = -\{\alpha n+\beta(n+1)\}p_n + \alpha(n+1)p_{n+1} + n\beta p_{n-1}.$$

With the aid of the step operator E, defined by $Ef(n) = f(n+1)$, this master equation may be written more shortly:

$$\dot{p}_n = \alpha(E-1)np_n + \beta(E^{-1}-1)(n+1)p_n.$$

The solution of this equation with the initial condition $p_n(0) = \delta_{n,n_0}$ can again be found explicitly, by means of the probability generating function.

The equation has one time-independent or stationary solution, which can be found directly from

$$0 = \alpha(E-1)n\, p_n^{st} + \beta(E^{-1}-1)(n+1)p_n^{st}$$

$$= (E-1)\left[\alpha n\, p_n^{st} - \beta E^{-1}(n+1)p_n^{st}\right].$$

The last line states that [ ] is independent of n:

$$\alpha n\, p_n^{st} - \beta n\, p_{n-1}^{st} = -J ,$$

where J (which is the net probability flow from n-1 to n) is constant. As $p_n$ vanishes at infinity one must have $J = 0$ and, therefore,

$$p_n^{st} = (\beta/\alpha)^n\, p_0^{st}.$$

Finally the normalization determines $p_0^{st} = 1 - \beta/\alpha$. If the radiation field is in thermal equilibrium this solution must be identical with the familiar equilibrium distribution:

$$p_n^{st} = p_n^{eq} = \text{const.}\ \exp\left[-n\frac{\hbar\omega}{kT}\right].$$

Then, without calculating any transition probability, we know the ratio $\beta/\alpha = \exp[-\hbar\omega/kT]$; which is Kirchhoff's law.

Markov processes with a discrete sequence of states n, in which only transitions between neighbouring states are possible, are called one-step processes or birth-and-death processes. Their stationary solution can always be found in the above manner. If the transition probabilities are linear functions of $n$ it is also possible to find their complete time-dependent solutions by means of the probability generating function. Examples I and II are such linear processes, but the following process is nonlinear. To be sure, all master equations are linear equations for the distribution $p_n$, but we call them "nonlinear" when the coefficients are nonlinear in $n$.

## 5. THE $\Omega$-EXPANSION

Example III. Dissociation of a diatomic gas:

$$AB \underset{\beta}{\overset{\alpha}{\rightleftharpoons}} A + B.$$

The total numbers $N_A$, $N_B$ of atoms A, B are fixed. If $n$ is the number of molecules AB, the probability per unit time for a dissociation is $\alpha n$. For an association to occur, one of the $N_A-n$ unattached atoms A must meet one of the $N_B-n$ atoms B; the probability per unit time is $\beta(N_A-n)(N_B-n)/\Omega$, where $\Omega$ is the volume of the reaction vessel. Hence the master equation is

$$\dot{p}_n = \alpha(E-1)np_n + (\beta/\Omega)(E^{-1}-1)(N_A-n)(N_B-n)p_n \quad .$$

The stationary solution can be found and identified with the known thermal-equilibrium solution to give

$$\frac{\beta}{\alpha} = \left\{2\pi kT \frac{m_A + m_B}{m_A\, m_B}\right\}^{3/2} e^{\chi/kT} \quad ,$$

where $\chi$ is the binding energy. The time-dependent solutions, however, cannot be found explicitly, owing to the non-linearity of the second coefficient.

It is essential for the Markov character (and thus for the validity of the master equation) that each molecule AB have a fixed dissociation chance $\alpha$, regardless of how and when it was formed. If the molecule can exist in different internal states this is no longer true. The obvious thing to do then is to introduce more variables $n_1$, $n_2$, ... for the number of molecules in internal states 1, 2, ..., and to include possible transitions between internal states. The multivariate process $\{n_1,n_2,...\}$ with probability distribution $p_{n_1n_2....}(t)$ may then again be Markovian. In general, whether or not a physical phenomenon constitutes a Markov process depends on the choice of variables. The task of the physicist is to find the appropriate set of variables in which the process is Markovian. Or at least approximately so, because no physical process is exactly a Markov process, unless all microscopic variables are included in the description, which would make the process deterministic but unmanageable.

We shall now obtain an approximate solution for the nonlinear master equation of Example III, valid when $\Omega$ is large and the fluctuations therefore relatively small. First it is necessary to exhibit the powers of $\Omega$ in the equation explicitly by setting $N_A = \Omega\rho_A$, $N_B = \Omega\rho_B$. For simplicity we take both equal, $\rho_A = \rho_B = \rho$. Next we expect that $p_n$ has the form of a sharp peak around some value of $n$, which is of order $\Omega$, having a width of order $\Omega^{\frac{1}{2}}$. Accordingly we get

$$n = \Omega\phi(t) + \Omega^{\frac{1}{2}}\xi, \qquad p_n = \Pi(\xi,t) \ .$$

This is a (time-dependent) transformation from $n$ to the new variable $\xi$; the function $\phi(t)$ will be chosen presently. Of course one is free to apply any transformation of variables; whether or not this particular one is useful will be decided by the result. The step operator $E$ shifts $\xi$ by an amount $\Omega^{-\frac{1}{2}}$, so that

$$E^{\pm 1}-1 = \pm\Omega^{-\frac{1}{2}}\frac{\partial}{\partial\xi} + \tfrac{1}{2}\Omega^{-1}\frac{\partial^2}{\partial\xi^2} \pm \cdots$$

Substitution of all this in the master equation yields

$$\begin{aligned}\frac{\partial\Pi(\xi,t)}{\partial t} - \Omega^{\frac{1}{2}}\frac{d\phi}{dt}\frac{\partial\Pi}{\partial\xi} = \\ &\alpha\left(\Omega^{\frac{1}{2}}\frac{\partial}{\partial\xi} + \tfrac{1}{2}\frac{\partial^2}{\partial\xi^2}\right)\left(\phi(t) + \Omega^{-\frac{1}{2}}\xi\right)\Pi \\ &+\beta\left(-\Omega^{\frac{1}{2}}\frac{\partial}{\partial\xi} + \tfrac{1}{2}\frac{\partial^2}{\partial\xi^2}\right)\left(\rho - \phi - \Omega^{-\frac{1}{2}}\xi\right)^2\Pi \ .\end{aligned}$$

The largest terms are those containing $\Omega^{\frac{1}{2}}$; they can be caused to cancel by choosing for $\phi(t)$ a solution of

$$\dot{\phi} = -\alpha\phi + \beta(\rho - \phi^2) \ .$$

This equation determines how the macroscopic part of $n$ varies with time. In fact, it is just the familiar rate equation for the densities or "concentrations" of the reactants. Thus, we have deduced the macroscopic equation from the master equation by singling out the largest terms and ignoring all terms that contain a lower power of $\Omega$.

Next collect the terms of order $\Omega^0$:

$$\frac{\partial\Pi}{\partial t} = \left[\alpha - 2\beta\rho + 2\beta\phi\right]\frac{\partial}{\partial\xi}\xi\Pi + \tfrac{1}{2}\left[\alpha\phi - \beta(\rho-\phi)^2\right]\frac{\partial^2\Pi}{\partial\xi^2} \ .$$

This is an equation of Fokker-Planck type, but the coefficients depend on t through $\phi(t)$. Yet its solution can be obtained in the form of a Gaussian, and one therefore only needs to determine the first and second moments of $\xi$. The first moment can be found directly by multiplying the equation by $\xi$ and integrating:

$$\partial_t\langle\xi\rangle = -[\alpha - 2\beta\rho + 2\beta\phi]\langle\xi\rangle \ .$$

Similarly, for the second moment multiply by $\xi^2$:

$$\partial_t\langle\xi^2\rangle = -2[\alpha - 2\beta\rho + 2\beta\phi]\langle\xi^2\rangle + [\alpha\phi + \beta(\rho-\phi)^2] \ .$$

When these equations are solved the entire distribution $\Pi(\xi,t)$ is known, but actually these two moments themselves are often all one wants to know.

As an application we compute the auto-correlation function of the fluctuations in n when the stationary state has been reached. The stationary value of $\phi$ is obtained from

$$\dot{\phi} = 0 = -\alpha\phi + \beta(\rho-\phi)^2 \ .$$

There are two roots, but one is larger than $\rho$ and therefore the other is $\phi^{st}$. On substituting $\phi^{st}$ in the equation for $\Pi$ the coefficients become independent of time, and therefore $\xi$ becomes an Ornstein-Uhlenbeck process. There is no need to invoke this fact, however, to find directly that

$$\langle\xi\rangle^{st} = 0, \qquad \langle\xi^2\rangle^{st} = \frac{\alpha\phi^{st}}{\alpha-2\beta\rho+2\beta\phi^{st}} \ .$$

Moreover, suppose that at the particular moment $t_1$ it happens that $\xi$ has the value $\xi_1$. Then the average of $\xi$ at a later time $t_2$, subject to this condition, is

$$\begin{aligned}\langle\xi(t)\rangle_{\xi_1} &= \xi_1 \exp\{-[\alpha-2\beta\rho+2\beta\phi^{st}]\,(t_2-t_1)\} \\ &= \xi_1\, e^{-\tau},\end{aligned}$$

where $\tau$ is an obvious abbreviation. Hence the autocorrelation function of $\xi$ is

$$\begin{aligned}\langle\xi(t_1)\xi(t_2)\rangle &= \iint\xi_1\xi_2\; P_2(\xi_1,t_1;\xi_2,t_2)d\xi_1 d\xi_2 \\ &= \iint\xi_1\xi_2\; P_1(\xi_1,t_1)\; P(\xi_2,t_2|\xi_1,t_1)d\xi_1 d\xi_2 \\ &= \int\xi_1 P^{st}(\xi_1)\langle\xi_2\rangle_{\xi_1} d\xi_1 = \langle\xi_1{}^2\rangle^{st}\, e^{-\tau} \\ &= \frac{\alpha\phi^{st}}{\alpha-2\beta\rho+2\beta\phi^{st}}\, e^{-[\alpha-2\beta\phi+2\beta\phi^{st}](t_2 - t_1)} \ .\end{aligned}$$

The autocorrelation of n is the same expression multiplied by $\Omega$. It is therefore a simple exponential and we have found the values of the two coefficients in it in terms of the parameters $\alpha$, $\beta$, $\rho$. The The Wiener-Khintchine theorem states that its Fourier transform is the spectral density of the fluctuations, which is often the quantity directly measured.

This is the "linear-noise approximation" for the fluctuations. If higher orders in $\Omega^{-\frac{1}{2}}$ are included, additional terms appear in the equation for $\Pi$, with the effect that the first coefficient is no longer linear in $\xi$, the second one no longer independent of $\xi$, and higher derivatives appear. The solution is no longer Gaussian, but it is still possible to determine the successive moments in any order desired. One consequence is that the autocorrelation function is supplemented with another exponential term, which decays about twice as fast as the one found in the linear-noise approximation.[5)]

## 6. LIMITATION OF THE $\Omega$-EXPANSION

The $\Omega$-expansion is based on the expectation that the fluctuations are small. The result confirmed that they are of relative order $\Omega^{-\frac{1}{2}}$, and higher terms are actually of order $\Omega^{-1}$, etc. Yet, if $\langle\xi\rangle$ or $\langle\xi^2\rangle$ grows in the course of time, these powers of $\Omega$ are only an appropriate measure of the magnitude of the fluctuations during a limited period. It is seen from the equations for $\langle\xi\rangle$ and $\langle\xi^2\rangle$ that they will remain finite or grow, according as

$$\alpha - 2\beta\rho + 2\beta\phi$$

is positive or negative. What is the meaning of this criterion?

Consider the macroscopic equation for $\phi$. Let $\phi(t)$ and $\phi(t) + \delta\phi(t)$ be two neighbouring solutions; then

$$\frac{d}{dt}\delta\phi = -[\alpha - 2\beta\rho + 2\beta\phi]\,\delta\phi + O(\delta\phi)^2 .$$

When the coefficient [ ] is positive the two solutions converge to one another; when [ ] is negative their distance grows. That is, the solution $\phi(t)$ is respectively stable or unstable for small perturbations. The conclusion is that fluctuations around a stable solution of the macroscopic equation remain small and can be handled by means of the $\Omega$-expansion, while for an unstable solution the fluctuations grow and the $\Omega$-expansion becomes spurious after an initial transient period.

A more precise investigation shows that for the unrestricted validity of the $\Omega$-expansion it is necessary (but not sufficient) that the system be globally asymptotically stable in the sense of Lyapounov, i.e., all solutions of the macroscopic equation tend to one and the same $\phi^{st}$. There is a variety of ways in which this condition can be violated, and in those cases no general method for obtaining approximate solutions of the master equation is available. It appears, however, that in many cases some quantities of interest can be computed even without solving the entire master equation. This will now be demonstrated for one particular case, namely, a bistable system. This situation occurs in many different fields, e.g., chemical reactions[6)], electronics[7)], and biology[8)].

Example IV.[9)] Our system consists of an even number $2N$ of spins, each of which has two possible positions: up and down. The number of up spins minus the number of down spins is the magnetization and will be denoted by $2n$. The interaction between the spins is supposed to take place through an overall field proportional to $n$, in such a way that the energy is $-\frac{1}{2}(g/N)n^2$. Here $g$ is an interaction constant, but the actual interaction strength is taken equal to $g/N$ to ensure that the thermodynamic limit $N \to \infty$ exists. (This is a frequently-used, though not quite honest, trick that makes it possible to incorporate the molecular-field approximation into the model itself.) The equilibrium distribution of $n$ is, taking into account the number of spin configurations belonging to each value of $n$,

$$p_n^{eq} = \text{const.} \binom{2N}{N+n} e^{(K/N)n^2} ,$$

where $K = g/2kT$. For large $N$, setting $n = Ns$,

$$p_n^{eq} = \text{const.}\ e^{-NU(s)}$$

$$U(s) = (1+s)\log(1+s) + (1-s)\log(1-s) - Ks^2 .$$

Of course this is an even function of $s$ in the interval $-1 < s < 1$. When $K < 1$ the distribution has a single maximum at $n=0$; when $K > 1$ it has a minimum at $n=0$ and two maxima at $n = \pm Ns_1$, with $s_1 = \tanh Ks_1$. Clearly, $K = 1$ is the Curie point.

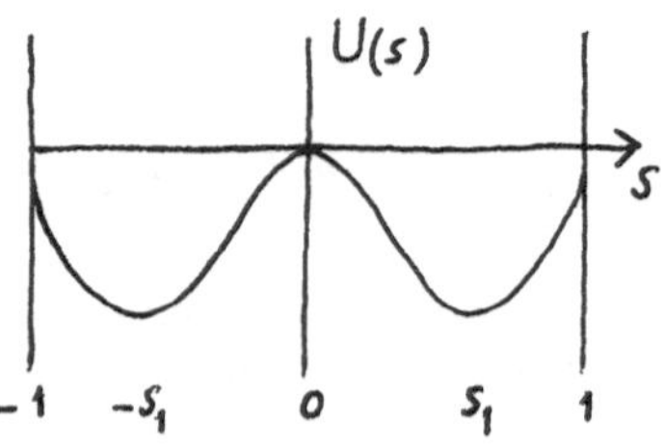


Now suppose that each of the up spins has a probability per unit time to flip to its down position equal to $e^{-(K/N)n}$; then the probability per unit time for $n$ to decrease by one unit is

$$\alpha_n = (N+n)e^{-(K/N)n} .$$

Similarly the probability per unit time for $n$ to increase by one unit is

$$\beta_n = (N-n)e^{(K/N)n} .$$

The master equation is

$$\dot{p}_n = (E-1)\alpha_n p_n + (E^{-1}-1)\beta_n p_n .$$

It is easily checked that the stationary solution of this equation is $p_n^{eq}$ .

The macroscopic equation is again obtained by setting $n = N\phi(t) + N^{\frac{1}{2}}\xi$ and extracting the largest term:

$$\dot{\phi} = -(1+\phi)e^{-K\phi} + (1-\phi)e^{K\phi} .$$

For convenience in interpretation we put it in the form

$$\dot{\phi} = -\frac{dV(\phi)}{d\phi}$$

$$V(\phi) = -\frac{2}{K}(1+\frac{1}{K})\cosh K\phi + \frac{2\phi}{K}\sinh K\phi .$$

$\phi$ can be visualized as the coordinate of some particle moving in a potential $V$ with so large a friction that its velocity is proportional to the force (as an electron in a conductor or semiconductor).

There is reason to emphasise that this potential function $V$, which governs the macroscopic equation, is not the same as the function $U$ connected with the equilibrium distribution (see note on p. 22). Yet it also has a single minimum when $K<1$; and when $K>1$ a maximum at $\phi=0$ and two minima at $\phi = \pm s_1$. Consequently, for $K<1$ all solutions of the macroscopic equation tend to $\phi=0$ and are therefore globally stable. The expansion in $N^{-\frac{1}{2}}$ is therefore valid without restrictions.

## 7. CALCULATION OF THE ESCAPE RATE

The two minima of $V(\phi)$ that exist when $K>1$ correspond to nonzero magnetization in the up or down direction and will be denoted by $\phi_u$ and $\phi_d$. They are stationary solutions of the macroscopic equation, locally stable but not globally. Every other solution tends to one or the other, depending on its initial value. This is the bistable situation for which we shall now investigate the effect of the fluctuations. Throughout we suppose that K is well above 1, so that the potential barrier between both stable stationary solutions is well developed.

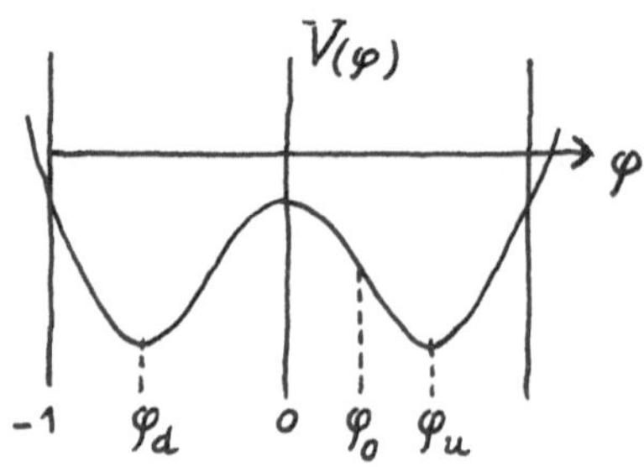


Consider the solution of the master equation determined by $p_n(0) = \delta_{n,n_0}$, where $n_0/N = \phi_0$ is located somewhere between 0 and $\phi_u$, but not too close to 0. The expansion in $N^{-\frac{1}{2}}$ yields

$$p_n(t) = \Pi\left(\frac{n - N\phi(t)}{N^{\frac{1}{2}}}, t\right)$$

Here $\phi(t)$ is the macroscopic solution with $\phi(0) = \phi_0$; and $\Pi(\xi, t)$ is the solution of the time-dependent Fokker-Planck equation having this $\phi(t)$ in the coefficients, and with initial condition $\Pi(\xi, 0) = \delta(\xi)$. Thanks to the local stability of $\phi_u$ the fluctuations do not grow and the distribution tends to a Gaussian located at $N\phi_u$.

This is not the complete story, however, because there is a tiny probability per unit time for a very large fluctuation which carries the system across the potential barrier, so that the magnetization flips from up to down. On a very long time scale, therefore, the single peak at $N\phi_u$ loses probability to the other potential minimum, until there are two peaks of equal height at both minima. The ultimate situation is the one described by $p_n^{eq}$. The rate at which probability escapes through the barrier will be shown to contain a factor $e^{-cN}$, which explains why this effect is not covered by the expansion in $N^{-\frac{1}{2}}$. For any macroscopic N the time scale involved may well be long compared to the age of the Universe, as demonstrated by the existence of paleomagnetism. Our problem is to compute this escape rate.

The picture is the following. Suppose one starts at $t=0$ with an arbitrary $p_n(0)$. This distribution evolves in time and develops two peaks, located at $\phi_u$ and $\phi_d$. They approach their local equilibrium, which in the linear-noise approximation (i.e. to order $N^{-\frac{1}{2}}$) consists of two Gaussian peaks at $\phi_u$ and $\phi_d$. The time needed for this is of the same order as the time needed for an initial $\phi$ to reach $\phi_u$ or $\phi_d$ according to the macroscopic equation. The probability is now divided into two parts $\pi_u$, $\pi_d$ ($\pi_u + \pi_d = 1$), one belonging to each peak. Subsequently, over a very long time scale $\tau$, these peaks exchange probability:

$$\dot{\pi}_u = -\dot{\pi}_d = -(\pi_u - \pi_d)/\tau \ .$$

We shall now compute $\tau$ following a method that was developed by Kramers[10)] for the case of diffusion in a continuous space.

After the local equilibrium has been reached $p_n$ varies very slowly. Hence one may take in the master equation $\dot{p}_n - 0$ and conclude as before (Example II) for this almost stationary solution that

$$\alpha_n p_n^{as} - \beta_{n-1} p_{n-1}^{as} = -J \ .$$

Divide by the factor $\alpha_n p_n^{eq} = \beta_{n-1} p_{n-1}^{eq}$ :

$$\frac{p_n^{as}}{p_n^{eq}} - \frac{p_{n-1}^{as}}{p_{n-1}^{eq}} = \frac{-J}{\alpha_n p_n^{eq}} \ .$$

Sum this identity from $n = N\phi_d + 1 \equiv d+1$ to $n = N\phi_u \equiv u$:

$$\frac{p_u^{as}}{p_u^{eq}} - \frac{p_d^{as}}{p_d^{eq}} = -J \sum_{d+1}^{u} \frac{1}{\alpha_n p_n^{eq}} \ .$$

Since the local peaks do not change in shape but only in height, the left-hand side is

$$\frac{\pi_u}{1/2} - \frac{\pi_d}{1/2} = 2(\pi_u - \pi_d) \ .$$

As $J$ is the flow of probability one has $\dot{\pi}_u = -\dot{\pi}_d = J$. Hence we have found

$$\tau = \tfrac{1}{2} \sum_{d+1}^{u} \frac{1}{\alpha_n p_n^{eq}} \ .$$

This remarkable formula expresses $\tau$ in quantities that are known and do not require that the master equation be solved. Obviously, the value of the sum is mainly determined by the terms around $n=0$ where $p_n^{eq}$ is very small. One may therefore put $\alpha_n = \alpha_0 = N$ and one finds, taking into account the normalization of $p_n^{eq}$,

$$\tau = \frac{1}{2N} \int_{-s_1}^{s_1} e^{NU(s)} N ds \int_{-1}^{1} e^{-NU(s)} N ds$$

$$= \frac{N}{2} \int_{-\infty}^{\infty} e^{NU(0) - \frac{1}{2}N|U''(0)|x^2} dx$$

$$\times \ 2 \int_{-\infty}^{\infty} e^{-NU(s_1) - \frac{1}{2}NU''(s_1)x^2} dx$$

$$= 2 \left\{ U''(s_1) \middle| U''(0) \right\}^{-\frac{1}{2}} e^{N\{U(0) - U(s_1)\}} .$$

It is easy to compute $\{\quad\}$, but the exponential is the dominant factor. It shows that the escape rate is, indeed, $e^{-cN}$. The coefficient $c$ is the difference of the maximum and minimum values of $U$ - not of the potential $V$! The conclusion is that, if we know at $t=0$ that our spin system is magnetized in the up-direction, then, after a time of order $\tau$, we can no longer be sure that it has not flipped to the down-direction. For magnetic memory elements and tapes it is of course necessary to make $\tau$ large by ensuring that $K$ is well above its critical value and that $N$ is large enough.

## 8. A FIRST-PASSAGE PROBLEM IN TWO DIMENSIONS

Example V. A more realistic picture of dissociation than the one in Example III is the following. The two atoms of the molecule AB vibrate with respect to one another. This vibrational mode gains and loses energy by successive collisions, until it happens to reach the energy $\varepsilon_a$ needed for dissociation. If we represent the vibration by a harmonic oscillator, and the effect of the collisions by a random force plus a damping, the equation that holds as long as the energy is less than $\varepsilon_a$ is

$$\ddot{x} + \beta\dot{x} + x = \alpha L(t).$$

We want to calculate how long it takes for the energy $\varepsilon = \frac{1}{2}(\dot{x}^2 + x^2)$ to reach the value $\varepsilon_a$. Admittedly this model is not very realistic, but it is good enough for an orientation, considering the many other uncertainties in the theory of chemical reactions. The theory of rotational relaxation of dipoles in a fluid provides another application.[11)]

The problem differs from the previous one in two respects. First, the range of the variable $x$ is continuous, but that is a minor modification. The second and more drastic difference, however, is that the equation is of second order. Kramers[10)] reduced it to first order by assuming that $\beta$ is large, so that the particle is overdamped and the inertial term $\ddot{x}$ may be neglected.[12)] We shall suppose, however, that $\beta$ is small, so that only a small fraction of the energy is dissipated during one period. It is then possible to average over the phase of the oscillation and reduce the problem to a diffusive motion in the one-dimensional energy scale.

The above Langevin equation for the harmonic oscillator is equivalent to a bivariate Fokker-Planck equation for the distribution in the space of x and the velocity v, namely

$$\frac{\partial P(x,v,t)}{\partial t} = -v\frac{\partial P}{\partial x} + x\frac{\partial P}{\partial v} + \beta\frac{\partial}{\partial v}vP + \frac{\alpha^2}{2}\frac{\partial^2 P}{\partial v^2} .$$

One knows that this equation must be satisfied by $P^{eq}(x,v) =$ const. $\exp[-(v^2 + x^2)/2kT]$, from which it follows that $\alpha^2 = 2\beta kT$.

Transform x, v to new variables $\varepsilon$, w, defined by

$$\varepsilon = \tfrac{1}{2}(v^2 + x^2), \qquad w = \arctan(x/v).$$

(They are the action-angle variables pertaining to the undamped, unfluctuating oscillator.) The transformed equation is, with the abbreviations $s = \sin w$, $c = \cos w$,

$$\begin{aligned}\frac{\partial P(w,\varepsilon,t)}{\partial t} = {} & -\frac{\partial P}{\partial w} + \beta\frac{\partial}{\partial\varepsilon}(2\varepsilon c^2 P) - \beta\frac{\partial}{\partial w}(scP) \\ & + \frac{\alpha^2}{2}\frac{\partial}{\partial\varepsilon}\left(2\varepsilon c^2\frac{\partial P}{\partial\varepsilon}\right) - \alpha^2\frac{\partial}{\partial w}\left(sc\frac{\partial P}{\partial\varepsilon}\right) \\ & + \frac{\alpha^2}{2}\frac{\partial}{\partial w}\left(s^2\frac{\partial P}{\partial w}\right) + \frac{\alpha^2}{2}\frac{\partial}{\partial\varepsilon}(c^2 - s^2)P .\end{aligned}$$

Average this equation over w and call

$$\frac{1}{2\pi}\int_0^{2\pi} P(w,\varepsilon,t)\,dw = \overline{P}(\varepsilon,t).$$

Then all terms with $\partial/\partial w$ disappear, but this does not yet reduce the equation to one for $\overline{P}(\varepsilon,t)$. The essential step to achieve that is that one sets

$$\overline{c^2 P} = \overline{c^2}\;\overline{P} = \tfrac{1}{2}\overline{P} .$$

This approximation is based on the idea that, regardless of the precise initial $P(w,\varepsilon,0)$, after a short while fluctuations will have spread out P over all values of w between 0 and $2\pi$. The equation then reduces to

$$\frac{\partial\overline{P}(\varepsilon,t)}{\partial t} = \beta\left[\frac{\partial}{\partial\varepsilon}\varepsilon P + kT\frac{\partial}{\partial\varepsilon}\varepsilon\frac{\partial P}{\partial\varepsilon}\right] .$$

This equation describes the diffusion in the energy scale.

Now cut off the energy scale at $\varepsilon_a$: when the oscillator reaches

$\varepsilon_a$ it escapes. That means that the equation for $\bar{P}(\varepsilon,t)$ has to be solved for $0 < \varepsilon < \varepsilon_a$ with boundary condition $\bar{P}(\varepsilon_a) = 0$. We suppose that $\varepsilon_a/kT$ is small. Then an almost stationary solution exists:

$$\bar{P}^{as}(\varepsilon) \approx \bar{P}^{eq}(\varepsilon) = \frac{1}{kT} e^{-\varepsilon/kT} \qquad (\varepsilon < \varepsilon_a).$$

Following Kramers' argument

$$\beta\left[\varepsilon \bar{P}^{as} + kT\,\varepsilon \frac{\partial \bar{P}^{as}}{\partial \varepsilon}\right] = -J \ .$$

Integrate between the limits $\varepsilon_a$ and some $\varepsilon_0 < \varepsilon_a$:

$$\beta e^{\varepsilon_0/kT} \bar{P}^{as}(\varepsilon_0) = \int_{\varepsilon_0}^{\varepsilon_a} e^{\varepsilon/kT} d\varepsilon/\varepsilon \ .$$

Take $\varepsilon_0$ sufficiently low for $\bar{P}^{as}$ to be practically equal to $\bar{P}^{eq}$ and approximate the integral on the right:

$$\frac{\beta}{kT} = \frac{J}{\varepsilon_a} kT\, e^{\varepsilon_a/kT} \ .$$

Hence the probability per unit time to escape is

$$J = \frac{\beta \varepsilon_a}{(kT)^2} e^{-\varepsilon_a/kT} \ .$$

Again this is very small because of the exponential factor. In chemistry it is called the Arrhenius factor and visualized as due to an intermediate "complex" with energy $\varepsilon_a$, which has to be formed before dissociation can occur. Of course that picture does not tell anything about the other factor, but it must be admitted that the value we found for it is somewhat unrealistic, because it depends sensitively on the precise model used.

Note to p. 17 : Rather one has for one-step processes

$$\frac{dU}{ds} = \log\left(1 + \frac{1}{\beta(s)} \frac{dV}{ds}\right) = -\log\left(1 - \frac{1}{\alpha(s)} \frac{dV}{ds}\right),$$

where we have set $\beta_n = \Omega\beta(n/\Omega)$, $\alpha_n = \Omega\alpha(n/\Omega)$.

## REFERENCES

1) Basic texts for physicists are: S. CHANDRASEKHAR, Rev. Mod. Phys. 15, 1 (1943) and M.C. WANG and G.E. UHLENBECK, Rev. Mod. Phys. 17, 323 (1945). (Both are reprinted in: N. WAX, ed., Selected Papers on Noise and Stochastic Processes, Dover, New York, 1954.) The best known textbooks are A.T. BHARUCHA-REID, Elements of the Theory of Markov Processes and Their Applications (McGraw-Hill, New York, 1960), and W.A. FELLER, An Introduction to Probability Theory and its Applications (two volumes)(Wiley, New York, 1968 and 1971).

2) A. KOLMOGOROFF, Grundbegriffe der Wahrscheinlichkeitsrechnung. (Ergebn. Mathem. Grenzgeb. 2, no. 3, Springer, Berlin, 1933) = Foundations of Probability Theory (Chelsea, New York, 1950).

3) And apart from one other, rather trivial process, J.L. DOOB, Annals of Mathem. 43, 351 (1942); also reprinted in WAX, reference 1.

4) K. ITÔ, Proc. Imp. Acad. Tokyo 20, 519 (1944); R.E. MORTENSEN, J. Statist. Phys. 1, 271 (1969).

5) For a recent survey, see N.G. VAN KAMPEN, Adv. Chem. Phys. 34, 245 (I. PRIGOGINE and S.A. RICE,eds., Wiley, New York, 1976).

6) The standard model is due to F. SCHLÖGL, Z. Physik 253, 147 (1972); see also K.H. JANSSEN, Z. Physik 270, 67 (1974) and I. MATHESON, D.F. WALLS and C. GARDINER, J. Statist. Phys. 12, 21 (1975).

7) R. LANDAUER, J. Statist. Phys. 13, 1 (1975).

8) G. NICOLIS and I. PRIGOGINE, Self-Organization in Nonequilibrium Systems (Wiley, Interscience, New York, 1977).

9) Th. W. RUIJGROK and J.A. TJON, Physica 65, 539 (1973).

10) H.A. KRAMERS, Physica 7, 284 (1940) = Collected Scientific Papers (North-Holland, Amsterdam 1956), p. 754. The escape rate has been explicitly calculated for special barriers by N.G. VAN KAMPEN, J. Statist. Phys. 17, 71 (1977) and M.R. PEAR and J.H. WEINER, J. Chem. Phys., to be published. For an application, see, e.g. P.G. DE GENNES, J. Statist. Phys. 12, 463 (1975).

11) E. PRAESTGAARD and N.G. VAN KAMPEN, to be published.

12) See also U.M. TITULAER, Physica 91A, 321 (1978).

# STOCHASTIC DIFFERENTIAL EQUATIONS WITH NON-MARKOV PROCESSES

Emilio Santos
Departamento de Fisica Teórica
Universidad de Santander
Santander (Spain)

## 1. INTRODUCTION

Roughly speaking, a stochastic process is any function, $X(t,\lambda)$, of time and chance. By chance we mean that there exists some measurable set, $\Lambda \equiv \{\lambda\}$, called sample space, where a probability measure is defined. Alternatively, we may say that a process is a probability measure defined in the set of functions of some parameter, which in physical applications is usually the time. A stochastic differential equation can be seen as a set of ordinary differential equations, one for each value of $\lambda$. Therefore, a stochastic equation establishes some relationship between stochastic processes. More precise definitions can be seen in standard books on probability theory (e.g., Papoulis, 1965; Feller, 1966; Prokhorov and Rozanov, 1969), stochastic processes (e.g., Barucha-Reid, 1960; Cox and Miller, 1965; Stratonovich, 1963, 1967, 1968), and stochastic differential equations (Arnold, 1974; Soong, 1973; Gikhman and Skorokhod, (1972).

A process is Markovian if it loses quickly its memory of the past. Markov processes have nice properties which allow us to deal with them rather easily. In contrast, non-Markov processes are difficult to handle, and many books on stochastic differential equations deal only with equations involving Markov processes. Consequently, the physicist is confronted with a real problem if he has to deal with an equation with non-Markov processes. The purpose of these lectures is to give a few ideas about the way in which this class of stochastic differential equations can be handled.

The plan of the lectures is to give first a summary of properties of Markov, Gaussian and stationary processes and an introduction to stochastic differential equations as they appear in physics. After that, we shall consider briefly the methods of solution.

## 2. MARKOV AND NON-MARKOV PROCESSES

The best known example of a stochastic process is the one of random steps. For instance, suppose that at times $\{0, 1, 2, \ldots, N\}$ a particle can be at any of the spatial points $\{0, \pm 1, \pm 2, \ldots\}$ in such a way that, starting at $x = 0$ at $t = 0$, if at time $t$ it is at the point $x$, at time $t + 1$ it is either at $x + 1$ or at $x - 1$ with equal probabilities. This rule can be expressed as

$$P[X(t+1)=x+1 / X(t)=x] = P[X(t+1)=x-1 / X(t)=x] = \tfrac{1}{2} . \qquad (2.1)$$

In this example, the sample space $\Lambda$ is set of all sets, $\{x_1, x_2, \ldots, x_N\}$, of N integers. Each element $\lambda \in \Lambda$ represents a possible history whose probability is $(1/2)^N$ if

$$|x_1| = |x_2-x_1| = |x_3-x_2| = \dots = |x_N-x_{N-1}| = 1 \;, \tag{2.2}$$

and zero otherwise. It is equivalent to define $\Lambda$ as the set of elements whose probability is non-zero; that is, any set of zero probability can be added to the sample space without changing the problem essentially. This is an example of a process, $X(t,\lambda)$, with a discrete parameter, i.e., a chain. It is a Markov chain because the probability for the position at time $t+1$ depends only on the position at time $t$, with independence of the previous history. That is,

$$P[X(t+1) = x\pm1 / X(t) = x;\ \text{anything}] = P[X(t+1) = x\pm1\ X(t) = x]. \tag{2.3}$$

It is not strictly correct to say that it does not depend on the previous history; it depends on it only through the position at time t.

As an example of a non-Markov chain, consider a motion in random steps such that, if the positions at times $t-1$ and $t$ were $x-1$ and $x$, respectively, then the position at time $t+1$ is either $x+1$ or $x-1$ with probabilities $p$ or $1-p$ respectively, and if at times $t-1$ and $t$ the positions were $x+1$ and $x$, then at time $t+1$ the position is $x+1$ or $x-1$ with probabilities $1-p$ or $p$. Also, we fix the initial positions as 0,1 for $t=0,1$. In the formulae, we have

$$\begin{aligned} &P[X(t+1)=x+1|X(t)=x;\ X(t-1)=x-1] = \\ &\qquad = P[X(t+1) = x-1|X(t) = x\ ;\ X(t-1) = x+1] = p\ , \end{aligned} \tag{2.4}$$

$$\begin{aligned} &P[X(t+1)=x+1/X(t)=x;\ X(t-1)=x+1] = \\ &\qquad = P[X(t+1) = x-1|X(t) = x;\ X(t-1) = x-1] = 1-p \end{aligned} \tag{2.5}$$

The fact that eqs. (2.4) and (2.5) do not coincide with eq. (2.3) shows that the random motion is not a Markov chain in this case. It is possible to define a new process $Y(t,\lambda)$ such that

$$Y(t+1,\lambda) \equiv X(t,\lambda)\ . \tag{2.6}$$

In terms of both processes $X$ and $Y$ it is possible to write eqs. (2.4), (2.5) and (2.6) as

$$\begin{aligned} &P[X(t+1)=x+1|X(t)=x;\ Y(t)=x-1] = \\ &\qquad = P[X(t+1) = x-1|X(t) = x;\ Y(t) = x+1] = p\ , \end{aligned} \tag{2.7}$$

$$P[X(t+1)=x+1/X(t)=x;\ Y(t)=x+1] =$$
$$= P[X(t+1) = x-1/X(t) = x;\ X(t-1) = x-1] = 1-p\ , \qquad (2.8)$$
$$P[Y(t+1)=x/X(t)=x] = 1\ . \qquad (2.9)$$

These equations are similar to eqs. (2.1) and (2.3) in that they involve only two times each, which is characteristic of a Markov process. In this way we have reduced a one-dimensional non-Markov process to a two-dimensional Markov one. The property is a general one; every non-Markov chain with finite memory can be reduced to a higher dimensional Markov chain. For processes with continuous time it is no longer possible to reduce a non-Markov process to a Markov one of higher dimensionality, in general.

For a continuous process $X(t, \lambda)$, $t \in [0,T]$, the sample space is the set of all functions $\{x(t)\}$. This set is badly defined and usually additional conditions are imposed on it, such as continuity, boundedness and so on, but we will not enter into these mathematical details here. Partial information on the process is obtained if we consider the related chain (process with discrete time) $X(t, \lambda)$, $t \in \{t_1, t_2, \ldots t_n\} \subset [0,T]$. The chain is completely defined by the probability

$$P[X(t_1) \in I_1,\ X(t_2) \in I_2,\ \ldots\ ,\ X(t_n) \in I_n]\ , \qquad (2.10)$$

where $I_1, \ldots I_n \subset R$ are intervals in the real line. As more and more probabilities of this type are known, the better defined is the process. In the limit it is considered that the process is completely defined if for every integer $n$ the probabilities (2.10) are known for all sets $\{t_1 \ldots t_n\}$ and $\{I_1 \ldots I_n\}$. This shows that a general stochastic process with a continuous parameter is a rather complex thing. For simplicity we shall assume that a probability density, $\rho(x_1,t_1\ ;\ x_2,t_2\ ;\ \ldots;x_n,t_n)$, exists which is related to the probability (2.10) by the integral

$$P[X(t_1) \in I_1, \ldots ] = \int_{I_1} dx_1 \ldots \int_{I_n} dx_n \rho(x_1,t_1\ ;\ldots\ x_n,t_n). \qquad (2.11)$$

From the definition of conditional probability, we have

$$\rho(x_1,t_1\ ;\ldots;\ x_n,t_n) =$$
$$\rho(x_1,t_1\ ;\ ..;x_{n-1},t_{n-1})\ \rho(x_n,t_n/x_1,t_1\ ;..;x_{n-1},t_{n-1})\ . \qquad (2.12)$$

Therefore, knowledge of the probability density at a single time, say $t = 0$ , and all conditional probabilities like the one in eq. (2.12), also defines the stochastic process.

A Markov process is such that the following relation, similar to eq. (2.3), holds

$$\rho(x_n,t_n|x_1,t_1;\ldots;\, x_{n-1},t_{n-1}) = \rho(x_n,t_n|x_{n-1},t_{n-1}),\ t_1 < t_2 \ldots < t_n. \tag{2.13}$$

Also, the two-time conditional probability, also called "transition" probability, fulfills the Chapman-Kolmogorov equation

$$\int_{-\infty}^{\infty} dx''\ \rho(x',t'|x'',t'')\ \rho(x'',t''|x,t) = \rho(x',t'|x,t). \tag{2.14}$$

As a consequence, Markov processes have the property that they are fully defined by giving just the probability density, $\rho(x,0)$, at the initial time, and the differential transition probability, $\rho(x+\Delta x,t+\Delta t|x,t)$. For instance, the stochastic process (called death process) associated with the decay of a set of radioactive atoms is a Markov process fully defined by

$$\rho(x,0) = \delta(x-N)\ ;\ \rho(x+\Delta x,t+\Delta t|x,t) =$$

$$(1-\lambda x\Delta t)\ \delta\ (\Delta x) + \lambda\, x\Delta t\ \delta(\Delta x+1) + O(\Delta t)\ ; \tag{2.15}$$

where the integer N is the number of atoms at the initial time. This is a process with a discrete variable ($x \in \{0,1..,N\}$) but we write eqs. (2.15) in continuous notation using Dirac´s deltas in order to unify the treatment.

A consequence of the Chapman-Kolmogorov equation is that the probability density $\rho(x,t)$ of a Markov process fulfills the following linear partial differential equation

$$\partial\rho(x,t)/\partial t = \sum_{k=1}^{\infty} (k!)^{-1}(-\partial/\partial x)^k\left[D_k(x,t)\ \rho\ (x,t)\right], \tag{2.16}$$

where

$$D_k(x,t) = \lim_{\Delta t\to 0} \langle \Delta x^k/\Delta t\rangle = \lim_{\Delta t\to 0} \int_{-\infty}^{\infty} \rho(x+\Delta x,t+\Delta t/x,t)\ \Delta x^k d\,\Delta x. \tag{2.17}$$

For a process in several dimensions there exists a straightforward generalization of this expression. For continuous processes, eq. (2.16) has no derivatives of order higher than two and is called the Fokker-Planck equation. For instance, this is not the case for the death process defined by eqs. (2.15) but it is true for a Wiener process. The Wiener process is a Markov process used in the theory of Brownian motion, defined by

$$\rho(x,0)=\delta(x)\ ;\ \rho(x+\Delta x,t+\Delta t/x,t)=(2\pi D\Delta t)^{-\frac{1}{2}}\exp[-(\Delta x)^2/2D\Delta t] \ . \tag{2.18}$$

As is well-known, it is the limit of the random-steps process defined by eq. (2.4) when the steps become shorter and more frequent. In this case, eq. (2.16) reduces to the Fokker-Planck equation that is the well-known diffusion equation.

## 3. GAUSSIAN AND STATIONARY PROCESSES

Besides the Markov property there are two other simplifying properties which stochastic processes may have: they may be Gaussian and/or stationary. Processes which are stationary and Gaussian, but non-Markov, may give us some hints about the ways of dealing with general non-Markov processes.

A process is called Gaussian (or normal) if every probability density like $\rho(x_1,t_1\ ;\dots;x_n,t_n)$ is a Gaussian function, i.e., the negative of its logarithm is a positive-definite quadratic function of $\{x_1,\dots,x_n\}$ . Defining the mean, $a(t)$, and the autocorrelation function, $C(t,t')$ , of the process, respectively, by

$$a(t) \equiv \langle X(t)\rangle\ ,\ C(t,t') \equiv \langle X(t)X(t')\rangle\ , \tag{3.1}$$

we can show that

$$\begin{aligned}&\rho(x_1,t_1\ ;\dots;\ x_n,t_n) = \\ &= (2\pi)^{-n}\int_{-\infty}^{\infty} d\eta_1\dots\int_{-\infty}^{\infty} d\eta_n \exp\Big[\sum_j\ i\ \eta_j(x_j-a(t_j)) - \\ &-\tfrac{1}{2}\sum_{j,k} C(t_j,t_k)\ \eta_j\ \eta_k\Big]\ .\end{aligned} \tag{3.2}$$

As a result, Gaussian processes are relatively simple; they are characterized by just two functions: the mean and the autocorrelation. For instance, the Wiener process, besides being Markovian, is a Gaussian process with

$$a(t) = 0\ ;\ C(t,t') = D\ \min(t,t')\ . \tag{3.3}$$

In contrast, the death process obeying eqs. (1.15) is Markovian but non-Gaussian. On the other hand, there are processes which are Gaussian but non-Markovian, as will be shown in a moment.

A process is said to be stationary in the strict sense if

$$\rho(x_1,t_1;\ldots x_n,t_n) = \rho(x_1,t_1+\tau;\ldots;x_n,t_n+\tau) , \tag{3.4}$$

for arbitrary $\tau$. It is said to be stationary in the broad sense if the mean is independent of $t$ and the autocorrelation depends only on the difference $t'-t$. It is easy to see that strict-sense stationarity implies broad-sense stationarity. The converse is not true, in general, but it is true for Gaussian processes, as is easy to see from eq. (3.2). There is a theorem by Doob which states that if a process is Markovian, Gaussian and stationary, it is an Ornstein-Uhlenbeck process, characterized by the autocorrelation function

$$C(t,t') = D \exp\left[-|t'-t|/t_c\right] , \tag{3.5}$$

where $D$ and $t_c$ are constants.

For a stationary process it is possible to define time averages as well as ensemble averages. For instance, the time-averaged autocorrelation of the process $X(t,\lambda)$ is given by

$$\bar{C}(s,\lambda) \equiv \lim_{\tau\to\infty} \tau^{-1} \int_0^{\tau} X(t,\lambda)\, X(t+s,\lambda)\, dt . \tag{3.6}$$

The time-averaged autocorrelation depends on $\lambda$, i.e., it is different in principle for each sample of the process. For some stationary processes time averages are independent of $\lambda$, or, more correctly, any time average has the same value for all samples $\lambda \in \Lambda$ except a set of probability zero. These processes are called ergodic. From this definition, it can be shown that, for an ergodic process, ensemble averages coincide with time averages almost certainly (i.e., except for a set of zero probability). In particular, the time-averaged autocorrelation, $\bar{C}(s,\lambda)$, coincides with the ensemble-averaged one, $C(s)$.

The concept of stationarity is too restrictive for some purposes, and it is possible to generalize it by introducing the concept of processes with stationary increments. We shall say that a process $X(t,\lambda)$ has stationary increments if, for any $\tau$, the process $X(t+\tau,\lambda) - X(t,\lambda)$ is stationary. This implies that the mean square increment,

$$D(t,t') \equiv \langle [x(t')-x(t)]^2 \rangle = C(t',t') + C(t,t) - 2C(t,t'), \tag{3.7}$$

depends only on the difference $t'-t$. If a process is stationary, then it has stationary increments, but the converse is not true. An example of a non-stationary process with stationary increments is the Wiener process defined by eq. (2.18).

For some processes it is possible to define the derivative. The derivative of the process $X(t,\lambda)$, if it exists, is another process, $\dot{X}(t,\lambda)$ , such that for each sample $\lambda \in \Lambda$ the function $\dot{X}(t)$ is the derivative of $X(t)$ . From the very definition, the derivative of a process with stationary increments is a stationary process and, conversely, the integral of a stationary process has stationary increments. It is also easy to deduce that if $C_x(t,t')$ is the autocorrelation of the process $X(t)$ , then the autocorrelation of $\dot{X}(t)$ is

$$C_{\dot{x}}(t,t') = \partial^2 C_x(t,t')/\partial_t \partial_{t'} \ . \tag{3.8}$$

Finally, the derivative of a Gaussian process is also Gaussian. It is interesting to consider the stationary process which is the derivative of the Wiener process. If it exists, its autocorrelation function must be, according to eqs. (3.3) and (3.8),

$$C(t,t') = D\ \delta(t'-t)\ . \tag{3.9}$$

But this means that, at any time, the process has a finite probability of not being finite, which is absurd. This is simply another way of saying that the Wiener process is not differentiable, which is well-known. In actual fact, there are processes which fulfill eq.(3.9) approximately. For reasons to be seen later, they are called white noise. An example of a process of this kind is the Ornstein-Uhlenbeck process for small enough $t_c$ (see eq. (3.5)). The integral of this process, with the condition that at time zero it takes the value zero, is a Gaussian process which has some similarity to the Wiener process whenever we are interested in its behaviour for time intervals much greater than $t_c$ . This is an example of a process which is non-Markovian but may be approximated by a Markov process, namely, the Wiener process.

The idea of approximating a non-Markov process by a Markov one has been used ever since the very early work on stochastic processes. Einstein, in his paper on Brownian motion in 1905, introduced what is now called the Wiener process by the following argument. We may assume that there is a time interval, $\Delta t$, which is so large that the Brownian particle has been subject to many collisions during $\Delta t$ and forgets its initial velocity completely. Nevertheless, $\Delta t$ is short enough to be treated as infinitesimal in comparison with the time of observation. This is equivalent to assuming that the velocity of a Brownian particle is a stationary process with an autocorrelation which can be approximated by eq. (3.9). This type of reasoning has led some people to believe that every stochastic process can be approximated by a Markov process, which is not true in general.

## 4. STOCHASTIC DIFFERENTIAL EQUATIONS

The usual way in which stochastic processes appear in physics is as follows. Any classical (i.e., non-quantal) physical system is described by a set of dynamical variables $\{q_1,\dots,q_r\}$ whose change with time determines the evolution of the system. For simplicity we shall assume that the set of variables is finite. The evolution is usually governed by a set of first-order differential equations of the type

$$\dot{q}_\nu = A_\nu(q_1,\dots,q_r), \qquad \nu = 1,\dots,r\ . \tag{4.1}$$

For instance, our q's may include the coordinates and momenta of a dynamical system and eqs. (4.1) may be Hamilton's equations. If it is possible to deal with all the variables together, then we are out of the domain of statistical physics and stochastic processes play no role. However, in many cases the set of variables is too large and, moreover, there is a small subset $\{q_1,\dots,q_s\}$ which contains the relevant parameters. In this case it is usually possible to divide the functions $A_\nu$ of eq. (4.1) into two parts, $H_\nu$ and $K_\nu$, so that the first s equations read

$$\dot{q}_\nu = H_\nu(q_1,\dots,q_s) + K_\nu(q_1,\dots,q_s,q_{s+1},\dots,q_r). \tag{4.2}$$

Next, we may replace the set of functions $\{q_{s+1}(t),\dots,q_r(t)\}$ by a suitable stochastic parameter $\lambda \in \Lambda$ , so that eqs. (4.2) are transformed into the set of stochastic differential equations

$$\dot{q}_\nu = H_\nu(q_1,\dots,q_s) + K_\nu(q_1,\dots,q_s;\lambda)\ . \tag{4.3}$$

The adequacy of the set of equations (4.3) depends on our ability to define a suitable probability measure in the sample space .

A typical example of this procedure is the Langevin formulation of Brownian motion. The motion of a Brownian particle in a liquid is an extremely complex problem which involves the coordinates and momenta of all molecules in the system. Nevertheless, we may assume that the only relevant parameters are the center-of-mass coordinates of the particle. Then, the motion can be described by the Langevin equation

$$\ddot{r} = -\gamma r + F(r,\dot{r},t;\lambda)\ , \tag{4.4}$$

which can be reduced easily to a set of six equations of the type (4.3).

The real problem is to give a suitable measure in the set of the functions $F(r,\dot{r},t)$ . For this, Langevin made two rather drastic assumptions :

a) For an adequate choice of $\gamma$ , the function $F$ depends only on $t$ (not on $r$ and $\dot{r}$ ), and

b) $F(t,\lambda)$ is a white-noise process.

The derivation of these assumptions from microscopic dynamics (in general, the derivation of eqs. (4.3) from eqs. (4.2)) is by no means trivial, and we shall not be concerned with this problem here. We shall consider, instead, the problem of solving stochastic differential equations of the type of eqs. (4.3). The solution of a stochastic differential equation consists in characterizing in full the stochastic process $q_\nu(t,\lambda)$ . That is, we must be able to calculate any probability like (2.10) with

$$\{q_1(t),\dots,q_s(t)\} \equiv X(t) \in R^s \quad \text{and} \quad I_k \subset R^s \ .$$

This is the complete solution of the stochastic equation but sometimes it is only possible and/or interesting to calculate the probability at a single time. If we assume that a probability density $\rho(q_1,\dots,q_s,t)$ exists, then the solution of the stochastic equation amounts to calculating it.

In accord with the title of this set of lectures, we shall exclude methods of solution which are specific for equations involving Markov processes. The methods adequate for non-Markov processes can be put in three classes, which we describe in the following. In the first place there are stochastic differential equations which are linear and have nonrandom coefficients. These can be solved by harmonic analysis. This method is well-known and can be found in standard textbooks, so that we shall give only a summary of it with the purpose of introducing some concepts (like Markov limit or Markov approximation) that will be considered later for nonlinear equations. Methods in the second class take advantage of the fact that the second term, $K_\nu$ , of eq. (4.3) is usually smaller than the first one, $H_\nu$ . As these methods involve some small parameter they might be called perturbation methods, but this name is usually reserved for cases in which the solution is obtained as an expansion in the small parameter. Finally, there is a very general, but frequently too involved, method resting upon integration in the functional sample space. It might be called the path-integral method, in view of the analogy with the Feynman formulation of quantum mechanics.

## 5. HARMONIC ANALYSIS

A way of studying linear equations involving stationary Gaussian processes (and, to some extent, nonstationary and/or non-Gaussian ones) is harmonic analysis, also called the Rice method (Rice, 1944, 1945). For any stationary process, in the broad sense, a function $S(\omega)$ is defined, called the spectral density or power spectrum, which is the Fourier transform of its autocorrelation function, i.e.,

$$C(s) \equiv \int_0^\infty S(\omega)\cos(\omega s)d\omega \ , \quad s \equiv t'-t \ . \tag{5.1}$$

If we assume that the process is defined in the time interval $[0, \infty]$, then $C(s)$ is defined for all positive values of $s$ and eq. (5.1) may be inverted to obtain $S(\omega)$ explicitly. Taking into account that a stationary Gaussian process with zero mean is fully characterized by the autocorrelation function, it is also fully defined by the spectrum. For a stationary ergodic process the spectrum is also related to the time-averaged autocorrelation function by eq. (5.1), which is called, in this case, the Wiener-Khintchine relation.

For a process with stationary increments it is also possible to define the spectrum by means of

$$D(s) = 2\int_0^\infty S(\omega)\,(1 - \cos\omega s)d\omega \ , \tag{5.2}$$

where $D(s=t'-t)$ was defined by eq. (3.7). It is easy to see that the previous definition of the spectrum, eq. (5.1), implies the new definition. Eq. (5.2) can be inverted by considering the inverse cosine transform of $D(\infty) - D(s)$. For instance, according to this definition, the spectrum of the Wiener process can be found to be

$$G_W(\omega) = 2D/\pi\,\omega^2 \ . \tag{5.3}$$

If a process $X(t)$ has stationary increments, its spectrum $S_X(\omega)$ and the spectrum $S_{\dot{X}}(\omega)$ of $\dot{X}(t)$ are related by

$$S_{\dot{X}}(\omega) = \omega^2 S(\omega)$$

For instance, the spectrum of the derivative of the Wiener process should be

$$S_{\dot{W}}(\omega) = 2D/\pi \ , \tag{5.4}$$

i.e., a white spectrum. We said that such a process does not exist, but a process with a white spectrum up to very high frequencies does exist, and is called white noise.

The concept of the spectrum is very useful in the solution of linear stochastic differential equations with constant nonrandom coefficients. Instead of giving the general theory of the Rice method, we shall illustrate it with an example. Consider the motion of a damped harmonic oscillator under the action of a random force $F(t)$. The equation of motion is

$$\ddot{x} = -\omega_0^2 x - \gamma\dot{x} + F(t) \tag{5.5}$$

which is a linear relation between a stochastic process, $F(t)$, whose properties are assumed known, and another process, $x(t)$, whose characteristic properties must be determined. In general, any stochastic differential equation connects some known processes with unknown ones. If the equation is linear, like eq. (5.5), and all the known processes are Gaussian, then all the unknown processes are also Gaussian because the Gaussian property is preserved by linear transformations. In this case the Rice method can give the general solution of the equation (if it exists). If the known processes are not Gaussian then the Rice method cannot give the general solution, but it provides some useful information.

If the known processes are stationary and we are searching for stationary solutions or solutions with stationary increments, then the Fourier transform of the equation provides relations between the spectra of the known and the unknown processes. For instance, from eq. (5.5) we obtain

$$(-\omega^2 + \omega_0^2 + i\gamma\omega)\,\tilde{X}(\omega) = \tilde{F}(\omega)\ . \tag{5.6}$$

Hence, taking into account that the spectra of processes are proportional to the square moduli of their Fourier transforms, we have

$$S_x(\omega) = \frac{S_F(\omega)}{(\omega^2 - \omega_0^2)^2 + \gamma^2\omega^2}\ , \tag{5.7}$$

which characterizes the stationary process $x(t)$ (assuming that $F(t)$ has zero mean).

If we want nonstationary solutions, we may use the Laplace transform instead of the Fourier transform. However, there is a more direct method, which consists in putting all known processes to the left of the equation and all unknown ones to the right and taking the two-time correlations on both sides. For instance, from eq. (5.5) we obtain

$$\langle [\ddot{x}(t)+\omega_0^2 x(t)+\gamma\dot{x}(t)]\,[\ddot{x}(t')+\omega_0^2 x(t')+\gamma\dot{x}(t')]\rangle = \langle F(t)F(t')\rangle . \qquad (5.8)$$

Hence, if we define $C_x(t,t')$, $C_F(t,t')$ as the autocorrelations of $x(t)$ and $F(t)$, it follows that

$$\left(\frac{\partial^2}{\partial t^2} + \omega_0^2 + \gamma\frac{\partial}{\partial t}\right)\left(\frac{\partial^2}{\partial t'^2} + \omega_0^2 + \gamma\frac{\partial}{\partial t'}\right) C_x(t,t') = C_F(t,t') . \qquad (5.9)$$

This is a linear partial differential equation which can be solved by standard methods to find the autocorrelation of $x(t)$ . On the other hand, the mean of $x(t)$ can be obtained from the equation

$$d^2\langle x\rangle/dt^2 = -\omega_0^2\langle x\rangle - \gamma d\langle x\rangle/dt + \langle F\rangle . \qquad (5.10)$$

In conclusion, knowledge of the mean and the autocorrelation of the process $F(t)$ allows the calculation of the mean and the autocorrelation of the process $x(t)$ . If $F(t)$ is Gaussian, then $x(t)$ is also Gaussian and is fully characterized by these two functions.

## 6. AN EXAMPLE OF THE MARKOV LIMIT

As an introduction to the perturbation methods to be considered later, it is interesting to study eq. (5.5) when the damping and the stochastic force are small. It is to be expected that some expansion in the small parameter $\gamma$ may give approximate solutions which are simpler than the exact one. If one tries to calculate the zeroth-order term naively one must take the limit of eq. (5.7), with the result

$$\lim_{\gamma\to 0} S_x(\omega) = (\pi/2\omega_0^2)\,\delta(\omega-\omega_0)\,\lim_{\gamma\to 0}[S_F(\omega_0)/\gamma] . \qquad (6.1)$$

This shows that a necessary condition for the existence of stationary solutions is that $S_F(\omega_0)/\gamma$ remain finite in the limit $\gamma\to 0$ . From eq. (6.1) we obtain the autocorrelation function

$$C_x(t,t') = \lim_{\gamma\to 0}\left[\pi S_F(\omega_0)/2\gamma\omega_0^2\right]\cos[\omega_0(t'-t)] , \qquad (6.2)$$

which describes the stationary solution as a set of undamped unperturbed oscillators whose mean square amplitude is given by $\lim_{\gamma\to 0}[\pi S_F(\omega_0)/\gamma\omega_0^2]$. If we want to know the approach to the stationary state, a somewhat more sophisticated method is needed.

In the first place, in order to reduce eq. (5.5) to the standard form of eqs. (4.3), it is convenient to write, instead of eq. (5.5), a system of two first-order equations. This can be done in several ways. A useful way is the following:

$$v(t) \equiv \dot{x}(t) + (\gamma/2)x(t); \ \dot{v}(t) = (\gamma^2/4-\omega_0^2)x(t) - (\gamma/2)v(t) + F(t). \tag{6.3}$$

The stochastic process $v(t)$ is essentially the velocity, because we will take the limit $\gamma \to 0$ afterwards; the definition of eqs. (6.3) has the purpose of simplifying the calculations. Eqs. (6.3) have the form of eqs. (4.3) with

$$q_1 = x \quad q_2 = v, \ H_1 = v, \ H_2 = -\alpha^2 x, \ K_1 = -(\gamma/2)x, \ K_2 = -(\gamma/2)v + F. \tag{6.4}$$

The way in which the various terms have been distributed amongst $H$ and $K$ is a matter of convenience, apart from the condition that both $K_1$ and $K_2$ must be small. Note that the constant

$$\alpha \equiv (\omega_0^2 - \gamma^2/4)^{\frac{1}{2}} , \tag{6.5}$$

which appears in $H_2$, is the imaginary part of the characteristic roots of eq. (5.5).

If $K_1$ and $K_2$ were zero, the solution of eqs. (6.3) would be

$$x(t) = x_0 \cos\alpha t + (v_0/\alpha) \sin\alpha t, \ v(t) = -x_0 \alpha \sin\alpha t + t_0 \cos\alpha t , \tag{6.6}$$

with $\{x_0, v_0\}$ some constants. This suggests that we introduce two new processes $\{x_0(t), v_0(t)\}$, related to $\{x(t), v(t)\}$ by eqs. (6.6) even if $K_1$ and $K_2$ are not zero. These stochastic processes change with time according to the equations

$$x_0(t) = \alpha^{-1}e^{-\gamma t/2}\int_0^t F(t')e^{\gamma t'/2}\sin(\alpha t')dt' + X e^{-\gamma t/2},$$

$$v_0(t) = e^{-\gamma t/2}\int_0^t F(t')e^{\gamma t'/2}\cos(\alpha t')dt' + V e^{-\gamma t/2} , \tag{6.7}$$

where $X$ and $V$ are the initial position and velocity. Now $x_0(t)$ and $v_0(t)$ can be interpreted physically as the position and velocity which the oscillator should have at time $t = 0$ in order to move deterministically (i.e., with $K_1 = K_2 = 0$) to the point to which it actually moves, starting from $x(0) = X$, $v(0) = V$, by the action of both the deterministic $(H_1,H_2)$ and the stochastic $(K_1,K_2)$ forces. Alternatively, we may say that $\{x_0(t), v_0(t)\}$ is the point in phase space at which the oscillator arrives if it moves from time $0$ to time $t$ under the action of all forces and then moves from $t$ to $2t$ under the action of the time-reversed deterministic forces. The advantage of the new processes $\{x_0(t), v_0(t)\}$ compared with the old ones $\{x(t), v(t)\}$ is that the new processes change only by the action of the stochastic forces, and, therefore, change slowly if these forces are small. The reader may realize the analogy with the interaction picture in quantum mechanics.

Eqs. (6.6) and (6.7), which are exact, give an explicit solution of eq. (5.5). Nevertheless, starting with a general system of stochastic equations of the type of eqs. (4.3), this is not possible, except in the limit $K_\nu \to 0$ . We shall perform this limit as an illustration, although in this particular case it is not really needed since we can calculate the exact solution from eqs. (6.7). We shall obtain the limit in which the damping constant $\gamma$ goes to zero with the condition that $\gamma^{-\frac{1}{2}}F(t)$ and $\gamma t \equiv \tau$ remain finite. This means that the stochastic forces go to zero, but at the same time they act during a time tending to infinity. (Note that, for convenience, the damping term $\gamma\dot{x}$ has been included in the stochastic part instead of in the deterministic part; it could be included in the deterministic part.)

In the first place, we shall obtain the averages of the position and velocity, which are, trivially,

$$\langle x_0(t)\rangle = X\, e^{-\tau/2},\ \langle v_0(t)\rangle = V\, e^{-\tau/2}\ , \tag{6.8}$$

where it has been assumed that the stochastic force $F(t)$ has zero mean. Next we should obtain the correlations of $x_0$ and $v_0$ , but it is simpler, and sufficient for our purposes, to calculate the mean square increment of $x_0$ and $v_0$ from time 0 to time $t$ . From eq. (6.7b) we obtain

$$\Delta v^2 \equiv \langle [v_0(t)-V]^2\rangle = V^2(e^{-\gamma t/2}-1)^2 +$$
$$e^{-\gamma t}\int_0^t dt' \int_0^t dt'' \langle F(t')\,F(t'')\rangle\, e^{\gamma(t'+t'')/2} \cos(\alpha t')\cos(\alpha t'').$$

By introducing the spectrum of $F(t)$ through eq. (5.1) we obtain for the second term, after changing the order of integration,

$$(1/4)e^{-\gamma t}\int_0^\infty d\omega\, S_F(\omega)\, \Big|\int_0^t dt'\, e^{(i\omega+\gamma/2)t'}\cos(\alpha t')\Big|^2 =$$

$$= (1/4)\int_0^\infty d\omega\, S_F(\omega)\, \left|\frac{e^{i(\omega-\alpha)t} - e^{-\gamma t/2}}{i(\omega-\alpha)+\gamma/2}\right|^2 ,$$

where the last equality has been written with the neglect of a term having $\omega+\alpha$ in place of $\omega-\alpha$ . This can be neglected because it would contribute only for $\omega$ near $-\alpha$, and the integral in $\omega$ goes only over positive values. (This approximation can be shown to be exact in the limit $t\to\infty$ ). Hence, we obtain

$$v^2 = (1/4)\int_0^\infty d\omega\, S_F(\omega)\, \frac{1+e^{-\tau}-2e^{-\tau/2}\cos[(\omega-\alpha)\tau/\gamma]}{(\omega-\alpha)^2+\gamma^2/4} +$$

$$+ V^2(1-e^{-\tau/2})^2 , \qquad \tau \equiv t\gamma . \tag{6.9}$$

At this point, we perform the limit $\gamma \to 0$ previously mentioned, with $\tau$ a constant and $S_F/\gamma$ remaining finite. To do this, we first realize that the limit of the expression under the integral sign is zero except for $\omega = \alpha$ . Next, we can take $S_F(\omega = \alpha)/\gamma$ outside the integral and put the lower limit equal to $-\infty$. After that, the calculation of the integral is straightforward and we obtain

$$\lim_{\gamma\to 0} \Delta v^2 = \lim_{\gamma \to 0} \left[\pi S_F(\omega_0)/2\gamma\right] \; (1-e^{-\tau}) + V^2(1-e^{-\tau/2})^2 , \tag{6.10}$$

where it has been taken into account that $\alpha$ goes to $\omega_0$ for $\gamma \to 0$ (eq. (6.5)). By a similar procedure it can be shown that

$$\lim_{\gamma\to 0} \Delta x^2 = \lim_{\gamma\to 0} \left[\pi S_F(\omega_0)/2\gamma\, \omega_0^2\right](1-e^{-\tau}) + X^2(1-e^{-\tau/2})^2 , \tag{6.11}$$

$$\lim_{\gamma \to 0} \langle x_0(t)\, v_0(t)\rangle = XVe^{-\tau} . \tag{6.12}$$

If the force $F(t)$ is Gaussian the processes $x_0(t)$ and $v_0(t)$ are also Gaussian, and eqs. (6.8), (6.10) and (6.11) can be recognized as defining two Ornstein-Uhlenbeck processes. In this case, it is easy to show, from eqs. (2.17), (6.8) and (6.10-12), that the probability density $\rho(x_0, v_0, t)$ fulfills the Fokker-Planck equation :

$$\frac{\partial \rho}{\partial \tau} = \frac{1}{2}\,\frac{\partial(x_0\rho)}{\partial x_0} + \frac{1}{2}\,\frac{\partial(v_0\rho)}{\partial v_0} + \beta\frac{\partial^2\rho}{\partial x_0^2} + \beta\,\omega_0^2\,\frac{\partial^2\rho}{\partial v_0^2} ,$$

$$\beta \equiv (\pi/4\,\omega_0^2) \lim_{\gamma\to 0} \left[S_F(\omega_0)/\gamma\right] . \tag{6.13}$$

The results obtained can be summarized as follows : The processes $x_0(t)$ and $v_0(t)$ , related by eqs. (6.6) to the position and the velocity of the harmonic oscillator, are stochastic non-Markov processes, except in special cases (e.g., if $F(t)$ is white noise). Nevertheless, they can be approximated by a Markov process if :

1) the damping and the random force are so small that they produce almost no change in $x_0$ and $v_0$ during a period of the deterministic motion, and
2) we are interested in time intervals at least of the order of the period of the deterministic motion. More precisely, we have in the

problem three characteristic times : $t_0 \equiv 1/\omega_0$ is the period of the deterministic motion, $t_s \simeq 1/\gamma$ is the time necessary for the stochastic forces to produce a significant change, and, finally, $t_c$ is the correlation time of the random force. These times are to be compared with the time interval $t$ in which we are interested. The conditions we have used in the derivation of eqs. (6.10-11) are the following :

$$t_c \lesssim t_0 \ll t_s \ , \quad t_0 \ll t \tag{6.14}$$

In fact, the condition that $S_F(\omega_0)/\gamma$ is finite means that the random force, besides being small enough, has a spectrum that contains, appreciably, frequencies of order $\omega_0$ or larger (if this were not so, either $S_F(\omega_0)/\gamma$ would be zero or $S_F(\omega/\omega \ll \omega_0)/\gamma$ would be too large). This means that the correlation time of the random force cannot be much larger than the period of the deterministic motion, which implies the first eq. (6.14). The second and third eqs. (6.14) were used in going from eq. (6.9) to eq. (6.10), where we assumed $\omega_0 \gg \gamma$. The existence of a Markov limit in these conditions can be understood on an intuitive basis. It means that every (non-Markov) process with memory loses its memory after a large-enough time; that is, the limit can exist only if $t \gg t_c$, which is clearly contained in eqs. (6.14). But this is not sufficient, because if we had $t_c \gg t_0$, there might be a kind of resonance between the stochastic and the deterministic motion, which, perhaps, would imply a much more protracted loss of memory. The condition $t_s \lesssim t_0$ might have the same effect.

A concept related to the Markov limit is that of the Markov approximation. For instance, we may use eq. (6.13), without taking the limit $\gamma \to 0$, as an approximation valid in the conditions expressed by eqs. (6.14). Then it is possible to return to the variables $x$ and $v$, related to $x_0$ and $v_0$ by eqs. (6.6). To do that we introduce the function $f(x,v,t)$ by

$$f(x(x_0,v_0,t), \quad v(x_0,v_0,t), \ t) = \rho(x_0,v_0,t) \ . \tag{6.15}$$

From eqs. (6.6), (6.13) and this one, it is straightforward to obtain

$$\frac{\partial f}{\partial t} = \frac{\partial}{\partial x}\left[\left(\frac{\gamma x}{2} - v\right) f\right] + \frac{\partial}{\partial v}\left[\left(\frac{\gamma v}{2} + \omega_0^2 x\right) f\right] + \frac{\pi S_F(\omega_0)}{4\,\omega_0^2}\left(\frac{\partial^2 f}{\partial x^2} + \omega_0^2 \frac{\partial^2 f}{\partial v^2}\right) \tag{6.16}$$

This Fokker-Planck equation defines the Markov approximation to the exact evolution equation for the probability density $f(x,v,t)$. This is a first-order (in $\gamma$ ) approximation, whilst the purely deterministic (Liouville) equation can be considered to be of zeroth order.

The concept of Markov limit has been the subject of active research in the last few years. Although the basic idea was used already by Einstein in his early work on Brownian motion, it seems that much attention was not paid to it in relation to stochastic differential equations until recently. The problem has been considered from a mathematical point of view by Lax (1966), Khas′minskiĭ (1966), Stratonovich (1968), Papanicolaou et al. (1971, 1972, 1973), and Coburn and Hersch (1973). In the domain of stochastic differential equations, the Markov limit may prove itself to be of an importance comparable to that of the "central limit theorem" in the domain of random variables. The Markov limit gives rise to a set of methods of approximation for the solution of stochastic differential equations, and these are considered in the following.

## 7. THE APPROXIMATE FOKKER-PLANCK EQUATION

We consider the solution of the stochastic differential equation

$$\dot{q} = H(q,t) + \varepsilon K(q,t;\lambda) , \tag{7.1}$$

which is of the type of eqs. (4.3). We insert the parameter $\varepsilon$ in front of $K$ in order to give a measure for the smallness of the stochastic term. This parameter plays the role of $\gamma^{\frac{1}{2}}$ in the example of the oscillator. In due time we shall consider the limit $\varepsilon \to 0$ . For the sake of simplicity we shall be concerned with a single variable, $q$ , although the generalization to several variables is straightforward. In some conditions (essentially those given in eqs. (6.15)), the solution of eq. (5.1) may be approximated by a continuous Markov process. Then, the probability density $f(q,t)$ fulfills a Fokker-Planck equation which we are going to derive. There are a number of methods for the solution of stochastic differential equations, which amount to deriving an approximate equation of the Fokker-Planck. Relevant papers are : Brissaud and Frisch (1973), Bourret (1962), Frisch (1968); Haken (1975), Keller (1962, 1964), Kubo (1963), 1969), Lax (1966), Terwiel (1974), and Zwanzig (1960, 1964). We do not attempt to give a review of these methods, partly because a recent paper, which reviews most of them, has been published by van Kampen (1976). We shall give only a few basic concepts in this section and the following one.

In order to obtain a Fokker-Planck equation similar to eq. (6.17), we begin by defining the stochastic process $q(t;\lambda)$ as the one fulfilling eq. (7.1) with the initial condition $q(0) = Q$ . Then we introduce the process $q_0(t)$ , which represents the value obtained at time 0 , by solving the deterministic equation

$$\dot{q} = H(q,t) \tag{7.2}$$

backwards with the final condition $q(t)$ at time $t$ . It can be shown that $q_0(t)$ fulfills the equation

$$\dot{q}_0(t;\lambda) = \varepsilon[dq_0(t)/dq(t)] \; K(q(q_0(t;\lambda),t),t;\lambda) \; , \tag{7.3}$$

where, to be clear enough, we write all functional dependences explicitly. We assume that the function $q_0(q,t)$ can be found by solving the deterministic eq. (7.2) backwards. The probability density $\rho(q_0,t)$ fulfills (approximately) the Fokker-Planck equation :

$$\frac{\partial\rho}{\partial t} = -\frac{\partial}{\partial q_0}(B\rho) + \frac{1}{2}\frac{\partial^2}{\partial q_0^2}(D\rho) \; , \tag{7.4}$$

where the drift coefficient, B , and the diffusion coefficient, D , must be found from the mean increment and the mean square increment of $q_0$ in an infinitesimal time interval t (see eq. (2.16)). By comparison with the example of the oscillator, we see that the Markov limit is obtained, if it exists, by first taking the limit $\varepsilon \to 0$ with $\tau \equiv \varepsilon^2 t$ a constant (which amounts to taking $t \to \infty$) , and then taking $\tau \to 0$ .

The mean change in $q_0$ is , from eq. (7.3) ,

$$\langle q_0(t)-Q\rangle = \varepsilon\int_0^t dt' \left\langle \frac{dq_0(t')}{dq(t')} \; K\,(q(q_0(t'),t'))\right\rangle_{q_0(t')\to Q} \; . \tag{7.5}$$

where, in putting $q_0(t') = Q$ , we neglect terms of order higher in $\varepsilon$. The meaning of the derivative $dq_0/dq$ is clear if we take into account that for a fixed time interval $t'$ the deterministic eq. (7.2) defines a mapping $q_0 \to q(q_0)$ , which we assume to be one-to-one. The derivative must be calculated as an explicit function of $q_0$ and $t'$ and then Q must be substituted for $q_0(t')$ . It should be remarked that a condition for the result of eq. (7.5) to be finite in the limit $\varepsilon \to 0$ , $\tau \equiv \varepsilon^2 t$ finite, is that the expression under the integral be of order $\varepsilon$ .

The mean square increment is

$$\langle[q_0(t)-Q]^2\rangle = \varepsilon^2\int_0^t dt'\int_0^t dt'' \; \langle J(t')K(t')J(t'')K(t'')\rangle_{q_0\to Q} \; , \tag{7.6}$$

where J is the derivative $dq_0/dq$ and only the dependence on time is written explicitly. Now we introduce new variables

$$s \equiv t'' - t' , \qquad \tau \equiv t\,\varepsilon^2 , \qquad \tau' \equiv t'\,\varepsilon^2 , \tag{7.7}$$

so that eq. (7.6) gives

$$\langle [q_0(\tau/\varepsilon) - Q]^2 \rangle = \tag{7.8}$$

$$\int_0^{\tau} d\tau' \langle J(\tau'/\varepsilon^2) \int_{-\tau'/\varepsilon^2}^{(\tau-\tau')/\varepsilon^2} ds\, K(\tau'/\varepsilon^2) K(\tau'/\varepsilon^2 + s) J(\tau'/\varepsilon^2 + s) \rangle_{q_0 \to Q} .$$

It is clear that the integral over $s$ goes from $-\infty$ to $\infty$ in the limit $\varepsilon \to 0$ . Besides this, if the expression under the first integral is a smooth function of $\tau$ in this limit, the diffusion coefficient is given by

$$D \equiv \lim_{\tau \to 0} \lim_{\varepsilon \to 0} \langle [q_0(\tau/\varepsilon) - Q]^2 \rangle / \tau = \int_{-\infty}^{\infty} ds \left[ J(s) \langle K(0) K(s) \rangle \right]_{q_0 \to Q} , \tag{7.9}$$

where it has been taken into account that $J(0) = 1$ . For the sake of completeness, we give the formula for the diffusion coefficients $D_{ij}$ in the case of several dimensions :

$$D_{ij} = \sum_k \int_{-\infty}^{\infty} ds \left[ \langle K_i(0) K_k(s) \rangle \; (\partial q_0^j(s) / \partial q^k(s)) \right]_{\vec{q}_0 \to \vec{Q}} . \tag{7.10}$$

This form of the diffusion coefficient was given by Lax (1966) (his eq. (5.41) ; note that our $D_{ij}$ is twice that of Lax). In practice, the limit $\varepsilon \to 0$ is not considered, and one uses the formula for finite $\varepsilon$ . In this case there are several expressions with different ranges of validity, although they have, presumably, the same limit for $\varepsilon \to 0$ , if this limit exists. It must be noted that eqs. (7.5) and (7.9) give the drift coefficient and diffusion coefficient at a definite time, which we have labelled $t=0$ . However, these coefficients may depend on time if the functions $H$ or $K$ depend explicitly on time.

The essential condition for the validity of eq. (7.10) is that the correlation time $t_c$ must be much shorter than the time $t_s$ needed for a significant change under the action of the stochastic terms alone. Finally, the Fokker-Planck equation for $f(q,t)$ can be found easily from eqs. (7.4) and (7.16).

## 8. EXPANSIONS IN THE CORRELATION TIME

In the previous two sections we have seen that the Markov approximation is valid when the correlation time $t_c$ of the stochastic forces is small compared with typical times involved in the problem. This suggests that the Markov limit may be a first-order approximation in a systematic expansion in the correlation time. There are several techniques for obtaining this expansion (see the references in section 7 -in particular, Frisch (1968) and van Kampen (1976)). We shall use the projection-operator technique (Zwanzig, 1960, 1964 ; Terwiel, 1974) , also called the smoothing method (Frisch, 1968).

We begin with eq. (7.1). For each value of $\lambda$ (7.1) can be considered to be a deterministic equation, so that the density $g(q,t;\lambda)$ of points in phase space fulfills the continuity equation :

$$\frac{\partial g(q,t;\lambda)}{\partial t} = -\frac{\partial}{\partial q}\left\{\left[H(q,t) + \varepsilon K(q,t;\lambda)\right] g(q,t;\lambda)\right\} . \qquad (8.1)$$

This is a stochastic (partial) differential equation, which in most cases should represent the Liouville equation. Hence the name "stochastic Liouville equation" for it (Kubo, 1963). We are not interested in the function $g(q,t;\lambda)$ itself, but in the averaged density

$$f(q,t) = \langle g(q,t;\lambda)\rangle \equiv \int dP(\lambda) g(q,t;\lambda) . \qquad (8.2a)$$

At the initial time we assume that $g$ coincides with $f$ for each $\lambda$ , i.e.,

$$f(q,0) = g(q,0;\lambda) = \langle g(q,0;\lambda)\rangle . \qquad (8.2b)$$

We may try naively to average eq. (8.1), so that

$$\frac{\partial f}{\partial t} = -\frac{\partial}{\partial q}(Hf) - \varepsilon\frac{\partial}{\partial q}\langle Kg\rangle \quad - \frac{\partial}{\partial q}(Hf) - \varepsilon\frac{\partial}{\partial q}\left[\langle K\rangle f\right] \equiv L_0 f . \qquad (8.3)$$

The first equation is exact, but useless. The second one is a closed equation for $f$ , but it is incorrect because it amounts to replacing the average of a product by the product of the averages. Methods like this which arbitrarily neglect correlations have been criticized by Keller (1964), who called them "dishonest methods". Nevertheless, eq. (8.3) is the zeroth-order approximation in the expansion we are seeking. After this, we write eq. (8.1) as

$$\frac{\partial g}{\partial t} = L_0 g + \varepsilon\frac{\partial}{\partial q}\left[(\langle K\rangle - K)g\right] \equiv L_0 g + L_1 g . \qquad (8.4)$$

We assume that it is possible to solve the deterministic part of eq.

(8.4). This is equivalent to finding the evolution operator $G_0$ such that

$$\frac{\partial}{\partial t} G_0(t,t') = L_0(t)\, G_0(t,t'), \quad G_0(t,t) = 1 . \tag{8.5}$$

If we call $G(t,t';\lambda)$ the total evolution operator, fulfilling

$$g(t;\lambda) = G(t,0;\lambda)\, g(0;\lambda) , \tag{8.6}$$

then eq. (8.4) implies the (Dyson) equation

$$G = G_0 + G_0 L_1 G . \tag{8.7}$$

We are not interested in $G$ , but in $\langle G \rangle$ , which is the evolution operator for the density function $f(q,t)$ (see eqs. (8.2)). In fact, by averaging eq. (8.6) we obtain

$$f(t) = \langle G(t,t';\lambda) \rangle \; f(0) . \tag{8.8}$$

In order to find $\langle G \rangle$ we might use the Keller (1964) expansion of eq. (8.7)

$$\langle G \rangle = G_0 + G_0 \langle L_1 \rangle G_0 + G_0 \langle L_1 G_0 L_1 \rangle G_0 + \ldots \tag{8.9}$$

Nevertheless, this is <u>not</u> an expansion in the correlation time, even if it is an expansion in the small parameter $\varepsilon$ (compare with eq. (8.4)). In fact, a typical term like

$$\int_0^{t_n} dt_{n-1} \ldots \int_0^{t_2} dt_1 \; \langle G_0(t_n,t_{n-1})\, L_1(t_{n-1}) \ldots G_0(t_1,0) \rangle \tag{8.10}$$

is of order $\varepsilon^{n-1} t^n$ , because it contains $n$ time intervals. Therefore eq. (8.9) is an expansion in $\varepsilon t$ , not in the correlation time. If we return to section 7, we see that in obtaining the Markov limit we assumed $t\,\varepsilon^2$ to be finite, which implies $\varepsilon t \to \infty$ . Therefore eq. (8.9) goes in the opposite direction to the Markov approximation.[+]

An expansion in powers of the correlation time $t_c$ can be obtained by the method of cumulants (see van Kampen, 1976). The projection-operator technique also provides a similar expansion. We define a projection operator $P$ which averages everything on its right. When we apply $P$ and $1-P$ to eq. (8.7) we obtain the following two equations

$$\begin{aligned} PG &= G_0 + G_0 P L_1\, PG + G_0 P L_1 (1-P)\, G = G_0 + G_0 P L_1 (1-P) G , \\ (1-P)G &= G_0 (1-P) L_1 PG + G_0 (1-P) L_1 (1-P) G , \end{aligned} \tag{8.11}$$

[+]See note 1 on p. 49

where we have taken into account that

$$PL_1P = \langle L_1 \rangle P = 0 ,$$

by eq. (8.4). Formal elimination of the operator $(1-P)G$ from eqs. (8.11) gives the desired expansion :

$$\langle G \rangle \equiv PG = G_0 + G_0PL_1 \left[1-G_0(1-P)L_1\right]^{-1} G_0(1-P)L_1PG$$

$$\equiv G_0 + G_0PL_1 \sum_{n=1}^{\infty} \left[G_0(1-P)L_1\right]^n PG . \qquad (8.12)$$

The important point is that this is an expansion in $(1-P)L_1$, not just in $L_1$ as in eq. (8.9). Now the operator $(1-P)$ has the property that it gives zero when inserted between two uncorrelated random variables with $P$ on the left and on the right, i.e., if $\{A,B\}$ are uncorrelated random variables,

$$P\,A(1-P)\,B\,P \equiv \left[\langle A\,B \rangle - \langle A \rangle \langle B \rangle\right] P = 0$$

This proves that a typical term of the expansion eq. (8.12), such as

$$\int_0^{t_n} dt_{n-1} \cdots \int_0^{t_2} dt_1 \langle G_0(t_n,t_{n-1})L_1(t_{n-1})G_0(t_{n-1},t_{n-2})(1-P)L_1(t_{n-1}) \ldots \rangle ,$$

is zero except when all the time intervals $t_{n-1} - t_{n-2}, \ldots, t_2 - t_1$ are smaller than $t_c$. Therefore, this term should be of order $\varepsilon^{n-1} t_c^{n-2} t^2$, which shows that eq. (8.12) is an expansion in powers of $\varepsilon t_c$.

Eq. (8.12) also has the form of a Dyson equation :

$$\langle G \rangle = G_0 + G_0 M \langle G \rangle , \qquad (8.13)$$

which corresponds to the integro-differential equation

$$\frac{\partial f(t)}{\partial t} = L_0(t)f(t) + \int_0^t dt' M(t,t')f(t') , \qquad (8.14)$$

where

$$M \equiv PL_1 \sum_{n=1}^{\infty} \left[G_0(1-P)L_1\right]^n . \qquad (8.15)$$

For instance, the second-order approximation is

$$\frac{\partial f(t)}{\partial t} = L_0(t)f(t) + \int_0^t \langle L_1(t)G_0(t,t')L_1(t') \rangle f(t')dt' . \qquad (8.16)$$

This is called Bourret's integral equation (Bourret, 1962) and it can be seen that it is related to eq. (7.9) for the Markov limit. It is clearly of order $\varepsilon^2 t_c t$ . The calculation of higher-order approximations is straightforward, though lengthy. (See notes 2 and 3 on p. .)

## 9. INTEGRATION IN FUNCTIONAL SPACES

The method of integration in functional spaces provides a general method for the solution of stochastic differential equations which has the advantage that it gives an intuitive picture of the problem involved. However, it is too complex in practice and, therefore, has limited application. Although the functional-integration approach to stochastic processes has its roots in the early work by Smoluchowski on Brownian motion, it was put in a firm mathematical basis by Wiener in 1923-30 (see, e.g., Kac, 1966; Gel´fand and Yaglom, 1960; Brush, 1961). This work was limited to Wiener processes, but the method can be extended to other processes as well. Functional-integration methods were given a strong push forward by the celebrated formulation of quantum mechanics by Feynman.

We shall consider only the application of the functional-integration method to the solution of first-order stochastic differential equations of the form

$$\dot{x}(t;\lambda) = F(x(t;\lambda), y(t;\lambda)) , \qquad (9.1)$$

where $y(t;\lambda)$ is a given stochastic process. We shall assume that we are interested in the calculation of the conditional probability density $\rho(x,t/x_0, 0)$ . To do this, we divide the time interval $[0,t]$ into $n$ parts of duration $\varepsilon = t/n$ each. We assume that we know the probability density

$$\rho(y_0,0;\, y_1,\varepsilon\, ;\ldots y_{n-1},\, t_{n-1}) \equiv \rho(y_0,y_1\ldots y_{n-1}) \, . \qquad (9.2)$$

For instance, if $y(t;\lambda)$ is Gaussian, knowledge of the mean and the autocorrelation is sufficient (see eq. (3.2)). Then we replace the differential equation (9.1) by the set of difference equations

$$x_{j+1} - x_j = \varepsilon\, F(x_j,y_j),\ j = 0,\ldots n \, . \qquad (9.3)$$

It should be noted that this is correct only if $x(t)$ is differentiable, which is not the case if $x(t)$ is a Wiener or an Ornstein-Uhlenbeck process. The probability density $\rho(x,t/x_0,0)$ is given by

$$\rho(x,t/x_0,0) = \lim_{\varepsilon\to 0} \int dy_1 \ldots \int dy_n\, \rho(y_0,\ldots,y_{n-1})\, \delta(x-x_n(x_0,y_0,\ldots,y_{n-1})), \qquad (9.4)$$

if we assume that the solution of eqs. (9.3) approaches that of eq. (9.1) for each $\lambda \in \Lambda$ in the limit $\varepsilon \to 0$ (remember that $\Lambda$ is the sample space). It is possible to change the integrals in eq. (9.4) from the variables $\{y_0,\ldots,y_{n-1}\}$ to the variables $\{x_1,\ldots,x_n\}$ and then perform the integral in $x_n$ by means of Dirac's delta. We then obtain

$$\rho(x,t|x_0,0) = \lim_{\varepsilon \to 0} \int_{-\infty}^{\infty} dx_1 \ldots \int_{-\infty}^{\infty} dx_{n-1}\, \rho(y_0,\ldots y_{n-1})\, |\partial x_j/\partial y_i| , \qquad (9.5)$$

where $|\partial x_j/\partial y_i|$ is the Jacobian of the transformation. In order to do the integrals in (9.5) it is convenient to have eqs. (9.3) written as

$$y_j = Y(x_j, \frac{x_{j+1} - x_j}{\varepsilon}) \qquad (9.6)$$

Eqs. (9.4) or (9.5) give, in principle, the solution of the stochastic differential equation (9.1). However, they are useless except when one is able to perform the integrals in a practical way, which is only possible in special cases.

As an example, let us suppose that eq. (9.1) has the particular form

$$\dot{x} = F(x) + y , \qquad (9.7)$$

and that $y(t;\lambda)$ is a Gaussian process with zero mean. In this case, the Jacobian which appears in eq. (9.5) is simply $\varepsilon^n$ and $\rho(y_0,..,y_{n-1})$ is given by eq. (3.2). Then eq. (9.5) can be written as

$$\rho(x,t|x_0,0) =$$

$$= \lim_{\varepsilon \to 0} (\varepsilon/2\pi)^n \int_{-\infty}^{\infty} d\eta_0 \ldots \int_{-\infty}^{\infty} d\eta_{n-1} \int_{-\infty}^{\infty} dx_1 \ldots \int_{-\infty}^{\infty} dx_{n-1}\, Z ,$$

$$Z \equiv \exp\left[\frac{1}{2}\sum_{j,k=0}^{n-1} \eta_j \eta_k C_{jk} - i \sum_{j=0}^{n-1} \eta_j (x_{j+1} - x_j - \varepsilon F(x_j))/\varepsilon\right] . \qquad (9.8)$$

Even in this example, the integrals involved cannot be done in practice except in certain particular cases. For instance, if the process $y(t)$ is white noise, then

$$C_{jk} = C\, \delta_{jk} \qquad (9.9)$$

and the integrals over $\eta$ can be done one at a time. Then it is easy to see that eq. (9.8) gives the Chapman-Kolmogorov equation. Nevertheless, in this case and, in general, whenever the process $x(t)$ is not differentiable, eqs. (9.3) do not give a good approximation to the stochastic differential equation, and some refinement is necessary. If $F(x)$ is a linear function, the integrals can be also evaluated. (For an application, see Santos, 1977). If $F(x)$ is a linear term plus a small perturbation, i.e.,

$$F(x) = a\,x + b + \alpha\, f(x) \,, \tag{9.10}$$

it is possible to obtain $\rho(x,t/x_0,0)$ by expanding

$$\exp[-i\,\eta_j\,\alpha f(x_j)] \simeq 1 - i\,\eta_j\,\alpha\, f(x_j) \tag{9.11}$$

and rearranging $\rho(x,t|x_0,0)$ as a series in powers of $\alpha$ . The calculation of the zeroth-order term is straightforward; the first-order term can be reduced to a definite integral, and so on.

Note 1. A more careful examination of eq. (8.10) shows that this term is of order $\varepsilon^{n-1} t_c^{(n-1)/2}\, t^{(n+1)/2}$ (provided that n is odd). Therefore, each term of eq. (8.9) is finite in the Markov limit $\varepsilon \to 0$ with $\varepsilon^2 t$ finite. Nevertheless, the essential point is that the Markov limit is not obtained with a finite number of terms, as is the case with the expansion eq. (8.12). I thank Dr. Spohn for this remark.

Note 2. Actually, Bourret proposed an equation similar to eq. (8.16) as the equation of motion for the average of any stochastic process or of any stochastic field that is governed by a stochastic equation of the type of eq. (8.4). That is, for us, eq. (8.4) is a continuity equation associated with the stochastic differential equation (7.1) and $f(q,t)$ is the probability density associated with the stochastic process $q(t)$. However, the expansion (8.14) can be applied to any linear stochastic (perhaps partial) differential equation of the type of eq. (8.4), where $g(t)$ or $g(q,t)$ is a stochastic process or a stochastic field. In this case, Bourret's eq. (8.16) is the equation for the average of g.

Note 3. The relation between eq. (8.16) and the Markov limit eq. (7.9) is not straightforward. In the Markov limit only the first term of eq. (8.15) is finite and all other terms go to zero. Therefore, eq. (8.16), which was derived from eqs. (8.14) and (8.15) by retaining the first term of the series only, should be sufficient in order to obtain the Markov limit. However, the first term of the right-hand side of eq. (8.16) divided by $\varepsilon^2$, goes to infinity when $\varepsilon \to 0$, whilst the other terms remain finite. Therefore, it is not possible to start from the equation (8.16) associated with eq. (7.1), but from a similar one associated with eq. (7.3), which should have $L_0=0$. It can be seen that, in the limit, eq. (8.16) becomes

$$\frac{\partial f}{\partial(\varepsilon^2 t)} = \frac{\partial}{\partial q_0}\int_{-\infty}^{0} ds \left\langle K(t) \frac{\partial}{\partial q_0} J(t+s)K(t+s)\right\rangle \rho(t) \qquad (10.1)$$

where the transformation from q to $q_0$ (see before eq. (7.2)) has been made so that $q(t) = q_0(t)$. It is seen that eq. (10.1) is similar to, but different from, the one which has eq. (7.9) as diffusion coefficient. Apparently, the connection between the Markov limit of Bourret's equation given by eq. (10.1) and the Markov limit given by eq. (7.9) remains an open problem. It can be checked that in the example of the harmonic oscillator considered in sections 5 and 6, eq. (10.1) gives a Fokker-Planck equation which is different from eq. (6.16) in that it does not contain the term with $\partial^2 f/\partial x^2$. This seems to indicate that eq. (10.1) is valid for times smaller than the deterministic period $t_0$, whilst eq. (6.16) is not. In fact, the stochastic force introduced in eq. (6.3) gives a diffusion in velocity space, but not in position space directly. The diffusion in position derives from the diffusion in velocity by the action of the deterministic motion, which shows that eq. (6.16) is correct only for times larger than the deterministic period. So, it seems that eq. (10.1) does not require the last inequality (6.14), but should require $t_c \ll t_0$ instead.

ARNOLD, L., (1974) Stochastic Differential Equations. Theory and Applications (John Wiley, NY).

BARUCHA-REID, A.T., (1960), Elements of the Theory of Markov Processes and their Applications (McGraw-Hill, NY).

BOURRET, R.C., (1962), Can. J. Phys. 40, 782.

BRISSAUD, A., and FRISCH, U., (1974), J. Math. Phys. 15, 524.

BRUSH, S.G.,(1961), Rev. Mod. Phys. 33, 79.

CLAVERIE, P.,and DINER, S.,(1978) Stochastic Electrodynamics and Quantum Theory, Intern. J. Quantum Chem. 13 (in press).

COBURN, R.,and HERSCH, R., (1973), Indiana Univ. Math. Journal (formerly J. Math. and Mechanics) 22, 1067.

COX, D.R., and MILLER, H.D., (1965), The Theory of Stochastic Processes (Chapman and Hall, London).

FELLER, W., (1966), An Introduction to Probability Theory and its Applications, (Wiley, NY)

FRISCH, U., (1968) in: Probabilistic Methods in Applied Mathematics I (ed. A.T. Barucha-Reid, Academic Press, NY).

GEL'FAND, I.M., and YAGLOM, A.M., (1960), J. Math. Phys. 1, 48.

GRIKHMAN, I.I., and SKOROKHOD, A.V., (1972), Stochastic Differential Equations (Springer-Verlag, Berlin).

HAKEN, H.,(1965), Rev. Mod. Phys. 47, 67.

KAC, M.,(1966) Bull. Am. Math. Soc. 72, 52.

KELLER, J.B.,(1962) in: Proc. Symp. Appl. Math. 13 (Amer. Math. Soc. Providence, R.I.) p. 227.

KELLER, J.B.,(1964) in: Proc. Symp. Appl. Math. 16 (Amer. Math. Soc. Providence, R.I.) p. 84.

KHAS'MINSKII, R.Z.,(1966), Theory Prob. Appl. 11, 211 and 390.

KUBO, R.,(1963), J. Math. Phys. 4, 174.

KUBO, R.,(1962) in: Stochastic Processes in Chemical Physics,(ed. K.E. Shuler, Interscience, NY.

PAPANICOLAOU,G.C., and HERSH, R., (1972), Indiana Univ. Math. J. 21, 815)

PAPANICOLAOU,G.C., and KELLER, J.B., (1971) SIAM J. Appl. Math. 21, 815

PAPANICOLAOU, G.C., McLAUGHLIN, D., and BURRIDGE, R. (1973), J. Math. Phys. 14, 84

PAPOULIS, A., (1965), Probability, Random Variables and Stochastic Processes (McGraw Hill, NY).

PROKHOROV, Yu. V., and ROZANOV, Yu. A.,(1969), Probability Theory (Springer, Berlin).

RICE, S.O., (1944, 1945), Bell Syst. Tech. J. 23, 282; 24, 46. (Reprinted in: Wax, N. 1954).

SANTOS, E., (1977), Lett. Nuovo Cimento 20, 308.

STRATONOVICH, R.L.,(1963, 1967), Topics in the Theory of Random Noise Vols. I and II (Gordon and Breach, NY)

STRATONOVICH, R.L.,(1968) Conditional Markov processes and their Application to the Theory of Optimal Control (American Elsevier, NY).

SOONG, T.T., (1973), Random Differential Equations in Science and Engineering (Academic Press, NY).

TERWIEL, R.H.,(1974), Physica 74, 248.

VAN KAMPEN, N.G.,(1976), Phys. Reports 24C, 171.

WAX, N., (Editor) (1954) Selected papers on Noise and Stochastic Processes (Dover, NY)

ZWANZIG, R.,(1960), J. Chem. Phys. 33, 1338; Boulder Lectures in Theoretical Physics 3, Eds. W.E. Brittin et al., Interscience 1961), p. 106.

ZWANZIG, R.,(1972) in: Statistical Mechanics, New Concepts, New Problems, New Applications (Proceedings of the 6th IUPAP Conference on Statistical Mechanics, eds. S.A. Rice et al., University of Chicago Press, Chicago).

# ASPECTS OF THE THEORY OF BROWNIAN MOTION

Robert M. Mazo
Chemistry Division, National Science Foundation
Washington, D.C. 20550

## 1. INTRODUCTION

In 1827, the well known botanist, Robert Brown, was examining the pollen of the plant Clarkia Pulchella through one of the, then new, achromatic objectives. He observed that the individual particles were in a very animated and irregular state of motion. This is now called the Brownian motion, and is the subject of this series of lectures.

Brown was, indeed, a renowned naturalist. As a young man, in 1801, he had sailed on an expedition to study the plant life of the new continent of Australia. The collection of flora which he brought back from the four-year voyage provided the raw materials for the researchers which established his reputation. His reputation became, in fact, immense. He received the appellation of princeps botanicorum. Among his discoveries was the existence of the nucleus of plant cells. The Brownian motion is regarded as one of his minor observations, at least in one history of biology which I have seen.

It is perhaps fitting that someone from the Pacific North-west of the United States lecture on Brownian motion, for Clarkia Pulchella is a common wild flower in that region. Indeed, it is quite possible that Brown's specimen was one brought back by the Lewis and Clark expedition sent to explore the Louisiana Purchase by President Thomas Jefferson. Interesting historical accounts of Brownian motion have been given by Brush (1968) and Kerker (1974).

Brown was not the first person to have observed the Brownian motion. Indeed, he mentioned a number of predecessors in his papers. However, he was apparently the first to study the phenomenon seriously. He showed that it was not limited to organic matter. Glass, minerals, even a fragment of the Sphinx, surely inorganic, exhibited the motion. It can be said that Brown removed Brownian motion from being a subject of biology, and made it a subject of physics. He also tried to ascertain the cause of this motion. This he was not able to do. But he was able to provide evidence against simple mechanical explanations, e.g., convection, capillarity, small air bubbles, etc.

In the years which followed, many explanations for the Brownian movement were given; most of these could have been ruled out by consideration of Brown's original experiments. The first worker to express views at all like the modern view was Christian Wiener (1826-1896) (not to be confused with Norbert Wiener (1894-1964) who, much later, created a rigorous mathematical theory of Brownian motion). Delsaux and Gouy were other natural philosophers whose speculations led to modern views, that the motion was caused by the unceasing bombardment of the moving particle by the particles of the surrounding medium. However, there was no experimental means to put these ideas to a definitive test, and they were not generally accepted. It was not until 1905 that a quantitative theory making predictions susceptible of experimental verification was put forward. It was due to Albert Einstein.

Einstein had always been interested in the problem of the size of atoms. Indeed, his doctoral dissertation was on this subject, and he continued to think about it well after he obtained his degree. In the course of these considerations he showed that the kinetic theory of matter required that small particles suspended in a fluid undergo an irregular motion. This motion was too chaotic to be described in any other way than statistically, and Einstein obtained some of the statistical characterizations of the motion (Einstein, 1905).

One statement in Einstein's first paper is, in retrospect, amusing. He writes "It is possible that the movements to be discussed here are identical to the so-called 'Brownian molecular motion'; however, the information available to me about the latter is so lacking in precision that I can make no judgement in the matter." His second paper, however, was called "On the Theory of Brownian Motion". It was published in 1906.

Einstein's theory concerned itself with the configuration-space aspect of Brownian motion. In these lectures we shall be primarily concerned with the phase-space aspects; that is, we shall consider both configuration space and velocity space. Consequently, we shall not pause to review Einstein's considerations in detail. We shall just note that (1) he concluded that the motion was so complex that it must be described statistically, and (2) he assumed that there was a time interval $\tau$, large compared to the time between collisions and small compared to macroscopic observation times, such that the motions executed by the particle in two consecutive intervals of length $\tau$ are mutually independent. From these hypotheses, he showed that the probability $\rho(\underset{\sim}{r},t)$ for finding the particle at $\underset{\sim}{r}$ at time t satisfied the diffusion equation

$$\frac{\partial \rho}{\partial t} = D \nabla^2 \rho . \tag{1.1}$$

He then showed that D, the diffusion coefficient, is related to the friction coefficient $\zeta$ by

$$D = kT/\zeta \quad . \tag{1.2}$$

The frictional coefficient is defined as the proportionality constant between the external force $\underset{\sim}{F}$ and the terminal velocity $\underset{\sim}{v}$ of a particle in a fluid:

$$\underset{\sim}{F} = \zeta \underset{\sim}{v}\,, \tag{1.3}$$

and k is Boltzmann's constant.

Because D and $\zeta$ can be measured, as can R, the universal gas constant, (1.2) enables one to determine Avogadro's number, $N_A = R/k$. This was an important prediction of the theory. Furthermore, Einstein showed that

$$\langle (\Delta \underset{\sim}{r})^2 \rangle = 6\,Dt\,, \tag{1.4}$$

another important verifiable result. Both (1.2) and (1.4) were, in fact, verified by J. Perrin within a few years of Einstein's work.

The dynamics of Brownian motion is associated with the names of Langevin, Fokker, and Planck. It is to these developments that we now turn.

## 2. DYNAMICAL THEORY - STOCHASTIC

A dynamical theory is a theory based on an equation of motion. The equation of motion in a statistical theory can be of either of two types: (a) an equation for the time development of the dynamical variables of the system, or (b) an equation for the time development of the probability distribution of the dynamical variables of the system. If both descriptions are available, they must be consistent. Even in nonprobabilistic classical mechanics, we have these two types of description. The first corresponds to Newton's equations, or Hamilton's equations. The second corresponds to Liouville's equation.

A description of type (a) in Brownian motion theory is associated with the name of Langevin. A description of type (b) is associated with the names of Fokker and Planck.

The Langevin description starts from considerations of immediate physical intuitive appeal, so let us consider this first. Our development will be a slight generalization of the classical development. Langevin's equation is just Newton's law of motion, force = mass times acceleration, with a specific prescription for the force. In particular, the force on a Brownian particle in a fluid is assumed to be expressible as the sum of two terms. The first term is a systematic frictional force, linearly related to the particle's velocity. The second term is a random force.

For the systematic friction, we shall write

$$F_{friction} = -m\int_0^t \beta(t-s)u(s)ds \tag{2.1}$$

where u(s) is the velocity of the Brownian particle at time s, m is the mass of the particle, and $\beta(t-s)$ is some function of the time, which we shall call the friction function. For steady motion in a viscous fluid, one is more used to writing

$$F_{friction} = -\zeta u , \tag{2.2}$$

all quantities being here taken at the same time. Of course, this is a special case of (2.1), i.e., $m\beta(t-s) = \zeta\delta(t-s)$. However, for non-steady motion in a viscous fluid it is known that the frictional force depends on the history of the motion, through an integral of the form (2.1). This is because the disturbances in the fluid caused by the moving particle take time to propagate; the fluid has inertia. The form of $\beta$ when macroscopic hydrodynamics applies is known. It is

$$\beta(t) = 6\pi\eta a\,\delta(t) - \tfrac{3}{2}(\pi a^4\rho\eta)^{\frac{1}{2}}\, t^{-\frac{3}{2}} \tag{2.3}$$

for a sphere of radius a in a fluid of density $\rho$ and viscosity $\eta$. However, we shall not need this specific form in the material to follow.

As concerns the random force, f, we shall assume that it is a second-order stationary Gaussian random process. The meaning of these terms has been given in Professor van Kampen's lectures. f is assumed to have zero mean

$$\langle f(t)\rangle = 0 \tag{2.4}$$

and covariance function

$$\langle f(t)f(t+s)\rangle = mkT\beta(s) \tag{2.5}$$

In fact, the precise form (2.5) is not an arbitrary hypothesis, but can be derived from a reasonable physical requirement. We shall see this in detail a bit later.

For this section of these lectures, we shall work in one dimension, as has already been foreshadowed by the lack of vector symbols in the notation. The reason is solely to keep the material as uncluttered as possible; the reader who has understood the ideas will have no difficulty in translating everything to three dimensions.

We shall furthermore assume that the random force is not correlated with the velocity of the particle at the initial time:

$$\langle u_0 f(t)\rangle = 0 . \tag{2.6}$$

This is one aspect of saying that the force f is indeed random.

With this background, Langevin's equation (generalized) can be written down. It is

$$\dot{u} + \int_0^t \beta(t-s)u(s)ds = m^{-1}f(t) \qquad (2.7)$$

$$\dot{x} = u$$

x being the particle's position. Now f(t) is a very wild random function, so it is not obvious that (2.7) possesses solutions; the usual existence theorems require a modicum of smoothness for the equation. Nevertheless, this can all be made completely respectable, mathematically (Doob, 1942). We shall take the physicist's approach. Namely, we shall ignore the possibility that our equations may be meaningless, but be careful to ask only questions of them for which they can be meaningfully interpreted.

Equation (2.7) can be solved by means of Laplace transformation. If we denote Laplace transforms by superior tildes

$$\int_0^\infty e^{-pt}g(t)dt = \tilde{g}(p) ,$$

then

$$p\,\tilde{u}(p) - u_0 + \tilde{\beta}(p)\tilde{u}(p) = m^{-1}\tilde{f}(p), \qquad (2.8)$$

$$p\tilde{x}(p) - x_0 = \tilde{u}(p) , \qquad (2.9)$$

where $u_0 = u(t = 0)$, and $x_0 = x(t = 0)$. Solving for $\tilde{u}(p)$, and performing the inverse Laplace transform, one gets

$$u(t) = u_0\,\chi(t) + m^{-1}\int_0^t \chi(t-s)f(s)ds , \qquad (2.10)$$

$$x(t) = x_0 + u_0\,\psi(t) + m^{-1}\int_0^t \psi(t-s)f(s)ds . \qquad (2.11)$$

Here, $\chi(t)$ and $\psi(t)$ are defined most simply by their Laplace transforms

$$\tilde{\chi}(p) = \left[p + \tilde{\beta}(p)\right]^{-1},$$

$$\tilde{\psi}(p) = \left[p(p + \tilde{\beta}(p))\right]^{-1} . \qquad (2.12)$$

Clearly,

$$d\psi/dt = \chi(t) \qquad (2.13)$$

The physical significance of $\chi$ and $\psi$ can be most easily seen as follows: multiply (2.10) and (2.11) by $u_0$ and average. By virtue of (2.6),

$$\langle u_0 u(t)\rangle \;/\; \langle u_0^2\rangle = \chi(t), \tag{2.14}$$

$$\langle u_0(x(t)-x_0)\rangle \;/\; \langle u_0^2\rangle = \psi(t). \tag{2.15}$$

That is, $\chi(t)$ measures how the velocity forgets its initial value, and $\psi(t)$ measures how the displacement forgets the inital velocity.

Let us now ask our first and main statistical question: what is the probability distribution of u and x? Since $u-u_0\chi$ and $x-x_0-u_0\psi$ are <u>linear</u> functionals of f, and since f is a <u>Gaussian</u> random process, it follows that $(u-u_0\chi\,,\ x-x_0-u_0\psi)$ is a bivariate Gaussian random process. This process has zero mean (because f has zero mean) and covariance matrix $\underset{\sim}{Q}$, given by

$$Q_{11} = \langle(u-u_0\chi)^2\rangle = (kT/m)(1-\chi^2(t)), \tag{2.16}$$

$$Q_{12} = Q_{21} = \langle(u-u_0\chi)(x-x_0-u_0\psi)\rangle = \left(\frac{kT}{m}\right)\psi(1-\chi), \tag{2.17}$$

$$Q_{22} = \langle(x-x_0-u_0\psi)^2\rangle = (kT/m)\left(2\int_0^t \psi(s)ds - \psi^2(t)\right). \tag{2.18}$$

That is, if, for the sake of brevity, we write $y_1 = u-u_0\chi$, $y_2 = x-x_0-u_0\psi$, then the probability density $P(y_1,y_2)$ is given by

$$P(y_1,y_2) = \pi^{-1}\,|\det \underset{\sim}{Q}|^{-\frac{1}{2}} \exp\,(-y^+\underset{\sim}{Q}^{-1}y). \tag{2.19}$$

Before going on, let us derive (2.16) to (2.18). This calculation will also enable us to derive (2.5). From (2.10), we see that

$$Q_{11} = \langle(u-u_0\chi)^2\rangle = \frac{1}{m^2}\int_0^t\int_0^t \chi(t-s)\,\chi(t-s')\,\langle f(s)f(s')\rangle\, ds ds'. \tag{2.20}$$

Since f is second-order stationary by assumption,

$$\langle f(s)f(s')\rangle = \phi\,(s-s'), \tag{2.21}$$

so that a simple change of variables gives

$$Q_{11} = \frac{1}{m^2}\int_0^t\int_0^t \chi(s)\,\chi(s')\,\phi\,(s-s')\,ds ds' \tag{2.22}$$

Now we have to put in a physical assumption. If $u_0$ has a Maxwellian distribution, then u will have a Maxwellian distribution for all later times. That is, thermal equilibrium is stable. Since then $\langle u^2 \rangle = \langle u_0^2 \rangle = kT/m$ in this special case, and the right-hand side of (2.22) is independent of the distribution of $u_0$, we have

$$Q_{11} = \langle u^2 \rangle - 2\langle u_0 u \rangle \chi + \langle u_0^2 \rangle \chi^2 = \frac{kT}{m}(1 - \chi^2(t)). \tag{2.16}$$

We can use this result to identify $\phi(s-s')$, for, from (2.16),

$$\frac{dQ_{11}}{dt} = -2\frac{kT}{m}\chi(t)d\chi/dt. \tag{2.23}$$

From the original (2.7) we have

$$\frac{d\chi}{dt} = -\int_0^t \chi(s)\beta(t-s)\,ds, \tag{2.24}$$

so that, finally,

$$\frac{dQ_{11}}{dt} = \frac{2kT}{m}\chi(t)\int_0^t \chi(s)\beta(t-s)\,ds. \tag{2.25}$$

However, from (2.22) we have

$$\frac{dQ_{11}}{dt} = \frac{2\chi(t)}{m^2}\int_0^t \chi(s)\,\phi\,(t-s)\,ds. \tag{2.26}$$

Comparing (2.25) and (2.26), we arrive at (2.5) as promised. This is a manifestation of the fluctuation-dissipation theorem.

Equations (2.17) and (2.18) can be derived in a similar manner. This will be left as an exercise for the reader.

Now that we know the distribution function for (u,x), we can ask: is there some simple equation satisfied by this distribution function? One could say, suppose there were! We have the solution already, so why search for the equation? The reason is that some problems, in finite geometries, are more simply treated by the governing differential equation supplemented by boundary conditions than by introducing additional forces into the Langevin equation to account for the walls. We are therefore leaving the Langevin description, and moving to the Fokker-Planck description, mentioned earlier.

The royal road for discussing a probability distribution is its characteristic function. The characteristic function of a distribution is its Fourier transform,

$$C(\lambda,\mu) = \langle \exp(i\lambda u + i\mu x) \rangle \tag{2.27}$$

Since the probability distribution implicit in the angular brackets $\langle \dots \rangle$ is Gaussian in our case, so is $C(\lambda, \mu)$, and it is easy to see that

$$C(\lambda, \mu) = \exp\Big\{ i\lambda u_0 \chi + i\mu (x_0 + u_0 \psi) - \\ - \tfrac{1}{2} (Q_{11}\lambda^2 + 2 Q_{12}\lambda\mu + Q_{22}\mu^2) \Big\} . \tag{2.27}$$

Therefore,

$$\frac{1}{C} \frac{\partial C}{\partial t} = \Big\{ i\lambda u_0 \dot{\chi} + i\mu u_0 \dot{\psi} - \tfrac{1}{2} (\lambda^2 \dot{Q}_{11} + 2\lambda\mu \dot{Q}_{12} + \mu^2 \dot{Q}_{22}) \Big\} . \tag{2.28}$$

We would like to eliminate the terms involving $u_0$, i.e., those which depend on initial conditions. Note that

$$\lambda \frac{\partial C}{\partial \lambda} = \Big\{ i\lambda u_0 \chi - Q_{11}\lambda^2 - Q_{12}\lambda\mu \Big\} C , \tag{2.29}$$

$$\mu \frac{\partial C}{\partial \lambda} = \Big\{ i\mu u_0 \chi - Q_{11}\lambda\mu - Q_{12}\mu^2 \Big\} C . \tag{2.30}$$

Some algebra, and the use of (2.16)-(2.18) yields

$$\frac{\partial C}{\partial t} - \mu \frac{\partial C}{\partial \lambda} = \frac{-\dot{\chi}}{\chi} \left( -\lambda \frac{\partial C}{\partial \lambda} - \frac{kT}{m} \lambda^2 C \right) \\ + \lambda\mu \left( \frac{\dot{\chi}}{\chi} \psi + 1 - \chi \right) C . \tag{2.31}$$

Returning from Fourier-transform space to phase space by the inverse transformation, the equation for the probability density becomes

$$\frac{\partial P}{\partial t} + u \frac{\partial P}{\partial x} = \zeta(t) \frac{\partial}{\partial u} \left( u + \frac{kT}{m} \frac{\partial}{\partial u} \right) P \\ + \frac{kT}{m} \sigma(t) \frac{\partial^2 P}{\partial x \partial u} , \tag{2.32}$$

where we have written

$$\zeta(t) = - \dot{\chi}/\chi ,$$

$$\sigma(t) = \zeta\psi - 1 + \chi . \tag{2.33}$$

We shall call eq. (2.31) the generalized Fokker-Planck equation. It is generalized because we allowed the friction to depend on history. If we had not done so, i.e., if $m\beta(t-s) = \zeta\delta(t-s)$ with $\zeta$ independent of time, then

$$\chi(t) = e^{-\zeta t/m},$$

$$\psi(t) = \frac{m}{\zeta}(1-e^{-\zeta t/m}). \tag{2.34}$$

Thus, in this special case, $\sigma = 0, \zeta(t) = \zeta/m$, a constant. This yields

$$\frac{\partial P}{\partial t} + u\frac{\partial P}{\partial x} = \frac{\zeta}{m}\frac{\partial}{\partial u}\left(u + \frac{kT}{m}\frac{\partial}{\partial u}\right)P, \tag{2.35}$$

which is the classical Fokker-Planck equation. It corresponds to the case where one considers the correlation time of the random force to be immeasurably short.

In this classical case, the stochastic velocity depends only on its own instantaneous value and the current value of the force, f. There is no history dependence; the classical case describes a Markov process. The derivation of the classical Fokker-Planck equation from the theory of Markov processes, in particular from the Chapman-Kolmogorov equation,can be found in a large number of easily accessible sources. The present derivation, when applied to the classical case, has a slight air of novelty, and is also more general, since if $\beta(t)$ is not a delta function, the process is not Markovian.

## 3. DIFFUSION

Another advantage of the approach we have taken in Section 2 is that it enables us to investigate the diffusion equation in a very economical manner, and to answer an old question posed by Uhlenbeck and Ornstein (Uhlenbeck and Ornstein, 1930).

The problem is essentially the following: The diffusion equation is

$$\frac{\partial F}{\partial t} = D\frac{\partial^2 F}{\partial x^2}, \tag{3.1}$$

where F(x,t) is the probability distribution in configuration space (we are continuing to speak in one-dimensional terms). It is known that (3.1) is a consequence of (2.35) for times long compared to $m\zeta^{-1}$. What is the exact equation satisfied by F(x,t)? This exact equation must go over into (3.1), of course, when $\zeta t/m \gg 1$.

This question can be attacked by means of eq. (2.27). Since

$$F(x,t) = \int P(x,u,t)\,du, \tag{3.2}$$

it follows that the characteristic function is $C(\lambda=0,\mu)$, which we shall call, with mild abuse of notation, $C(\mu)$. Therefore,

$$\frac{\partial C(\mu)}{\partial t} = (i\mu u_0 \chi - \tfrac{1}{2}\mu^2 \dot{Q}_{22})\, C(\mu). \tag{3.3}$$

Taking inverse Fourier transforms,

$$\frac{\partial F}{\partial t} = \frac{kT}{m}\, \psi(1-\chi)\, \frac{\partial^2 F}{\partial x^2} - u_0\, \chi\, \frac{\partial F}{\partial x}. \tag{3.4}$$

This clearly goes over into (3.1) as $t \to \infty$, since $\chi(t) \to 0$, and one expects $\psi(t) \to$ constant. This is certainly the case when $m\beta = \zeta\delta(t)$, since then (2.34) holds. It should not be surprising that the equation for F depends on the initial velocity $u_0$, since one would expect, at short times, that the distribution of displacements would, in fact, be distorted in the direction of the initial velocity, the more so the higher the initial velocity.

Equation (3.4) answers a long-standing question of Ornstein and Uhlenbeck. In their classic paper on Brownian motion (Uhlenbeck and Ornstein, 1930), they find $P(x,u,t)$ by solving the Fokker-Planck equation, and then evaluate $F(x,t)$ by integrating over u, eq. (3.2). They then state: "It seems impossible to derive from (19) the rigorous differential equation for $F(x_0,x,t)$, which, for $t \gg \beta^{-1}$, would become the diffusion equation, and of which (24) would be the fundamental solution". In this quotation, (19) is the same as our eq. (2.35) and $\beta$ is what we have called $\zeta/m$. (24) is the result of integrating the solution of (2.35) over the velocity, namely,

$$F(x,t) = \left[\zeta^2/2\pi mkT(2\zeta t/m - 3 + 4e^{-\zeta t/m} - e^{-2\zeta t/m})\right]^{\frac{1}{2}} \times$$

$$\times \exp\left\{-\frac{\zeta^2}{2kTm}\; \frac{\left[x - x_0 - mu_0(1-e^{-\zeta t/m})/\zeta\right]^2}{2\zeta t/m - 3 + 4e^{-\zeta t/m} - e^{-2\zeta t/m}}\right\}. \tag{3.5}$$

Eq. (3.5) is a solution of (3.4) in the classical case.

In normal circumstances, however, we are not given this particular problem. Rather, we are given a distribution, usually a Maxwellian distribution, of the initial velocity $u_0$. Let this distribution function be $\phi(u_0)$. Returning to eq. (2.27) and writing $\overline{C}(\mu) = \int C(\mu)\phi(u_0)du_0$, we have

$$\overline{C}(\mu) = \exp(-\tfrac{1}{2} Q_{22}\mu^2) \int e^{i\mu(x_0 + u_0\psi)} \phi(u_0)du_0 . \tag{3.6}$$

If $\phi$ is Maxwellian

$$\phi_M = \left(\frac{m}{2\pi kT}\right)^{\frac{1}{2}} \exp\left(-mu_0^2/2kT\right), \tag{3.6}$$

then

$$\bar{\sigma}(\mu) = e^{i\mu x_0} e^{-\frac{1}{2}\mu^2(Q_{22} + \psi^2 kT/m)} \tag{3.7}$$

Using (2.18), we finally have

$$\frac{\partial f}{\partial t} = \frac{kT}{m} \psi \frac{\partial^2 f}{\partial x^2}, \tag{3.8}$$

where $f = \int F(x,t;u_0)\, \phi_m(u_0) du_0$.

In the classical case, (2.34), one can see at once how fast $\psi(t) \to m/\zeta$. For a general $\beta(t)$, one must compute $\psi$ explicitly for each case. Note that, for the classical case, as $t \to \infty$ the coefficient of $\partial^2 f/\partial x^2$ becomes $kT/\zeta$, which is Einstein's relation (1.2). Thus our result is completely consistent with well-known, tried and true, results, but extends these to the realm of shorter times. Of course, one cannot trust this type of analysis for extremely short times. For then the governing equations are Newton's laws, where the force cannot be split into a systematic friction and a random force. Rather, one has to integrate the equations of motion of a conservative dynamical system. This is the problem to which we shall address ourselves in the remainder of these lectures.

Let us close this section by acknowledging that the results in sections 2 and 3 were first obtained by Adelman and Dufty, respectively (Adelman, 1976, Dufty, 1974), and independently by the present author (unpublished).

## 4. PROJECTION-OPERATOR METHODS

The obvious next topic to discuss is the molecular basis for the phenomenological description of Brownian motion which has been occupying us. Several methods have been applied to investigate this problem. The one which we shall use is the projection-operator technique. In this section we shall explain the technique itself, and reserve the application to Brownian motion for Section 5.

The projection-operator technique is merely a way, or rather a family of ways, to rewrite an equation of motion in a form which one hopes will be more convenient for approximation than the original form. Before approximations are made, the rewritten equations are rarely any more tractable than the original ones.

In statistical mechanics there are two main classes of projection operators. One class consists of operators defined on the space of distribution functions for a many-body physical system. We call these Zwanzig projection operators (Zwanzig, 1960). The second class is defined on the space of dynamical variables. We call these Mori projection operators (1965). They will never appear in the same problem, and it will always be quite obvious which is under discussion.

Let us first discuss Zwanzig projection operators. Let $\rho$ be a function of the dynamical variables of a system, and the time. Suppose $\rho$ satisfies an evolution equation of the form

$$\frac{\partial \rho}{\partial t} = -i\, L\rho\,, \tag{4.1}$$

where L is some linear operator. Very often, $\rho$ contains more information than we really want to know. Suppose there is a projection operator $\hat{P}$, which projects only the desired information out of $\rho$ , i.e.,

$$f \equiv \hat{P}\rho \tag{4.2}$$

is the function containing what we want to know. To say that $\hat{P}$ is a projection operator is to say that it is idempotent, i.e., $\hat{P}^2 = \hat{P}$ . f is usually called the "relevant part" of $\rho$.

Furthermore, the "irrelevant part" of $\rho$ is

$$\rho - f = (1-\hat{P})\,\rho \equiv g\,. \tag{4.3}$$

We shall often write $1 - \hat{P} = \hat{Q}$. From (4.1) we get

$$\frac{\partial f}{\partial t} = -i\,\hat{P}L\rho = -i\hat{P}Lf - i\hat{P}Lg\,, \tag{4.4a}$$

$$\frac{\partial g}{\partial t} = -i\,\hat{Q}L\rho = -i\hat{Q}Lf - i\hat{Q}Lg\,. \tag{4.4b}$$

Now, one can easily solve (4.4b) for g(t) treating $i\hat{Q}Lf$ as an inhomogeneous term:

$$g = \int_0^t e^{-i\hat{Q}L(t-\tau)}\,\hat{Q}(-iL)f(\tau)d\tau + e^{-i\hat{Q}Lt}g(0)\,. \tag{4.5}$$

One has merely to insert (4.5) in (4.4a) to obtain an equation for f:

$$\frac{\partial f}{\partial t} = \hat{P}\,(-iL)f + \hat{P}\,(-iL)\int_0^t e^{-i\hat{Q}L(t-\tau)}\,\hat{Q}(-iL)f(\tau)d\tau + \hat{P}(-iL)\,e^{-i\hat{Q}Lt}g(0) \tag{4.6}$$

Note that eq. (4.6) does not depend on g(t) but only on the initial value of g, through the last term. This is sometimes called the initial-value term. Thus, eq. (4.1) has been transformed into an equation for f(t), the relevant part of $\rho$, only. But a price has been paid for this. The equation for f is an integro-differential equation. It depends on the history of f in the time interval (0,t). Also, it has rather complicated operators, $\hat{Q}L$, in the exponentials. Nevertheless, (4.6) is often more convenient than (4.1).

We go on now to discuss Mori operators. Suppose $\underset{\sim}{A}$ is a set of dynamical variables (we use vector notation to indicate that there may be more than one variable under consideration). Suppose $\underset{\sim}{A}$ satisfies

$$\dot{\underset{\sim}{A}} = iL\underset{\sim}{A}(t), \tag{4.7}$$

where L is a linear operator. We note the operator identity

$$e^{i(B+C)t} = e^{iBt} + \int_0^t e^{i(B+C)(t-\tau)}\, iCe^{iB}\, d\tau, \tag{4.8}$$

which can easily be verified by differentiation. We apply this to (4.7) as follows: From (4.7),

$$\underset{\sim}{A}(t) = e^{iLt}\,\underset{\sim}{A}(0). \tag{4.9}$$

Now consider projection operators, $\hat{P}$ and $\hat{Q} = 1-\hat{P}$, as before, and set, in (4.8), $B = \hat{Q}L$, $C = \hat{P}L$. Then, combining (4.7), (4.8) and (4.9), one has

$$\dot{\underset{\sim}{A}}(t) = e^{iLt}\,\hat{P}\,iLA(0) + \underset{\sim}{F}^{+}(t) + \int_0^t d\tau\, e^{iL(t-\tau)}\hat{P}iL\underset{\sim}{F}^{+}(\tau). \tag{4.10}$$

where we have written

$$\underset{\sim}{F}^{+}(t) = e^{i\hat{Q}Lt}\,\hat{Q}iL\underset{\sim}{A}(0), \tag{4.11}$$

which is the equation of motion for $\underset{\sim}{A}$ rewritten in projection operator form.

It is quite obvious that all of this operator manipulation has been adapted from that developed for quantum-mechanical perturbation calculus. In particular, the difference in the sign of i in (4.1) and (4.7) was chosen to conform to that in the equations of motion for density matrices and Heisenberg-representation operators in quantum theory. However, it should be clearly borne in mind that the work to follow will be based exclusively on classical mechanics. We have used operator techniques made familiar by quantum theory, but no quantum physics.

It is not obvious, at this stage, what advantages (4.6) and (4.10) have over (4.1) and (4.7) respectively. Indeed, in some cases, there need be no advantage. For the problem of Brownian motion, we shall try to show the advantage in the next section.

## 5. DYNAMICAL THEORY - MOLECULAR

From a molecular point of view, a particle undergoing Brownian motion can be thought of as a heavy particle of mass M. The fluid is made up of light particles of mass m. The Hamiltonian of the system is

$$H = P^2/2m + \sum_i p_i{}^2/2M + U(\underset{\sim}{r}^N, \underset{\sim}{R}) . \tag{5.1}$$

Here, $\underset{\sim}{P}$ is the momentum of the heavy particle, $\underset{\sim}{p}$ that of the light particles, and U the interaction potential between them. $\underset{\sim}{r}^N$ is shorthand for the position vectors of the N light particles. $\underset{\sim}{R}$ is the position of the heavy particle.

The Liouville operator governing the microscopic dynamics is therefore

$$iL = \frac{\underset{\sim}{P}}{M} \cdot \frac{\partial}{\partial \underset{\sim}{R}} + \underset{\sim}{F} \cdot \frac{\partial}{\partial \underset{\sim}{P}} + iL_0 . \tag{5.2}$$

$\underset{\sim}{F} = - \partial U/\partial \underset{\sim}{R}$ is the force exerted by the light particles (hereafter called the bath, for short) on the Brownian particle. $L_0$ is the Liouville operator of the bath particles, considering the Brownian particle as fixed, i.e.

$$iL_0 = \sum_i \frac{p_i}{m} \cdot \frac{\partial}{\partial \underset{\sim}{r}_i} + \sum_i \underset{\sim}{F}_i \cdot \frac{\partial}{\partial \underset{\sim}{p}_i} ,$$
$$\underset{\sim}{F}_i = - \partial U/\partial \underset{\sim}{r}_i . \tag{5.3}$$

We are now almost in a position to try to derive the Fokker-Planck equation from statistical mechanics, using the projection-operator method. The question remaining is what projection operator to use. We adopt the following (Lebowitz and Résibois, 1965). Let $h(\underset{\sim}{R},\underset{\sim}{P},\underset{\sim}{r}^N,\underset{\sim}{p}^N)$ be a function in the phase space of the N + 1 particles; capital and lower-case letters have the same meaning as in (5.3) and (5.2). Then we set

$$\hat{P}\, h = \phi_0\, (\underset{\sim}{r}^N, \underset{\sim}{p}^N;\ \underset{\sim}{R}) \int h \; d\underset{\sim}{r}^N \, d\underset{\sim}{p}^N . \tag{5.4}$$

That is, we average h over the bath variables, and then multiply by $\phi_0$ to return from 1-body phase space to (N+1)-body phase space. $\phi_0$ is the Gibbs distribution for the bath alone, i.e.,

$$\phi_0 = \frac{1}{Z_0} \exp(-H_0/kT), \tag{5.5}$$

where $H_0$ is the Hamiltonian of eq. (5.1) with the $P^2/2M$ (kinetic energy of the Brownian particle) omitted.

The rationale for this choice of projection operator is that (a) it does project out the one-body distribution function of the Brownian particle and (b) the function $\phi_0$, thrown in apparently ad hoc, does describe, roughly, the state of the bath. The reader should verify that $\hat{P}$ is actually a projection operator.

From now on, we shall call the distribution function of the Brownian particle $\rho_1(\underset{\sim}{R},\underset{\sim}{P},t)$ instead of P as in Section 2; we have already used the symbol P for too many different things. We shall let $\rho$ without a subscript be the (N+1)-body distribution function (bath plus Brownian particle).

We can apply eq. (4.6) directly, remembering that $f = \phi_0 \rho_1$, while $g = \rho - \phi_0 \rho_1$. Now we have an initial-value problem to solve, and $\rho(t=0)$, the initial value, is at our disposal. Let us choose

$$\rho(0) = \phi_0 \rho_1(0). \tag{5.6}$$

This choice has the pleasant effect of causing the last term in (4.6), the initial-value term, to vanish identically. But, of course, the choice (5.6) was arbitrary. Why make it? It is a reasonable initial condition. It says that the bath, at zero time, is in equilibrium with the Brownian particle. Assuming one knows nothing else about the bath, what else could one assume? One also suspects that, given any other initial condition that is not too wild, intermolecular interactions would drive the initial-value term to zero very rapidly on the time scale on which $\rho_1$ changes. This expectation was one of the original arguments for choosing (5.6), but models are now known for which the hypothesis of rapid decay is false (Hynes, 1973, Chang, Mazo and Hynes, 1974).

One can easily verify by integration by parts that

$$\hat{P}\, i\, L_0 = 0, \tag{5.7}$$

$$\hat{P}\left(\frac{\underset{\sim}{P}}{M} \cdot \frac{\partial}{\partial \underset{\sim}{R}}\right)\hat{Q} = 0. \tag{5.8}$$

Remembering that the function f in (4.6) is given by $f = \phi_0 \rho_1$, a certain amount of tedious algebra yields

$$\frac{\partial \rho_1}{\partial t} + \frac{\underset{\sim}{P}}{M} \cdot \frac{\partial \rho_1}{\partial \underset{\sim}{R}} =$$

$$\frac{\partial}{\partial \underset{\sim}{P}} \cdot \int_0^t ds \int d\underset{\sim}{r}^N \, d\underset{\sim}{p}^N \, \underset{\sim}{F} \, e^{-i\hat{Q}L(t-s)} \, \underset{\sim}{F} \cdot \phi_0 \left( \frac{\partial}{\partial \underset{\sim}{P}} + \frac{\underset{\sim}{P}}{MkT} \right) \rho_1(s). \qquad (5.9)$$

Of course, if we had not assumed the special initial condition (5.6) we would get an extra term, arising from the last term in (4.6).

It is important to realize that (5.9) is exact; no approximations have yet been introduced. It is now time to approximate! The key observation here is that, for $M \gg m$, the momenta of a bath particle, $\underset{\sim}{p}$, and of the Brownian particle, $\underset{\sim}{P}$, have quite different magnitudes. The ratio is of order $(m/M)^{\frac{1}{2}}$. Of course, both $\underset{\sim}{p}$ and $\underset{\sim}{P}$ are variables, and do not have unique fixed magnitudes. What this statement means is that the important ranges of $\underset{\sim}{p}$ and $\underset{\sim}{P}$ have roughly the above ratio. The argument here is that the important range - the range in which the distribution functions have sensible magnitude - is expected to cluster roughly around the equilibrium root-mean-square values of p and $\underset{\sim}{P}$, i.e., $(mkT)^{\frac{1}{2}}$ and $(MkT)^{\frac{1}{2}}$, respectively.

If this is granted, then, to lowest order,

$$i\,L \approx i\,L_0, \qquad (5.10)$$

for the terms involving $\underset{\sim}{P}$ in L are a factor $(m/M)^{\frac{1}{2}}$ smaller than those involving $\underset{\sim}{p}$, i.e., than $iL_0$.

If we then replace L by $L_0$ in the exponential in (5.9), from (5.7) we have $\hat{Q}L_0 = L_0$. Furthermore, $\phi_0$ commutes with $L_0$ ($\phi_0$ is an eigenfunction of $L_0$ with eigenvalue zero), and $L_0$ does not operate on the $\underset{\sim}{P}$ and $\underset{\sim}{R}$ variables. Hence the integral in (5.9) becomes

$$\int d\underset{\sim}{r}^N d\underset{\sim}{p}^N \phi_0 \underset{\sim}{F} \, e^{-iL_0(t-s)} \, \underset{\sim}{F} = \langle \underset{\sim}{F}(0) \underset{\sim}{F}_0(t-s) \rangle_{eq}, \qquad (5.11)$$

where $F_0(t-s)$ is the force on the Brownian particle at time t-s, if the Brownian particle were stationary, and eq denotes an equilibrium average. This comes about as follows: From eq. (4.7), $\exp[-iL_0(t-s)]\,\underset{\sim}{F} = \underset{\sim}{F}_0(s-t)$. The equilibrium average of a dynamical variable is symmetric in time: $\langle A(0)A(-s)\rangle_{eq} = \langle A(0)A(s)\rangle_{eq}$. Hence, we get (5.11).

Finally then, (5.9) becomes

$$\frac{\partial \rho_1}{\partial t} + \frac{\underset{\sim}{P}}{M} \cdot \frac{\partial \rho_1}{\partial \underset{\sim}{R}} = \int_0^t ds \, \underset{\sim}{G}(t-s) : \frac{\partial}{\partial \underset{\sim}{P}} \left( \frac{\partial}{\partial \underset{\sim}{P}} + \frac{\underset{\sim}{P}}{MkT} \right) \rho_1(s)$$

$$\underset{\sim}{G}(s) = \langle \underset{\sim}{F}(0)\, \underset{\sim}{F}_0(s) \rangle_{eq} \qquad (5.12)$$

Except for the fact that the friction factor in eq. (2.35) is a constant, while that in (5.12) is an integral kernel, we have succeeded in deriving the Fokker-Planck equation from the basic equation of molecular dynamics, the Liouville equation. We must, of course, discuss further the approximations involved, but it will be more convenient to do this after we have looked at the complementary problem, that of the Langevin equation.

To investigate the Langevin equation, it is clear that we should consider the variable $\underset{\sim}{A}$ in (4.7) to be the Brownian particle's momentum, $\underset{\sim}{P}$. As projection operator in the space of dynamical variables we choose

$$\hat{P}\,\underset{\sim}{A} = \int d\underset{\sim}{r}^N d\underset{\sim}{p}^N\, \phi_0\, \underset{\sim}{A}\,, \tag{5.13}$$

i.e., a partial average over the bath variables only. Let us note that (5.7) holds for $\hat{P}$ as given by (5.13) also. The equation of motion for $\underset{\sim}{P}$ appropriate to the projection operator (5.13) is then (4.10). First, we note that $iL\underset{\sim}{P}(0) = \underset{\sim}{F}(0)$, and that $\langle F\rangle_{eq} = 0$: the mean force on a particle in equilibrium is zero. Hence, the first term in (4.10) vanishes, and $\underset{\sim}{F}^+$ becomes

$$\underset{\sim}{F}^+(t) = e^{i\hat{Q}Lt}\, iL\underset{\sim}{P}(0) = e^{i\hat{Q}Lt}\,\underset{\sim}{F}(0). \tag{5.14}$$

Note that

$$\hat{P}\,\underset{\sim}{F}^+\,(t) = \langle \underset{\sim}{F}^+(t)\rangle_{eq} = 0\,, \tag{5.15}$$

because $\hat{P}\,\hat{Q} = 0$. Thus $\underset{\sim}{F}^+(t)$ has one of the attributes of a random force: its average over the bath variables is zero.

A bit of algebra then yields

$$\dot{\underset{\sim}{P}} = \underset{\sim}{F}^+(t) + \int_0^t ds e^{iL(t-s)}\left(\frac{\partial}{\partial \underset{\sim}{P}} - \frac{1}{MkT}\,\underset{\sim}{P}\right)\cdot\langle \underset{\sim}{F}(0)\,\underset{\sim}{F}^+(s)\rangle\,. \tag{5.16}$$

This equation is exact. To proceed further towards the Langevin equation, we must make approximations. Here too the physical idea will be that the important range of $\underset{\sim}{P}$ is near $(MkT)^{\frac{1}{2}}$, while the important range of the bath momenta is a factor $(m/M)^{\frac{1}{2}}$ smaller.

Now $\underset{\sim}{F}^+(s)$ is given by eq. (5.14). Again, we make the approximation that, to order $(m/M)^{\frac{1}{2}}$, we may replace $iL$ by $iL_0$. But then $\langle \underset{\sim}{F}(0)\underset{\sim}{F}^+(s)\rangle$ becomes $\langle \underset{\sim}{F}(0)\underset{\sim}{F}_0(s)\rangle$ , which is independent of $\underset{\sim}{P}$, since the subscript zero means the Brownian particle is held stationary. Thus, (5.16) becomes

$$\dot{\underset{\sim}{P}} = \underset{\sim}{F}^+(t) - \frac{1}{MkT}\int_0^t ds\, \underset{\sim}{P}(t-s)\cdot\langle \underset{\sim}{F}(0)\underset{\sim}{F}_0(s)\rangle \tag{5.17}$$

which is precisely of the form of the Langevin equation we started with, in eq. (2.7) of our phenomenological discussion.

Now, it is quite clear that eq. (5.17) and eq. (5.12) cannot both be correct! For we have seen in Section 2 that a Langevin equation of the form (5.17) gives rise to a Fokker-Planck equation of the type (2.32) rather than (5.12). In the former case, the friction is time-dependent, but occurs only at the same time as the distribution function. In the latter case, the friction is retarded. We make essentially the same approximation in both cases, so how could this arise? Of course, it is well known that the same approximation, made at different stages of a calculation, may yield quite different results. Granted this, which of (5.12) or (5.17) is right, or is neither?

The answer to this is not known (at least to the writer). If I were forced to choose between (5.17) and (5.12), I would choose (5.17), the Langevin equation. The reason for this is that, in one special case, the Langevin equation, (5.17), is known to be *exact*. This case is the Rubin model, a heavy particle in a one-dimensional lattice of light particles all harmonically bound (Rubin, 1960). Rubin found many of the properties of this model by direct integration of Hamilton's equations for the system; the equations are linear. Deutch and Silbey showed that, for this special system, the Langevin equation is exact (Deutch and Silbey, 1971). Mazo, Chang and Hynes considered the Fokker-Planck equation (5.12) for the model, and showed that the solutions differed from Rubin's exact answer (Mazo, Chang and Hynes, 1974).

There is a saying, "for instance is not proof", and that certainly applies here. Nevertheless, this example gives a rationale for making a guess. In the harmonic-lattice case, what spoils the Fokker-Planck equation is that terms of higher order in the small parameter $(m/M)^{\ell}$ have long lifetimes, of order $(M/m)^{\ell}$. Since the approximated operator occurs under a time integral, one has no *a priori* right to neglect small terms in the mass ratio for all times. Put otherwise, the approximation is not uniform in time. It is not known whether the same reasoning applies for a particle in a fluid, but it is very likely that this is indeed the case.

Of course, if one adopts (5.17), then one knows the proper equation for the distribution function. It is (2.32), suitably written for three space dimensions.

The advantage of the discussion of the molecular basis for Brownian-motion theory does not seem very great. We are still left with a problem, namely, to justify the approximations which yielded the known phenomenological answer. Yet the advantages are there.

There are at least three: (1) the theory itself points out what has to be done to justify the approximations; (2) the theory gives an explicit algorithm for the friction factor, which was a phenomenological constant in the older theory; (3) the molecular theory points the way to generalization of Brownian-motion theory to new domains. It is this last point to which we now wish to turn.

## 6. CONCENTRATION DEPENDENCE

So far, our considerations have been restricted to a single Brownian particle in a medium. In practice, there is always more than one particle present. When the concentration of Brownian particles is high enough, one might suspect that the friction factor will depend on the Brownian-particle number density, and thus the diffusion coefficient will become concentration-dependent. It is well known experimentally that this occurs. We now want to develop a molecular theory for the phenomenon (Mazo, 1969).

Instead of having N bath particles and one Brownian particle, we now have N bath particles and n + 1 Brownian particles. We are interested in the one-body distribution function for the heavy particles. Therefore, we single out one of them; call it particle zero. The other n Brownian particles can be considered part of the solvent (there is no reason why all the solvent molecules must be identical).

The Liouville operator is given, as before, by eq. (5.2), but now $iL_0$ contains terms referring to the n Brownian particles. We choose as projection operator, $\hat{P}$, the same operator as given in (5.4); now, however, $\phi_0$ contains the kinetic and potential energies of the n Brownian particles.

The arguments which we have given before hold here too. That is, $\underset{\sim}{P}_0 \cdot (\partial/\partial \underset{\sim}{R}_0) + \underset{\sim}{F}_0 \cdot \partial/\partial \underset{\sim}{P}_0$ is $\mathcal{O}(m/M)^{\frac{1}{2}}$ smaller than $iL_0$. The fact that there are now some small terms in $iL_0$ - those involving the Brownian particles - does not invalidate this remark.

One can now go through the same arguments we went through earlier, and obtain again eq. (5.12). In the present case,

$$\underset{\sim}{G}(s) = \langle \underset{\sim}{F}_0(0)\, \underset{\sim}{F}_0(s) \rangle \tag{6.1}$$

where $\underset{\sim}{F}_0$ is the force on particle zero. Here, however, the propagator which produces $\underset{\sim}{F}_0(s)$ from $\underset{\sim}{F}_0(0)$ is $\exp(-iL_0 s)$, and $iL_0$ permits, in fact demands, that the n Brownian particles, which we are considering part of the bath, move. Of course, we know they will move more slowly than the solvent particles, but, nevertheless, they move.

Thus, our expansion in $(m/M)^{\frac{1}{2}}$ is not a systematic expansion in this parameter.

There is, however, no requirement that when one has a small parameter one must expand everything available in terms of that parameter. In fact, there are known cases where this is a very bad thing to do. The examples of nonlinear oscillations and singular perturbation theory come to mind. In the former example, for instance, the lowest-order effect of a small nonlinear force is to renormalize the oscillation frequency. If one tried to treat the problems by straightforward expansion in the small nonlinearity, one would obtain a first-order solution valid only for very short times (one which increases without bound as time increases).

In the present case, if we had expanded everything available in powers of $(m/M)^{\frac{1}{2}}$ then we would superficially have got the same equation (5.12), exoept that the meaning of $\underset{\sim}{G}$ would be different. The force correlation would be computed with <u>all</u> of the Brownian particles held fixed, and then averaged over the configurations of the n + 1 Brownian particles. This would yield a constant term in the force correlation due to forces between the Brownian particles themselves. Since $\rho_1$, the Brownian-particle distribution function, is a slowly varying function, a constant term in $\underset{\sim}{G}$ would lead to secular behaviour on the right-hand side of (5.12) because of the time integral. This is analogous to the secular behaviour in the more familiar nonlinear oscillation problem, to which we alluded above. By not expanding the bath propagator in the mass ratio we have, at least, avoided this problem.

Thus, the Fokker-Planck equation for a solution of Brownian particles which is not infinitely dilute has exactly the same formal structure as in the infinitely dilute case. The sole difference lies in the fact that the friction function, $\underset{\sim}{G}$, is concentration-dependent. Let us see how far we can analyse this concentration dependence.

When the concentration is not too high, a power-series expansion in the Brownian-particle concentration is possible. The derivation of this expansion has been given previously (Mazo, 1969). We shall not repeat the derivation here, since the techniques are not especially relevant to the main theme of these lectures, irreversible processes. They belong, rather, to the theory of cluster expansions.

Let us just state the result, to first order in the concentration $n/\Omega$ of Brownian particles, $\Omega$ being the volume of the system:

$$\langle \underset{\sim}{F}_0 (0)\, \underset{\sim}{F}_0(t) \rangle_{n+1} =$$

$$\langle \underset{\sim}{F}_0(0)\, \underset{\sim}{F}_0(t) \rangle_1 + \frac{n}{\Omega} \int d\underset{\sim}{R} d\underset{\sim}{P} d\underset{\sim}{r}^N \, d\underset{\sim}{p}^N \times$$

$$\times \left[ \frac{\exp(-\beta H_2)}{Z_2} \underset{\sim}{F}_0^{(2)}(0)\, \underset{\sim}{F}_0^{(2)}(t) - \frac{\exp(-\beta H_1)}{\Omega \lambda Z_1} \underset{\sim}{F}_0^{(1)}(0) \underset{\sim}{F}_0^{(1)}(t) \right]. \quad (6.2)$$

We must explain the notation. The subscripts on the angular brackets $\langle \cdots \rangle_{n+1}$, etc., denote the number of Brownian particles present in the averaging. The subscripts on the Hamiltonians $H_2$, etc., denote the same. $Z_2$ is the partition function for a system of two Brownian particles (and, of course, N solvent particles), and $\lambda = (2\pi MkT)^{3/2}$. The superscripts on the F's also denote the number of Brownian particles (the subscript positions have already been usurped).

We would like to indicate how one might give a reasonable estimate of this first-order term. In the system with two heavy particles present, the force correlation will, at first, decay rather rapidly, i.e., on the scale of decay for a single Brownian particle. This is because the heavy particles move slowly, and the early stages of decay are governed by the solvent-particle motions, just as in the single-body case. But the force correlation will not decay to zero. Rather, it will decay to the correlation function of the solvent-averaged mean force between Brownian particles. This function will then decay on the much longer time scale of the Brownian-particle motion.

The integrand in eq. (6.2) is then, effectively, the force correlation function of this mean force, the initial rapid decay being essentially the same for both terms and largely cancelling. The average force depends on time because the position of the particles depends on time. Thus, we apparently still have to solve the equations of motion for two heavy particles in the solvent, a task even more impossible than for a single heavy particle. However, there is a saving feature.

We already have a theory of how individual Brownian particles move in a fluid, on the average, namely, the entire theory of Section V. Therefore, if we confine ourselves to first order in the density $n/\Omega$ we may use the zeroth-order results to determine the time dependence of the correlation function. If one knows the mean-force law, this reduces the evaluation of the complicated integral (6.2) to a low-dimensional quadrature - in fact, to a one-dimensional integral (Mazo, 1965). Of course, the answer one obtains depends on a parameter, namely, the friction constant at infinite dilution. This parameter enters the dynamics that determines the time dependence. So this method does not completely determine the initial slope (as a function of concentration) of the force correlation from first principles. Rather, it gives the slope in terms of the infinite-dilution intercept. But this is a worthwhile step. It is an enormous simplication of the problem.

One should now ask, does this make sense? We have asserted, but not derived here, that we have a density expansion of a force correlation function. It has been well known in kinetic theory since the mid 1960's that such density expansions for gases do not exist. The individual terms diverge! These divergences are caused by the dynamics of small numbers of molecules, which allows certain repeated collisions involving delicate initial correlations in separated parts of phase space. In reality these collisions do not take place. They are an artifact of the expansion method. There are always other particles present (mean-free-path effects) to destroy these correlations before they cause trouble.

In the present case we also have small numbers of Brownian particles in the early terms of the density series - two in the term we have written in (6.1). But there are always a large number of solvent molecules present. These will dominate the motion of the Brownian particle, and prevent the kind of divergence which occurs in the gas case. There may possibly be as-yet undiscovered pathologies in the expansion we have derived. But, if so, they will have a different origin from that in the more familiar case of dense gases.

Let us end this discussion by stating that there is no reason to expect our considerations to be valid when there are long-range forces between the Brownian particles. Long-range forces have pathologies of their own which are not removed simply by having a solvent present. On the other hand, the case of long-range forces does not seem to be terribly important physically. As far as I know it has not been treated.

## 7. NONUNIFORM SYSTEMS

So far, we have been considering a heavy particle in a uniform, homogeneous medium of light particles, and the generalization of this situation to finite concentrations. We now want to ask: what is the equation of motion of the heavy particle when the surrounding medium is nonuniform? What happens when there are gradients of temperature, velocity, or, in the case of multicomponent solvents, composition?

This problem can be treated using the same methods we have used up to now in these lectures. Our basic starting point is, again, eq. (4.6). However, the choice of projection operator which we have so far used is clearly no longer appropriate. The formal equation (4.6) is invariant to the choice of projection operator. But, if we are to make approximations, and stop at some finite order of approximation, it behoves us to use a projection operator which gives a result as close as possible to the exact one in lowest order. The projection operator (5.4) does not do this, for it essentially projects on to a description of a particle moving in a uniform fluid.

The simplest projection operator which takes nonuniformity into account seems to be

$$\hat{P}h = \phi(\underset{\sim}{r}^N, \underset{\sim}{p}^N; \underset{\sim}{R}) \int h \, d\underset{\sim}{r}^N d\underset{\sim}{p}^N , \tag{7.1}$$

where $\phi$, in contrast to $\phi_0$, is not an equilibrium distribution function, but a local-equilibrium distribution function. That is

$$\phi = Z^{-1}\exp\Big\{- \sum_{\alpha} \sum_{j=1}^{N} \beta_j (p_j - mu_j)^2/2m_\alpha - \frac{1}{2} \sum_{\alpha,\beta} \sum_{j,l} U_{\alpha\beta}(r_{jl})\beta_j$$

$$+ \sum_{\alpha} \sum_{j} \nu_\alpha(r_j) . \tag{7.2}$$

Here, Z is a normalization factor; the Greek subscripts label species, the Latin, particles. $\beta_j$ is $(kT(r_j))^{-1}$, i.e., the inverse of the local temperature. $U_{\alpha\beta}$ is the potential of the interaction between a particle of species $\alpha$ and one of species $\beta$. Finally $\nu_\alpha(r_j)$ is an abbreviation for $\beta(r_j)\,\mu_\alpha(r_j)$, where $\mu_\alpha(r_j)$ is the chemical potential of species $\alpha$ at position $r_j$, and $u_\alpha(r_j)$ is the mean velocity of species $\alpha$ at position $r_j$.

One now need only go through exactly the same manipulations we have already gone through in obtaining eq. (4.12). They are more complex because the Liouville operator operates on $u_\alpha$, $\beta$ and $\nu_\alpha$, since these are functions of particle position. Nevertheless, modulo some extra algebra, the procedure is the same (Mazo, 1969) The results, however, look quite complicated, and we only write them down for the case of small gradients, i.e., when we linearize in $\nabla\beta$, $\nabla u$ and $\nabla\nu$. Let us just display the final result:

$$\frac{\partial \rho_1}{\partial t} + \underset{\sim}{P}_0 \cdot \frac{\partial \rho_1}{\partial \underset{\sim}{R}_0} + \langle \underset{\sim}{F}_0 \rangle \cdot \frac{\partial \rho_1}{\partial \underset{\sim}{P}_0} = \frac{\partial}{\partial \underset{\sim}{P}_0} \cdot \Big[ \zeta' \Big( \frac{\partial}{\partial \underset{\sim}{P}_0} + \frac{P_0 - Mu_0}{MkT} \Big)$$

$$+ \eta \frac{\nabla\beta}{\beta} + \sum_{\alpha} \gamma_\alpha \nabla_T \nu_\alpha \Big] \rho_1 \tag{7.3}$$

In addition to linearization, we have also made the assumption that $\rho_1$ varies slowly with respect to the scale of time variation of the correlation functions, so that it may be taken outside the integral i.e., for example,

$$\int_0^t G(t-s)\, \rho_1(s)ds \approx \rho_1(t) \int_0^t G(t-s)\, ds$$
$$\approx \rho_1(t) \int_0^{\infty} G(s)ds \tag{7.4}$$

The zero subscripts, as before, refer to the Brownian particle, which we call particle zero. $\langle F_0 \rangle_\ell$ es the average force on the Brownian particle in the local equilibrium environment. The other quantities in (7.3) are defined as follows:

$$\zeta' = \int_0^{\infty} \langle \underset{\sim}{F}_0(0)\, \underset{\sim}{F}_0(s) \rangle\, ds + \frac{1}{kT} \int_0^{\infty} \langle \underset{\sim}{F}_0(0)\, \underset{\sim}{F}_0(s) \sum_j \frac{\underset{\sim}{r}_j}{m_j} (\underset{\sim}{p}_j - m_j \underset{\sim}{u}_j) \rangle : \nabla \underset{\sim}{u} \tag{7.5a}$$

$$\eta = \frac{1}{kT} \int_0^{\infty} \langle \underset{\sim}{F}_0(s)\, \underset{\sim}{Q}(0) \rangle\, ds \tag{7.5b}$$

$$\gamma_\alpha = \int_0^{\infty} \langle \underset{\sim}{F}_0(s)\quad \underset{\sim}{j}_\alpha(0) \rangle\, ds \tag{7.5c}$$

Here $\underset{\sim}{Q}$ is the microscopic heat-flow vector:

$$\underset{\sim}{Q} = \sum_j \frac{\underset{\sim}{p}_j}{m_j} \Big[ \{ p_j^2/2m_j - h\,(\underset{\sim}{r}_j) \}\, \underset{\sim}{1}$$
$$+ \frac{1}{2} \sum_{j \neq k} (U_{jk}\, \underset{\sim}{1} - \underset{\sim}{r}_{jk} \cdot \frac{\partial}{\partial \underset{\sim}{r}_k} U_{jk})$$
$$+ U_{0k}\, \underset{\sim}{1} - \underset{\sim}{r}_{0k} \cdot \frac{\partial}{\partial \underset{\sim}{r}_{0k}} U_{0k} \Big]\quad \delta(\underset{\sim}{r}_j - \underset{\sim}{r}) \tag{7.6a}$$

and $\underset{\sim}{j}_\alpha$ is the microscopic current of species $\alpha$ :

$$\underset{\sim}{j}_\alpha = \sum \frac{\underset{\sim}{p}_j}{m_j}\, \delta(\underset{\sim}{r}_j - \underset{\sim}{r}) \tag{7.6b}$$

h being the local enthalpy density.
(We have written these formulae for the case of no mean mass flow.)
The symbol $\nabla_T$ means $\nabla - (\nabla T) \partial/\partial T$ .

Thus, the Fokker-Planck equation has two new terms, one involving the temperature gradient, and one involving the chemical-potential, or concentration, gradient. These terms will describe thermal diffusion and mutual diffusion, whereas the ordinary Fokker-Planck equation describes self-diffusion. The new terms have their own new friction constants $\eta$ and $\gamma_2$ . Furthermore, the expression for the ordinary friction constant has an additional term. This term has never been thoroughly analysed, as far as I know, but is likely to be small, for it is proportional to $\nabla \underset{\sim}{u}$, which is small by hypothesis. Equation (7.3) has been derived by other techniques by Nicolis (1965) and by Zubarev and Bashkirov (1968).

We shall leave the subject here, and not enter into a discussion of physical situations in which one might want to apply the equation. In any event, the consequences of eq. (7.3) have not been studied very much.

## 8. CONCLUDING REMARKS

The aspects of Brownian motion theory which we have considered in these lectures are, of course, only a subset of the topics which could have been discussed. Let us mention some of these other topics briefly.

1. Rotational Brownian Motion and Rotational Diffusion

Our considerations have been restricted to the translational case, but the case of rotation is also interesting and important. Clearly, angular momentum takes the place of linear momentum, and Euler angles the place of cartesian coordinates. Otherwise the theory is rather similar to the theory we have been treating in these lectures. The differences are technical, not conceptual.

Rotational Brownian motion occurs in a number of contexts. One of the earliest was the dielectric dispersion theory of Debye, where rotational Brownian motion acts to counterbalance the aligning effect of an external electric field on polar molecules. Other areas are fluorescence depolarization and flow birefringence. In the former, molecular rotation changes the orientation of the transition dipole between the absorption and emission steps of a fluorescence process, leading to depolarization. In the latter, rotational Brownian motion opposes the aligning effect of hydrodynamic torques on optically anisotropic molecules.

2. Internal Degrees of Freedom

We have treated our Brownian particles as if they were point masses. Real Brownian particles are composed of many molecules, and clearly have internal degrees of freedom. They can collide inelastically with solvent molecules. In the normal course of events, i.e., thermal equilibrium, this is not likely to produce anything interesting. But one can envisage a situation in which the Brownian particle has a different internal temperature from that of the ambient medium. Consider, for example, a soot particle bathed in sunlight in an ambient medium which is transparent to the radiation. This case has been analyzed (Slinn et al, 1970; Mazo, 1974a; Mazo, 1974b). Indeed, the diffusion coefficient depends on both the particle internal temperature and the bath temperature, in the case of a gas. In the case of a dense fluid, only the particle temperature is relevant.

### 3. Polymer Dynamics

The transport properties of high-polymer solutions depend on the configurations of the polymer molecules. For flexible molecules, these configurations undergo a sort of internal Brownian motion. This is complicated by the fact that there are strong geometrical constraints on the possible configurations, due to the chemical bonds holding the molecule together, and by the long-range hydrodynamic interaction between polymer segments. It is usually treated in the diffusion approximation, where the momentum distribution of the segments is assumed to have relaxed to its equilibrium value (Kirkwood,1967; Zwanzig, 1969).

### 4. The Critical Region

When the solvent is near its critical point, density fluctuations become very large and the compressibility becomes very large. It is then no longer possible to treat the solvent as an incompressible fluid for which Stokes' law holds. It turns out, though, that linearized hydrodynamics predicts no critical anomalies. The numerical value of the friction constant changes, but that is all. Nonlinear effects and fluctuations do lead to changes near the critical point. Near the critical point there is a slowing down of Brownian diffusion – an example of a very general critical slowing-down phenomenon.

Very close to the critical point the correlation length of the fluctuations becomes large compared to the size of the Brownian particle, and an analysis predicts a rather sharp decrease in the diffusion coefficient. This subject has been reviewed recently by Gitterman (1978).

### 5. The Problem of the Higher-order Terms

In our discussion of both Fokker-Planck and Langevin equations from the molecular point of view we talked about an expansion of the general formal equations in powers of $(m/M)^{\frac{1}{2}}$, and stopped at the lowest-order term. In essence, we stopped because the lowest-order term gave us what we expected to get, i.e., the known phenomenological equations. It is really necessary to look seriously at what is left over in order to justify stopping. The smallness of the expansion parameter is, by itself, not adequate justification. This is emphasised in the present case by the inequivalent results in the Fokker-Planck and Langevin cases that we have previously mentioned.

Hynes, Kapral and Weinberg (Hynes et al., 1975) have studied both Fokker-Planck and Langevin equations by introducing Langevin equations for higher powers of the momentum. These have their own random forces and damping, and are all coupled. The couplings between the momentum itself and the nonlinear modes is claimed to have no parts which decay on the long time scale of the momentum decay. This eliminates the major conceptual problem of the earlier theory at the expense of considerable technical complications due to the simultaneous consideration of many coupled modes. It is found that, in lowest order, the usual equations are valid. To higher order in m/M there is no closed set of equations, but a hierarchy for all of the modes.

With these brief and fragmentary remarks we close. There are doubtless other subjects that we could have listed as further topics — quantum Brownian motion, Brownian motion as a natural limit to measurements, etc. But we shall stop here. Our aim has been to take an old, established problem in the field of stochastic processes, to discuss some of its more recent developments, and to connect it with the molecular theory of many-body systems.

## REFERENCES


ADELMAN, S., (1976) J. Chem. Phys. 64, 124

BRUSH, S., (1968) Arch. Hist. Exact Sci. 5, 1

CHANG, E.L., MAZO, R.M., and HYNES, J.T., (1974) Mol. Phys. 28, 997

DEUTCH, J., and SILBEY, R., (1971) Phys. Rev. A3, 2049

DOOB, J.L., (1942) Ann. Math. 43, 351

DUFTY, J., (1974) Phys. Fluids, 17, 328

EINSTEIN, A., (1905) Ann. der Physik, 17, 549

GITTERMAN, M., (1978) Rev. Mod. Phys. 50, 85

HYNES, J.T.,(1973) J. Chem. Phys. 59, 3459

HYNES, J.T., KAPRAL, R., and WEINBERG, M., (1975a) Physica 81A, 485

HYNES, J.T., KAPRAL, R., and WEINBERG, M., (1975b) Physica 81A, 509

KERKER, M., (1974) J. Chem. Ed. 51, 764

KIRKWOOD, J.G., (1967), Macromolecules, ed. P.L. AUER, Gordon and Breach, N.Y. (a collection of reprints of original papers)

LEBOWITZ, J., and RESIBOIS, P., (1965) Phys. Rev. 139A, 1101

MAZO, R.M., (1965) J. Chem. Phys., 43, 2873

MAZO, R.M., (1969a) J. Stat. Phys., 1, 89

MAZO, R.M., (1969b) J. Stat. Phys., 1, 101

MAZO, R.M., (1974a) J. Stat. Phys., 12, 427

MAZO, R.M., (1974b) J. Chem. Phys., 60, 2634

MORI, H., (1965) Progr. Theor. Phys. (Japan), 33, 423

NICOLIS, G., (1965), J. Chem. Phys. 43, 1110

RUBIN, R.J., (1960) J. Math. Phys. 1, 309

SLINN, W.G.N., SHEN, S.F., and MAZO, R.M., (1970) J. Stat. Phys. 2,251

UHLENBECK, G.E., and ORNSTEIN, L.S., (1930) Phys. Rev. 36, 823

ZUBAREV, D.N., and BASHKIROV, A.G., (1968) Physica 39, 334

ZWANZIG, R.W., (1960) J. Chem. Phys. 33, 1388

ZWANZIG, R.W., (1969) in: Stochastic Processes in Chemical Physics, ed. K.E. SHULER, Interscience, N.Y.

# PATH-INTEGRAL METHODS IN NONEQUILIBRIUM THERMODYNAMICS AND STATISTICS

R. Graham
Universität Essen GHS
West Germany

## 1. INTRODUCTION

Nonequilibrium thermodynamics and statistics is a broad field of physics[1-4]. It is the general theory of physical systems that can be characterized by a set of macroscopic parameters but which are not in thermodynamic equilibrium. The phenomena one wants to understand and describe are very diverse and include the approach to thermodynamic equilibrium states, the approach to, and behaviour in, steady states, fluctuations in steady states, instabilities of steady states, and related phenomena such as symmetry changes and self-organization in steady states.

There have been many different approaches to nonequilibrium thermodynamics, which differ in their scope and domain of validity. One well-known approach is based on linear-response theory[5], which is applicable for small departures from thermodynamic equilibrium; another approach is based on a certain general form of the "nonequilibrium statistical operator"[6]. Yet other approaches make use of the "master equation", or equations derived from it, like the Boltzmann equation or the Fokker-Planck equation.

There have also been phenomenological approaches to this field, most notably those put forward by Onsager[7,8] and Prigogine[9-12] and their respective schools. The central aim of these works is to establish new potentials from which the properties of macroscopic systems away from equilibrium can be derived. The two phenomenological approaches mentioned have also been related to fluctuation theory in some restricted cases, in the work of Onsager and Machlup[8] and of Prigogine and Glansdorff[11].

Surprisingly, the precise relations between these two major phenomenological approaches have hardly been studied. (An exception is the book by Gyarmati[2].) One reason seems to be the fact that neither the Onsager-Machlup nor the Prigogine-Glansdorff theory has been formulated in sufficient generality to be readily comparable with each other.

In these lectures, we will try to develop a general theory from which both the Onsager-Machlup theory and the Prigogine-Glansdorff theory may be obtained as special cases. The relation to Schlögl's[13-15] statistical formulation and to the generalization of the Prigogine-Glansdorff theory is also considered. The theory we present is an extension of our earlier approaches to the subject[16-18].

In section 2 we start by considering the basic postulates and equations that apply to macroscopic systems. We show that the description can be based on a Fokker-Planck equation. Since this equation has also been derived from a more fundamental microscopic description in many cases[19,20], in the following I shall accept the Fokker-Planck equation as a sufficiently secure starting point.

Onsager and Machlup[8] require more: they restrict their analysis to linear Langevin equations of a special form that holds near thermodynamic equilibrium. It is this assumption which allows them to derive their results, but it makes a comparison with the Prigogine-Glansdorff approach, which describes small fluctuations in steady states far from thermodynamic equilibrium, impossible.

The central theme of these lectures is to avoid the linearity assumption of the Onsager-Machlup theory and the assumption that one is close to the thermodynamic equilibrium state. Nevertheless, the language introduced by Onsager and Machlup is found to be very useful. It is the language of probability densities for paths in the space of macroscopic variables. Averages and correlation functions reduce to path integrals[21] in this language. In fact, since Wiener originally invented integrals of this type for the description of Brownian motion[2] - the simplest problem studied in nonequilibrium thermodynamics and statistics - this is the classic field of their application.

The beauty of path integrals in this context derives from the resulting pronounced formal analogy to equilibrium thermodynamics and statistics. For example, nonequilibrium potentials are related to probability-density functionals of paths, much as equilibrium potentials are related to probability distributions in equilibrium.

In section 3 we consider the path-integral formulation for small fluctuations in steady states. This analysis is related in section 4 to various variational principles. The Onsager-Machlup theory near equilibrium and the Prigogine-Glansdorff theory for fluctuations in nonequilibrium steady states are discussed as special cases. In section 5 the linearity assumption is lifted and the path-integral formulation of the non-linear theory is given[23]. In section 6 the relation of the probability-density functional to dissipative potentials is discussed. In section 7 the variational principles for the nonlinear case are formulated and the relation of our theory to Schlögl's[13] statistical formulation of the Prigogine-Glansdorff theory is obtained. In section 8 we discuss the relation between the path-integral solution of the Fokker-Planck equation and the variational principles for the nonlinear case.

We shall not have time to consider in detail specific applications of the formalism we develop. In particular, we shall not exploit here the new approximation schemes (like variational principles) offered by the path-integral approach for the solution of nonlinear statistical problems[21], nor shall we discuss the diagrammatic perturbation theory following from the path integral. Concerning these topics, the interested reader is referred to the literature[(24-28)]. Instead we have chosen to emphasise the general aspects of the theory in these lectures. The only application that we consider, in section 9, is to a very well-studied system which shows the simplest kind of instability and self-organization that one can conceive – the Bénard convection cell.

## 2. BASIC POSTULATES AND EQUATIONS OF STATISTICAL NONEQUILIBRIUM THERMODYNAMICS

Let us consider a system described by a set of macroscopic variables $q^\nu$ ($\nu = 1, 2, \ldots, n$) and let us assume that the set $q^\nu$ is "complete", i.e., that all other variables of the system have much shorter relaxation times. The variables $q^\nu$ as functions of time then define an n-dimensional Markovian stochastic process which completely describes the macroscopic dynamics of the system. Let us assume, furthermore, that the system is "aged", i.e., that it is in a state where all the fast variables had sufficient time to relax, and let the boundary conditions of the system be stationary, i.e., time-independent, when expressed in the $q^\nu(t)$. Then the n-dimensional Markov process $q^\nu(t)$ is stationary. Its conditional probability density

$$P = P(qt|q_0t_0) \tag{2.1}$$

then only depends on $t - t_0$.

The equation of motion satisfied by $P$ can be obtained from the microscopic equations by a number of techniques, most notably by projector techniques[20]. However, without entering into a discussion of the microscopic properties of the system, one may deduce many of the properties of the equation of motion satisfied by $P$ from the basic postulates made above.

The Markov property implies that the conditional probability density satisfies the Chapman-Kolmogorov equation

$$P(qt|q_0t_0) = \int dq_1\, P(qt|q_1t_1)P(q_1t_1|q_0t_0) \quad (t > t_1 > t_0) \tag{2.2}$$

The macroscopic nature of the variables $q^\nu$ implies that they refer to a collective property of a sufficiently large number of the microscopic constituents. Therefore, they take on continuous values to a good approximation. Furthermore, their change in time is continuous to a good approximation, since it is brought about by the local interaction of the microscopic constituents. Equation (2.2) can then be transformed into a differential equation by using the Kramers-Moyal expansion

$$\frac{\partial P}{\partial t} = \sum_{s=1} \frac{(-1)^s}{s!} \left[ \frac{\partial^s Q^{\nu_1 \nu_2 \cdots \nu_s}(q)\, P}{\partial q^{\nu_1} \cdots \partial q^{\nu_s}} \right] \tag{2.3}$$

where

$$\begin{aligned} Q^{\nu}(q_0) &= \lim_{\varepsilon \to 0} \frac{1}{\varepsilon} \int dq (q^{\nu} - q_0^{\nu}) P(qt+\varepsilon \mid q_0 t) \equiv K^{\nu}(q_0) \\ Q^{\nu_1 \cdots \nu_s}(q_0) &= \lim_{\varepsilon \to 0} \frac{1}{\varepsilon} \int dq (q^{\nu_1} - q_0^{\nu_1}) \cdots (q^{\nu_s} - q_0^{\nu_s}) P(qt+\varepsilon \mid q_0 t) \end{aligned} \tag{2.4}$$

We now express the continuity of the process $q^{\nu}(t)$ by the requirement that, for $\varepsilon \to 0$,

$$|q^{\nu}(t+\varepsilon) - q^{\nu}(t)| \leq K \varepsilon^{\alpha} \tag{2.5}$$

with probability 1, where $K$ is some positive constant and $\alpha > 0$. As a consequence of (2.5), the coefficients $Q^{\nu_1 \cdots \nu_s}(q)$ vanish for all $s > 2$.

In order to see this, one introduces the quantities

$$\Delta q = (q^{\nu} - q_0^{\nu}) \xi_{\nu} \tag{2.6}$$

with the arbitrary real parameters $\xi_{\nu}$ and

$$Q^{(s)} = Q^{\nu_1 \cdots \nu_s} \xi_{\nu_1} \cdots \xi_{\nu_s}, \quad Q^{(0)} = 1 . \tag{2.7}$$

Then one has

$$Q^{(s)} = \lim_{\varepsilon \to 0} \frac{1}{\varepsilon} \langle (\Delta q)^s \rangle \tag{2.8}$$

where the angular brackets denote the average over $q$ with $P(qt+\varepsilon \mid q_0 t)$. Since $P$ is non-negative it can be used as a weight function to define a scalar product of all functions of $q$ that are square-integrable with the weight $P$. With the Schwartz inequality for this scalar product, one proves

$$\langle (\Delta q)^{r+s} \rangle^2 \leq \langle (\Delta q)^{2r} \rangle \langle (\Delta q)^{2s} \rangle \tag{2.9}$$

which is equivalent to

$$(Q^{(r+s)})^2 \leq Q^{(2r)}\, Q^{(2s)} \quad . \tag{2.10}$$

Because of (2.5) we have

$$Q^{(s)} = 0 \qquad s \geq N + 1 > \frac{1}{\alpha} \tag{2.11}$$

where N is some integer larger than $1/\alpha$ . Applying the inequality (2.10) for r = 1, s = N - 1, we obtain

$$\left[Q^{(N)}\right]^2 \leq Q^{(2)} Q^{(2N-2)} = 0 \text{ for } N > 2 . \tag{2.12}$$

The right-hand side of (2.12) vanishes because of (2.11), since $2N - 2 \geq N + 1$ for $N > 2$. Hence $Q^{(N+1)} = 0$ implies $Q^{(N)} = 0$, etc., down to N = 2 in (2.12).

Equation (2.3) is therefore reduced by the requirement of continuity (2.5) to the Fokker-Planck equation

$$\frac{\partial P}{\partial t} = -\frac{\partial}{\partial q^\nu}(K^\nu(q)P) + \tfrac{1}{2}\frac{\partial^2}{\partial q^\nu \partial q^\mu}(Q^{\nu\mu}(q)P). \tag{2.13}$$

It is interesting to note that the positivity of P has been used in an essential way in going from (2.3) to (2.13). If P is a quasi-probability density of a quantum system, the positivity of P is not guaranteed, and one may have continuous stochastic quantum processes involving higher than second-order derivatives in (2.13).

The boundary conditions that have to be used with (2.4) depend, of course, on the specific system and the specific questions one wants to ask. The conservation of the normalization requires that

$$\frac{\partial}{\partial t}\int P dq = \oint df_\nu \left[K^\nu(q)P - \tfrac{1}{2}\frac{\partial}{\partial q^\mu} Q^{\nu\mu}(q)P\right] = 0 \tag{2.14}$$

where the integral on the right-hand side is over the boundaries of the q-space. Generally, we shall assume the integrand on the right-hand side of (2.14) to vanish at the boundaries or to satisfy cyclic boundary conditions in angular variables.

The matrix of functions $Q^{\nu\mu}(q)$ is non-negative, owing to its definition. If we make the stronger assumption that $Q^{\nu\mu}(q)$ is positive-definite, the solutions W(qt) of eq. (2.13) can be shown to satisfy the H-theorem. We shall formulate this theorem under the assumption that a time-dependent solution $W_0(q)$ of (2.13) exists. Then this theorem states[29] that the the quantity

$$K(t) = \int dq\, W(qt) \ln \frac{W(qt)}{W_0(q)} \tag{2.15}$$

is positive for $W(qt) \neq W_0(q)$ and vanishes for $W(qt) = W_0(q)$:

$$K(t) \begin{cases} > 0 & \text{for} \quad W(qt) \neq W_0(q) \\ = 0 & \text{for} \quad W(qt) = W_0(q) \end{cases} \tag{2.16}$$

and is decreasing in time until $W(qt) = W_0(q)$:

$$\dot{K}(t) \begin{cases} < 0 & \text{for} \quad W(qt) \neq W_0(q) \\ = 0 & \text{for} \quad W(qt) = W_0(q) \end{cases} \quad . \tag{2.17}$$

Equation (2.16) is proved, with the help of the inequality $x - 1 + \ln \frac{1}{x} > 0$ for all real $x > 0$, $x \neq 1$, by rewriting eq. (2.15) in the form

$$K(t) = \int dq\, W(qt) \left( \ln \frac{W(qt)}{W_0(q)} - 1 + \frac{W_0(q)}{W(qt)} \right) \quad . \tag{2.17}$$

Equation (2.17) is proved by first making the rearrangements

$$\begin{aligned} \dot{K}(t) &= \int dq (\ln W - \ln W_0 + 1) \dot{W} \qquad (2.18) \\ &= \int dq (\ln W - \ln W_0 + 1) \left[ -\frac{\partial}{\partial q^\nu} (K^\nu W) + \tfrac{1}{2} \frac{\partial^2}{\partial q^\nu \partial q^\mu} (Q^{\nu\mu} W) \right], \end{aligned}$$

performing a partial integration, in which the boundary terms vanish:

$$\dot{K}(t) = \int dq\, W \left[ \frac{\partial}{\partial q^\nu} \left( \ln \frac{W}{W_0} \right) \right] \left[ K^\nu - \tfrac{1}{2} \frac{\partial Q^{\nu\mu}}{\partial q^\mu} - \tfrac{1}{2} Q^{\nu\mu} \frac{\partial \ln W}{\partial q^\mu} \right] \tag{2.19}$$

then adding an integral, which vanishes because of $\dot{W}_0 = 0$,

$$0 = - \int dq \frac{W}{W_0} \left[ \frac{\partial}{\partial q^\nu} \left( \ln \frac{W}{W_0} \right) \right] \left[ \left( K^\nu - \tfrac{1}{2} \frac{\partial Q^{\nu\mu}}{\partial q^\mu} \right) W_0 - \tfrac{1}{2} Q^{\nu\mu} \frac{\partial W_0}{\partial q^\mu} \right] \tag{2.20}$$

to obtain, finally,

$$\dot{K}(t) = -\tfrac{1}{2} \int dq\, W Q^{\nu\mu} \left[ \frac{\partial}{\partial q^\mu} \left( \ln \frac{W}{W_0} \right) \right] \left[ \frac{\partial}{\partial q^\nu} \left( \ln \frac{W}{W_0} \right) \right] \leq 0 \quad . \tag{2.21}$$

For positive-definite $Q^{\nu\mu}$ eq. (2.21) vanishes only if $\ln \frac{W}{W_0} = \text{const}$, and is negative otherwise. If there are no mutually disconnected parts of q-space, $\ln \frac{W}{W_0}$ must be constant throughout q-space and equal 0 by normalization.

The H-theorem implies under quite general conditions that the steady-state distribution $W_0(q)$ is unique. Moreover, there is no periodic solution of (2.13) under the same general conditions. These conditions are: (i) that the system is finite; (ii) that the phase space spanned by the $q^\nu$ is irreducible in the sense that all its points are accessible, starting from any point[29].

If the system is infinite, spontaneous symmetry-breaking is a possibility. Then $W_0$ need no longer be unique. If the phase space may be reduced into several disconnected parts, $\ln(W/W_0) = \text{const}$ may be satisfied with different constants in different parts, and $W_0$ is also not unique. In the opposite case, the normalization condition implies const $= 0$.

The time-independent solution $W_0(q)$ defines a macroscopic potential $S(q)$ by

$$W_0(q) \quad = \quad \exp S(q) \quad . \tag{2.22}$$

If the time-independent state described by $W_0(q)$ is an equilibrium state, $S(q)$ is the appropriate thermodynamic potential of the system provided the values of the $q^\nu$ are fixed by coupling the system to additional reservoirs. Strictly, $S(q)$ is then the entropy of the total system, including these additional reservoirs. If the time-independent state is not an equilibrium state but a steady state, $S(q)$ is a generalization of the usual thermodynamic potential.

To see this we rewrite eq. (2.13) in terms of $S(q)$:

$$\frac{\partial}{\partial q^\nu}\left(r^\nu(q)e^{S(q)}\right) \quad = \quad 0 \tag{2.23}$$

$$K^\nu(q) - \tfrac{1}{2}\frac{\partial Q^{\nu\mu}}{\partial q^\mu} \quad = \quad \tfrac{1}{2}\, Q^{\nu\mu}(q)\,\frac{\partial S(q)}{\partial q^\mu} + r^\nu(q) \quad . \tag{2.24}$$

Equation (2.24) gives the decomposition of the total drift velocity in phase space $(K^\nu(q) - \tfrac{1}{2}\partial Q^{\nu\mu}/\partial q^\mu)$ into a part derived from the potential $S(q)$ and a residual part $r^\nu(q)$. Actually eq. (2.24) is to be considered as the definition of $r^\nu(q)$. In eq. (2.23), $r^\nu(q)$ reappears as the drift velocity of the probability current in the steady state. Equations (2.23) and (2.24) are well-known equations if $S(q)$ describes an equilibrium state. Equation (2.23) is then the Liouville equation for the equilibrium distribution. Equation (2.24) expresses the total drift of the macroscopic variables as a superposition of the drift which appears in the Liouville equation, describing reversible conservative motion, and the dissipative drift, which is proportional to the generalized thermodynamic forces $\partial S(q)/\partial q^\nu$. These are just the constitutive equations, which are the basis of Onsager's approach to nonequilibrium thermodynamics.

In the nonequilibrium steady state the total drift is a similar superposition; however, it is no longer possible, in general, to assign reversible processes to $r^\nu(q)$ and irreversible processes to $\frac{1}{2}Q^{\nu\mu}(q)\frac{\partial S}{\partial q^\mu}$, a fact which will be considered in more detail later. Thus, in nonequilibrium steady states irreversible processes still take place. This is the most important difference between nonequilibrium steady states and equilibrium states from a general point of view.

However, we want to point out that, even in nonequilibrium steady states, not all degrees of freedom of the system need to be involved in irreversible drift processes in phase space. Sometimes there exist some variables in such systems which in the steady state have $r^\nu(q) = 0$. The most striking example is a single-mode laser near threshold. Sufficiently close to the threshold, the photon number is much slower than all other variables and is the only macroscopic variable which has to be considered. In the laser, photons are emitted and reabsorbed by the atoms by being radiated through or absorbed by the mirrors. This is clearly an irreversible process and the laser can be in a steady state but not in an equilibrium state. Nevertheless, the photon number in the steady state satisfies $r(n) = 0$, since the alternative allowed by (2.23) is $r(n) = \text{const}\ e^{-S(n)}$ and diverges for $n \to \infty$. For such peculiar cases, eqs. (2.23), (2.24) for the steady state have the same formal properties as for equilibrium states. In particular, $S(q)$ then has all the formal properties of a thermodynamic potential. Another example of $r^\nu(q) = 0$ occurs at the Bénard instability on long time scales, which is considered in section 9.

If $r^\nu(q)$ is not zero for reasons of symmetry and the boundary conditions, it is very hard to determine for nonequilibrium steady states. It is necessary to obtain the steady-state distribution $W_0(q)$ first.

## 3. LINEAR PROCESSES

We first consider systems with distribution functions that peak sharply around their maximum and thus exhibit small fluctuations, either in nonequilibrium steady states or in equilibrium states. The latter case was considered by Onsager and Machlup using the properties of Gaussian path integrals, and we shall also make use of this language here[8].

Let us take $q^\nu = 0$ at the most probable state. Then we have the expansion

$$S(q) = -\tfrac{1}{2}\,\sigma^{-1}_{\mu\nu}\, q^\nu q^\mu + S(0) \tag{3.1}$$

$$K^\nu(q) = B^\nu_\mu\, q^\mu \tag{3.2}$$

$$Q^{\nu\mu}(q) = \text{const.} \tag{3.3}$$

Since $S(q)$ is defined by the time-independent solution of the Fokker-Planck equation, $\sigma^{-1}_{\mu\nu}$ is related to $B^{\nu}_{\mu}$ and $Q^{\nu\mu}$ by

$$-\sigma^{\nu\lambda}B_{\lambda}{}^{\mu} - B^{\nu}{}_{\lambda}\sigma^{\lambda\nu} = Q^{\nu\mu} \; ; \; \sigma^{-1}_{\mu\lambda}\sigma^{\lambda\nu} = \delta_{\mu}{}^{\nu} \tag{3.4}$$

from which $\sigma_{\mu\nu}$ has to be determined for given $B^{\mu}{}_{\nu}$, $Q^{\mu\nu}$. The drift rate in the steady state, $r^{\nu}(q)$, is defined by

$$B^{\nu}{}_{\mu}q^{\mu} = -\tfrac{1}{2}Q^{\nu\mu}\sigma^{-1}_{\mu\lambda}q^{\lambda} + r^{\nu}(q) \tag{3.5}$$

The flow described by $r^{\nu}(q)$ turns out to be incompressible, owing to eq. (3.4):

$$\frac{\partial r^{\nu}(q)}{\partial q^{\nu}} = B^{\nu}{}_{\nu} + \tfrac{1}{2}Q^{\nu\mu}\sigma^{-1}_{\mu\nu} = 0 \tag{3.6}$$

and conserves the potential $S(q)$:

$$r^{\nu}(q)\frac{\partial S(q)}{\partial q^{\nu}} = -(B^{\nu}{}_{\mu} + \tfrac{1}{2}Q^{\nu\lambda}\sigma^{-1}_{\lambda\mu})\sigma^{-1}_{\nu\varkappa}q^{\mu}q^{\varkappa} = 0, \tag{3.7}$$

again owing to eq. (3.4). Equations (3.6), (3.7) together ensure that eq. (2.23) is satisfied.

It is remarkable how the properties of $r^{\nu}(q)$ generalize well-known properties of reversible flows in equilibrium states, which are incompressible due to Liouville's theorem and conserve the entropy of the total system. In the linearized theory of steady states, $r^{\nu}(q)$ has the same properties but need not be reversible.

For the following it will be useful to consider the Langevin equation, which is stochastically equivalent to the Fokker-Planck equation. It is given by [+)]

$$\dot{q}^{\nu}(t) = B^{\nu}_{\mu}q^{\mu}(t) + g^{\nu}{}_{a}\xi^{a}(t) \tag{3.8}$$

with

$$g^{\nu}{}_{a}g^{\mu}{}_{a}\delta^{ab} = Q^{\nu\mu} \tag{3.9}$$

---

+) Equation (3.8) is a stochastic differential equation. Since it is linear, we need not enter here into a discussion of whether the Itô calculus or the Stratonovich calculus is used in (3.8).

$$\langle \xi^a(t) \rangle = 0, \quad \langle \xi^a(t)\, \xi^b(0) \rangle = \delta^{ab}\,(t) \tag{3.10}$$

$$\xi^a(t) \quad \text{Gaussian}$$

$$\langle q^\mu(t-\varepsilon)\, \xi^a(t) \rangle = 0 \quad \text{if} \quad \varepsilon > 0 \tag{3.11}$$

with $\delta^{ab} = \delta_{ab}$ = Kronecker symbols (a, b = 1, ..., n). The properties (2.34) of $\xi^a(t)$ can be combined in one expression for the probability density of the variables

$$\xi_i^a = \frac{1}{\varepsilon} \int_{t_i}^{t_{i+1}} \xi^a(\tau)\, d\tau \tag{3.12}$$

where we have introduced on the time axis an equidistant lattice with lattice constant $\varepsilon$. The probability density equivalent to (3.10) is the Gaussian

$$W_\varepsilon(\{\xi_i\}) = \prod_i \left(\frac{\varepsilon}{2\pi}\right)^{n/2} \exp(-\tfrac{1}{2}\varepsilon\, \xi_i^a \xi_i^b\, \delta_{ab}) \quad . \tag{3.13}$$

It is often useful to take formally the limit $\varepsilon \to 0$ in this expression:

$$W(\{\xi(\tau)\})\, D\mu(\{\xi(\tau)\}) = \exp\left[-\tfrac{1}{2}\int d\tau\, \xi^a(\tau)\, \xi^b(\tau)\, \delta_{ab}\right] \times D\mu(\{\xi(\tau)\}) \tag{3.14}$$

with the "volume element" in $\{\xi(\tau)\}$-space

$$D\mu(\{\xi(\tau)\}) = \lim_{\varepsilon \to 0} \left[\prod_i \left(\frac{\varepsilon}{2\pi}\right)^{n/2} d\,\xi_i\right] \quad . \tag{3.15}$$

To be sure, eqs. (3.14) and (3.15) have a formal meaning only, and can lead to ambiguities if handled too carelessly. Such ambiguities have to be resolved by going back to eq. (3.13).

Eqs. (3.14) and (3.15) can be used to obtain the probability density for paths $q^\nu(\tau)$ in the $\{q(\tau)\}$-space. We have only to insert eq. (3.8) to obtain[17)]

$$\begin{aligned} &W(\{q(\tau)\})\, D\mu(\{q(\tau)\}) \\ &= \exp\left[-\tfrac{1}{2}\int d\tau \left\{ Q_{\nu\mu}\, (\dot{q}^\nu(\tau) - B^\nu{}_\lambda\, q^\lambda(\tau))(\dot{q}^\mu(\tau) - B^\mu{}_\kappa\, q^\kappa(\tau))\right\}\right] J D\mu(\{q\}) \end{aligned} \tag{3.16}$$

$J$ is the Jacobian of the transformation (3.8). Since (3.8) is linear, $J$ is independent of $q$ and can be absorbed in the normalization of the probability density. In eq. (3.16) we have introduced the matrix

$Q_{\nu\mu}$ , which is the inverse of $Q^{\nu\mu}$:

$$Q_{\nu\mu} Q^{\mu\kappa} = \delta_\nu^{\kappa} = Q^{\kappa\mu} Q_{\mu\nu} .$$

Thus, we assume at this point that $Q^{\nu\mu}$ has an inverse.

This assumption simplifies the formal expressions considerably. It is not always satisfied in practice, however. Already in the simple example of the Brownian motion of a particle, described by the equations

$$\begin{aligned} \dot{q} &= \frac{p}{m} \\ \dot{p} &= -\frac{\partial V(q)}{\partial q} - \gamma p + \sqrt{2mkT\gamma}\, \xi(t), \end{aligned} \tag{3.17}$$

we have
$$Q^{\nu\mu} = \begin{pmatrix} 0 & 0 \\ 0 & 2mkT\gamma \end{pmatrix} . \tag{3.18}$$

In such oases one can either introduce additional parameters to make $\|Q^{\nu\mu}\|$ finite in all intermediate steps of the calculation (this is the procedure which we will assume in the following) or else increase the order of the equations, e.g., by eliminating $p$ in (3.17), which is the way Onsager and Machlup handled this problem. A third possibility is to consider instead of (3.16) its functional Fourier transform with respect to $\dot{q}$, which only contains $Q^{\nu\mu}$ and not its inverse[30].

We proceed with the assumption that $Q_{\nu\mu}$ exists. Then $Q^{\nu\mu}$ and $Q_{\nu\mu}$ are necessarily positive-definite and it is useful to define a metric form

$$ds^2 = Q_{\nu\mu}\, dq^\nu\, dq^\mu > 0$$

and to use $Q^{\nu\mu}$ and $Q_{\nu\mu}$ to raise and lower indices. We define the "volume element" in q-space as

$$D\mu(\{q\}) = \lim_{\varepsilon\to 0} \prod_i \frac{dq(\tau_i)}{\sqrt{(2\pi\varepsilon)^n \|Q^{\nu\mu}\|}} . \tag{3.19}$$

Of course, the expressions (3.16), (3.19) are just as formal as (3.8), (3.14), (3.15), from which they are derived. However, these expressions have remarkable formal properties and are very useful for formal manipulations. These properties derive from the fact that (3.16) is a multi-dimensional Gaussian. For Gaussian functions, integrating over its variables under given constraints (e.g., keeping some variables fixed) gives, apart from a constant factor, the same result as maximizing the Gaussian under the same constraints.

We make use of this property to derive from (3.16) the conditional probability density $P(qt|q_0 0)$ of the process. It is obtained from (3.16) by the path integral

$$P(qt|q_0 0) = \int_{q(0)=q_0}^{q(t)=q} D\mu(\{q\}) \exp\left[-\int_0^t d\tau \, \mathcal{L}(\dot{q}(\tau), q(\tau))\right] \tag{3.20}$$

with

$$\mathcal{L}(\dot{q}, q) = \tfrac{1}{2} Q_{\nu\mu} (\dot{q}^\nu - B^\nu_\lambda q^\lambda)(\dot{q}^\mu - B^\mu_\varkappa q^\varkappa) + \text{const.} \tag{3.21}$$

The function $\mathcal{L}(\dot{q}, q)$ is called the Onsager-Machlup function. We "evaluate" this integral by maximizing the Gaussian integrand under the given constraints

$$q^\nu(0) = q_0^\nu \quad , \quad q^\nu(t) = q^\nu . \tag{3.22}$$

With these boundary conditions we have to satisfy the Euler-Lagrange equations

$$\frac{d}{d\tau} \frac{\partial \mathcal{L}}{\partial \dot{q}^\nu} - \frac{\partial \mathcal{L}}{\partial q^\nu} = 0 \quad . \tag{3.23}$$

In matrix notation they read explicitly

$$(1 \frac{d}{d\tau} + B^T) \cdot Q^{-1} \cdot (1 \frac{d}{d\tau} - B) \cdot \underline{q} = 0 \tag{3.24}$$

where $\quad (B^T)_\nu{}^\mu = B^\nu{}_\mu \quad .$

Putting

$$Q^{-1} \cdot (1 \frac{d}{d\tau} - B) \cdot q = \lambda \tag{3.25}$$

we have

$$\lambda(\tau) = e^{-B^T \tau} \lambda_0 \quad . \tag{3.26}$$

The boundary condition at $\tau = t$ yields

$$e^{-Bt} \cdot q - q_0 = \int_0^t d\tau \; e^{-B\tau} \cdot Q \cdot e^{-B^T \tau} \cdot \lambda_0 \quad . \tag{3.27}$$

If we insert the solution (3.25), (3.26) into (3.20), we obtain

$$\begin{aligned} P(qt/q_0 0) &\sim \exp\left[-\tfrac{1}{2} \int_0^t d\tau \; \lambda_0 e^{-B\tau} . Q . e^{-B^T \tau} \cdot \lambda_0\right] \\ &= \exp\left[-\tfrac{1}{2} \lambda_0 \cdot (e^{-Bt} q - q_0)\right] \end{aligned} \tag{3.28}$$

If we write

$$\lambda_0 = (q - e^{Bt} . q_0) . \sigma^{-1}(t) . e^{Bt} \tag{3.29}$$

we obtain from (3.27) a matrix differential equation for $\sigma(t)$:

$$\dot{\sigma}(t) = B.\sigma(t) + \sigma(t).B^T + Q; \quad \sigma(0) = 0 \tag{3.30}$$

from which $\sigma$ has to be determined. The normalized conditional probability density then is

$$P(qt|q_0 0) = \frac{1}{\sqrt{(2\pi)^n \, ||\sigma(t)||}} \exp\left[-\tfrac{1}{2}(q-e^{Bt}.q_0).\sigma^{-1}(t).(q-e^{Bt}.q_0)\right] \tag{3.31}$$

The same solution could, of course, be obtained directly from the Fokker-Planck equation. For $t \to \infty$ we have $e^{Bt}.q_0 \to 0$, $\sigma(t) \to \sigma$. Thus $P$ approaches the steady-state solution $W_0 \sim e^{S(q)}$ with $S(q)$ given by (2.25), as we know it must.

The conditional probability density (2.51) can be used to reconstruct the probability density of paths by

$$W(\{q(\tau)\})D\mu(\{q(\tau)\}) = \lim_{\varepsilon \to 0} \prod_j \{P(q_j t_j / q_{j-1} t_{j-1}) dq_j\} \tag{3.32}$$

with

$$t_{j+1} - t_j = \varepsilon \tag{3.33}$$

$$\sigma(\varepsilon) = Q\varepsilon + O(\varepsilon^2) \tag{3.34}$$

$$||\sigma(\varepsilon)|| = \varepsilon^n ||Q|| (1 + \varepsilon B^\nu_\nu + O(\varepsilon^2)) \tag{3.35}$$

$$q_j - e^{B\varepsilon}.q_{j-1} = q_j - q_{j-1} - \varepsilon B.q_{j-1}, \tag{3.36}$$

and putting

$$q_j - q_{j-1} = \varepsilon \dot{q}(\tau), \qquad q_j = q(\tau) \tag{3.37}$$

we obtain, taking $\varepsilon \to 0$ formally,

$$W(\{q(\tau)\})D\mu(\{q\}) =$$

$$= \lim_{\varepsilon \to 0} \left\{ \prod_j \frac{dq_j}{\sqrt{(2\pi\varepsilon)^n \, ||Q||}} \right\} \exp\left[-\tfrac{1}{2} \int d\tau \left[(\dot{q}-B.q).Q^{-1}.(\dot{q}-B.q) + B^\nu_\nu\right]\right]$$

Comparing with eqs. (3.16), (3.19), we see that we have determined the Jacobian in (3.16) :

$$J = \exp\left[-\tfrac{1}{2} \int d\tau \; B^\nu_\nu\right] \tag{3.39}$$

$J$ is, of course, independent of $q$, as promised. We could also have obtained this Jacobian directly from the transformation formula (3.8)[17]. The uniqueness of the factor $\frac{1}{2}$ in the exponent of (3.39) has been the subject of some debate in the literature. However, the way we derived it from the conditional probability density (3.31) leaves no other choice (cf. also the discussion in section 5).

The replacement (3.37) and the appearance of $\dot{q}$ in (3.38) needs some further comment. Let us first consider the analytical properties of the paths that contribute to the probabilities computed from (3.38). The probability $p_<(\varepsilon, \lambda)$ that

$$|q_j^\nu - q_{j-1}^\nu| < K\varepsilon^\lambda \qquad (K > 0 \text{ real}, \ \lambda \text{ real}) \tag{3.40}$$

is obtained from (2.51) in the limit $\varepsilon \to 0$ as

$$p_<(\varepsilon, \lambda) = 2\,\phi\left(\frac{K \quad \varepsilon^{\lambda - \frac{1}{2}}}{\sqrt{Q^{\nu\nu}}}\right) \tag{3.41}$$

with

$$\phi(x) = \frac{1}{\sqrt{2\pi}} \int_0^x dy\; e^{-y^2/2} \ . \tag{3.42}$$

Since

$$\lim_{\varepsilon \to 0} p_< = \begin{cases} 1 & \text{for} \quad \lambda < \frac{1}{2} \\ 0 & \text{for} \quad \lambda > \frac{1}{2} \end{cases} \tag{3.43}$$

we have, with probability 1,

$$\begin{aligned} |q_j^\nu - q_{j-1}^\nu| &< K\,\varepsilon^{\frac{1}{2}-\alpha} \\ & \qquad\qquad\qquad \alpha > 0 \\ |q_j^\nu - q_{j-1}^\nu| &> K\,\varepsilon^{\frac{1}{2}+\alpha} \end{aligned} \tag{3.45}$$

i.e., $q^\nu(t)$ is continuous but not differentiable at $t = t_j$.

Let us next ask for the probability $P_<(\varepsilon, \lambda)$ that inequality (3.40) holds <u>everywhere</u> in a fixed finite time interval $T$. Since the influence of $B$ has dropped out for $\varepsilon \to 0$ and the increments of $q^\nu$ due to diffusion in each time interval are independent, we have

$$P_<(\varepsilon, \lambda) = (p_<(\varepsilon, \lambda))^{T/\varepsilon} \tag{3.46}$$

and thus

$$\lim_{\varepsilon \to 0} P_<(\varepsilon, \lambda) = \begin{cases} 1 & \\ & \text{for} \\ 0 & \end{cases} \begin{matrix} \lambda < \frac{1}{2} \\ \\ \lambda \geqslant \frac{1}{2} \end{matrix} \tag{3.47}$$

Thus, the paths in any given finite time interval are everywhere continuous simultaneously. Finally, we consider the probability $P_>(\varepsilon, \lambda)$ that

$$|q_j^\nu - q_{j-1}^\nu| > K\varepsilon^\lambda \tag{3.48}$$

holds everywhere in a fixed finite time interval T. It is given by

$$P_{>}(\varepsilon,\lambda) = (1 - p_{<}(\varepsilon,\lambda))^{T/\varepsilon} \tag{3.49}$$

and thus

$$\lim_{\varepsilon\to 0} P_{>}(\varepsilon,\lambda) = 0 \qquad \text{for} \qquad \lambda \le \tfrac{1}{2}$$

$$\lim_{\varepsilon\to 0} P_{>}(\varepsilon,\lambda) = \lim_{\varepsilon\to 0} \exp\left[-\frac{2T}{\varepsilon}\frac{K\,\varepsilon^{\lambda-\frac{1}{2}}}{\sqrt{Q^{\nu\nu}2\pi}}\right] = \begin{cases} 0 & \text{for} \quad \tfrac{1}{2} \le \lambda < \tfrac{3}{2} \\ 1 & \phantom{\text{for}} \quad \lambda > \tfrac{3}{2} \end{cases} \tag{3.50}$$

Thus, in any fixed finite time interval T the trajectories are differentiable somewhere with probability 1, although the probability that the paths are differentiable at any given point is zero. The points on the time axis where the trajectories are differentiable therefore have measure zero, but they lie dense on the time axis (just as the rational numbers lie dense on the real axis). The second-order time derivative does not exist anywhere in any given finite time interval.

Frequently, the statement is made that

$$\frac{|q_j^{\nu} - q_{j-1}^{\nu}|}{\varepsilon} = K\,\varepsilon^{-\frac{1}{2}}\,. \tag{3.51}$$

This is correct as a rule of thumb, but has to be used with some care, since it is not satisfied for the dense set of points on the time axis, where the paths are differentiable. At any given point the probability that the left-hand side of this relation is actually smaller is given by $p_{<}(\varepsilon,\tfrac{1}{2}) = 2\,\phi(K/\sqrt{Q^{\nu\nu}})$ (cf. eq. (3.41)). In any arbitrarily small but finite time interval T, the left-hand side is actually of order $\varepsilon^0$ with probability 1 at an arbitrarily large number of points (cf. eq. (3.50)).

Let us return now to eq. (3.37). We require that, at those points where the trajectories are differentiable, the left-hand side of eq. (3.37) coincides with the time derivatives. At all other points we are free to define what we mean by $\dot{q}$. We fix this definition by extending all the formal rules of calculus that apply to ordinary derivatives and hold on the dense set of points on the time axis so that they hold also at all other points on the time axis. This requires, in particular, that if the limit $\varepsilon \to 0$ is to exist it must be possible to replace eq. (3.37) by

$$q_{j+1} - q_j = \varepsilon\,\dot{q}(\tau), \qquad q_j = q(\tau) \tag{3.52}$$

without changing the limit. It can easily be checked that this replacement is in fact possible, with no change in the result, when we go from (3.32) to (3.38).

Before going further, we want to discuss here briefly the relation of the formulation of nonequilibrium thermodynamics in terms of a denumerable number of discrete variables and the formulation necessary for continua. For simplicity we only consider the example of heat conduction in a solid crystal, described by energy conservation

$$\frac{\partial \varepsilon}{\partial t} = - \operatorname{div} \vec{j}_\varepsilon \tag{3.53}$$

and an equation for the energy-current density

$$j_{\varepsilon,i} = \sum_j \varkappa_{ij}(T) \nabla_j \frac{1}{T} + g_i = \sum_j \varkappa_{ij}(T) \nabla_j \left(\frac{\partial s}{\partial \varepsilon}\right) + g_i \tag{3.54}$$

where $\vec{g}$ is a fluctuating, Gaussian, $\delta$-correlated contribution to $\vec{j}_\varepsilon$, $s$ is the local entropy density in the solid, and $\varkappa_{ij}$ its heat-conductivity tensor.

One way of making contact between eqs. (3.53), (3.54) and our description is to relate

$$\begin{aligned} &q^\nu \to \varepsilon(\vec{r}t), && \dot{q}^\nu \to \dot{\varepsilon}(\vec{r}t) \\ &\frac{\partial}{\partial q^\nu} \to \frac{\delta}{\delta \varepsilon(\vec{r})}, && \sum_\nu \to \int d^3r \\ &S \to \int d^3r\, s(\vec{r}), && Q^{\nu\mu} \to 2 \sum_{ij} \nabla_i \varkappa^s_{ij} \nabla'_j \delta(\vec{r} - \vec{r}') \end{aligned} \tag{3.55}$$

from which

$$\sum_{ij} \langle\langle (\nabla_i g_i(\vec{r}\, t))(\nabla'_j g_j(\vec{r}' t')) \rangle\rangle = 2 \sum_{ij} \nabla_i \varkappa^s_{ij} \nabla'_j \delta(\vec{r} - \vec{r}')\, \delta(t - t') \tag{3.56}$$

immediately follows, where

$$\varkappa^s_{ij} = \tfrac{1}{2}(\varkappa_{ij} + \varkappa_{ji}).$$

$K^\nu$ in the continuous formulation has a bulk and a surface contribution,

$$\left(K^\nu - \tfrac{1}{2} \frac{\partial Q^{\nu\mu}}{\partial q^\mu}\right)\Big|_{\text{bulk}} \longrightarrow \sum_{ij} \left[\nabla_i \varkappa^s_{ij} \nabla_j \frac{\partial s}{\partial \varepsilon}\right] + r^\nu\Big|_{\text{bulk}} \tag{3.57}$$

$$\left(K^\nu - \tfrac{1}{2} \frac{\partial Q^{\nu\mu}}{\partial q^\mu}\right)\Big|_{\text{surface}} \longrightarrow \sum_{ij} \left[n_i \varkappa^s_{ij} \nabla_j \frac{\partial s}{\partial \varepsilon}\right] + r^\nu\Big|_{\text{surface}} \tag{3.58}$$

where $\vec{n}$ is the normal vector on the surface, and

$$\mathcal{H}^{a}_{ij} = \tfrac{1}{2}(\mathcal{H}_{ij} - \mathcal{H}_{ji}) \tag{3.59}$$

As a consequence, the steady-state drift rate is given by+)

$$r^{\nu}\Big|_{\text{bulk}} \longrightarrow - \sum_{ij} \left[ \nabla_i \mathcal{H}^{a}_{ij} \nabla_j \frac{\partial s}{\partial \varepsilon} \right] \tag{3.60}$$

$$r^{\nu}\Big|_{\text{surface}} \longrightarrow \sum_{ij} n_i \left[ \mathcal{H}^{a}_{ij} \nabla_j \frac{\partial s}{\partial \varepsilon} \right] \tag{3.61}$$

Equation (2.23) is then satisfied. Near thermodynamic equilibrium Onsager's symmetry relations imply

$$\mathcal{H}^{a}_{ij} = 0 \,, \qquad \text{i.e.} \quad r^{\nu} = 0 \,. \tag{3.62}$$

Far from thermodynamic equilibrium one could have $\mathcal{H}^{a}_{ij} \neq 0$ and hence $r^{\nu} \neq 0$, in principle.

## 4. VARIATIONAL PRINCIPLES AND PROBABILITY DENSITIES FOR LINEAR PROCESSES.

Now let us take a closer look at the Onsager-Machlup function that we have obtained. We recall that the theory is also valid for small fluctuations in steady states that are far from thermodynamic-equilibrium states. Inserting eq. (3.5) in eq. (3.21) and using the properties (3.6), (3.7) we obtain

---

+) There are also contributions to $r^{\nu}$ containing terms (e.g., in the bulk, $\int d^3r' \frac{\delta}{\delta \varepsilon(\vec{r}')} ( \nabla_i \mathcal{H}^{a}_{ij} \nabla_j' \delta(\vec{r} - \vec{r}'))$, which are ill-defined unless the spatial $\delta$-function is replaced by a function with a symmetrical wavenumber cut-off, in which case these contributions vanish. They are therefore omitted in (3.60), (3.61). In the expression for $\frac{\partial r^{\nu} e^{S}}{\partial q^{\nu}}$ these terms just cancel in any case.

$$\mathcal{L} = \tfrac{1}{2}\left[\phi(\dot{q}-r,\ \dot{q}-r) + \psi\left(\frac{\partial S}{\partial q}, \frac{\partial S}{\partial q}\right) - \dot{S}\left(\dot{q}, \frac{\partial S}{\partial q}\right)\right] + \text{const} \tag{4.1}$$

where the dissipative potentials $\phi$ and $\psi$ are defined by

$$\phi(\dot{q}, \dot{q}) = Q_{\mu\nu}\,\dot{q}^{\mu}\,\dot{q}^{\nu} \tag{4.2}$$

$$\psi\left(\frac{\partial S}{\partial q}, \frac{\partial S}{\partial q}\right) = \tfrac{1}{4}Q^{\mu\nu}\,\frac{\partial S}{\partial q^{\mu}}\,\frac{\partial S}{\partial q^{\nu}} \tag{4.3}$$

and the production of $S$ has the usual form

$$\dot{S}\left(\dot{q}, \frac{\partial S}{\partial q}\right) = \dot{q}^{\nu}\,\frac{\partial S}{\partial q^{\nu}} \tag{4.4}$$

Both $\phi$ and $\psi$ are positive within our assumptions about $Q^{\mu\nu}$.

The form (4.1) of $\mathcal{L}$ was derived by Onsager and Machlup for systems near thermodynamic equilibrium. The latter assumption was made by these authors in order to be able to use the reciprocity relations, which hold near thermodynamic equilibrium. However the reciprocity relations do not really have to be used, and we have obtained their result here without invoking microscopic reversibility at all. The reason that this was possible is that eqs. (3.6), (3.7) are valid for general steady states. We have obtained an additional generalization as compared with the Onsager-Machlup case, by including the drift velocity in phase space in the steady state, $r^{\nu}$. In the presence of such a drift in the steady state the only change is that $\dot{q}^{\nu} - r^{\nu}$ enters the dissipative potential $\phi$, while all other results remain unchanged.

However, there is one important drawback: In order to find $r^{\nu}$ for a general steady state the time-independent distribution $\sim e^{S(q)}$ has to be determined first, i.e., an important part of the problem must be solved first. This is not necessary for systems near thermodynamic equilibrium. In the latter case the decomposition

$$K^{\nu}(q) = \tfrac{1}{2}\,Q^{\nu\mu}\,\frac{\partial S(q)}{\partial q^{\mu}} + r^{\nu}(q) \tag{4.5}$$

coincides with the decomposition of $K^{\nu}(q)$ into two parts with definite parity under time reversal. Below, we shall prove this statement in a much more general case than the one considered here, and we therefore omit a more detailed explanation for the moment. Suffice it to say that $r^{\nu}(q)$, the drift rate in phase space, describes all purely reversible processes in thermodynamic equilibrium and therefore transforms like $\dot{q}^{\nu}$ under time reversal; $\tfrac{1}{2}Q^{\nu\mu}\,\partial S(q)/\partial q^{\mu}$ is the purely irreversible part of the drift through phase space near thermodynamic equilibrium and does not contribute to the probability current through phase space in equilibrium.

Having determined $\mathcal{L}(\dot{q},q)$ in the form

$$\mathcal{L}(\dot{q},q) = \tfrac{1}{2}\left[\phi(\dot{q}-r,\ \dot{q}-r) + \psi(\partial S/\partial q, \partial S/\partial q) - \dot{S}(\dot{q}, \partial S/\partial q)\right] + \text{const} \tag{4.6}$$

we can now write the probability density in the steady state as

$$W_0(q)dq = \int_{q(t)=q} D\mu(\{q\})\exp\left[-\int_{-\infty}^{+\infty}\mathcal{L}(\dot{q},q)d\tau\right]. \tag{4.7}$$

and, similarly, the simultaneous probability densities in the steady state as

$$W_\ell(q_\ell t_\ell; q_{\ell-1}t_{\ell-1}; \ldots; q_1t_1)dq_\ell \ldots dq_1 = \int_{(C_\ell)} D\mu(\{q\})\exp\left[-\int_{-\infty}^{+\infty}\mathcal{L}(\dot{q},q)d\tau\right] \tag{4.8}$$

where the constraint $C_\ell$ is

$$C_\ell : \quad q_1 = q(t_1), \ldots, q_\ell = q(t_\ell) \quad . \tag{4.9}$$

Because of the Gaussian form of the probability density functional $W(\{q(\tau)\})$, which is a consequence of the linearity of the equations, the path integrals in eqs. (4.7), (4.8) are easily carried out. Apart from normalization constants, they are given by the maximum of the integrand under the corresponding constraint, a property which we have already used above. Thus, in self-explanatory notation we have

$$W_0(q) \sim \exp\left[-\int_{-\infty}^{+\infty}\mathcal{L}(\dot{q},q)d\tau\right]^{q(t)=q}_{max.} \tag{4.10}$$

and

$$W_\ell(q_\ell t_\ell, \ldots, q_1t_1) \sim \exp\left[-\int_{-\infty}^{+\infty}\mathcal{L}(\dot{q},q)\,d\tau .\right]^{C_\ell}_{max} \tag{4.11}$$

As another consequence of the Gaussian nature of the probability densities, the maximizing paths $q^\nu = q^\nu_{max}(\tau, \{C_\ell\})$ are identical to average paths under the constraint $C_\ell$ .

This argument can be generalized immediately to obtain the probability that any constraint C is satisfied by the macroscopic variables $q^\nu$ in the steady state:

$$W(\{C\}) \sim \exp\left[-\int_{-\infty}^{+\infty}\mathcal{L}(\dot{q},q)d\tau\right]^{C}_{max} \tag{4.12}$$

with the average paths

$$\langle q^{\nu}(\tau)\rangle_C \quad = \quad q^{\nu}_{max}(\tau, \{C\}) \tag{4.13}$$

being given by the maximizing paths satisfying the constraint C. The only condition on C is the requirement that it be compatible with the system's being in the steady state. Thus, it is not possible to constrain the system at a given time more closely than fixing its macroscopic variables does.

Let us define for each constraint C a macroscopic potential $S(\{C\})$ by

$$W(\{C\}) \quad = \quad \exp S(\{C\}) \quad . \tag{4.14}$$

Then we have, in the steady state,

$$S(\{C\}) \quad = \left[-\int_{-\infty}^{+\infty} d\tau \;\; \mathcal{L}(\dot{q},q)\right]^{C}_{max} \quad + \quad \text{const} \quad . \tag{4.15}$$

Since $S(\{C\})$ is again a bilinear form in the $q^{\nu}$ the probability (4.14) is again Gaussian, and further averaging over $W(\{C\})$ is again equivalent to further maximization.

A special case of the very remarkable result (4.15) is an expression for the macroscopic potential $S(q)$:

$$S(q) \quad = \quad \left[-\int_{-\infty}^{+\infty} d\tau \;\; \mathcal{L}(\dot{q},q)\right]^{q(t)=q}_{max} \quad + \text{ const} \; . \tag{4.16}$$

Equation (4.15) reveals that $\mathcal{L}(\dot{q},q)$ is the most general nonequilibrium thermodynamic potential of the system in the steady state. It corresponds to the most complete constraint possible, i.e., to complete specification of the paths $q^{\nu}(\tau)$, $\nu = 1, \ldots, n$, $-\infty < \tau < +\infty$. The potential for any less-specific constraint is obtained by minimizing $\int_{-\infty}^{+\infty} \mathcal{L}(\dot{q},q)d\tau$ under this constraint. In this sense the latter quantity is the true generalization of the well-known equilibrium thermodynamic potentials to the case of time-dependent properties and nonequilibrium steady states.

Let us consider now some alternative formulations of the results above. First, it is not necessary to take the time integral in eq. (4.16) to infinity. The potential $S(q)$ may also be derived from

$$S(q) \quad = \quad \left[-\int_{-\infty}^{t} d\tau \;\; \mathcal{L}(\dot{q},q)\right]^{q(t)=q}_{max} \quad + \quad \text{const.} \tag{4.17}$$

Assuming that $S(q(t\to -\infty)) = 0$, we may simplify (4.17) by doing the integral over $S$ to obtain

$$S(q) = \Big[-\int_{-\infty}^{t} d\tau\, (\phi(\dot{q}-r,\, \dot{q}-r) + \psi(\frac{\partial S}{\partial q}, \frac{\partial S}{\partial q}))\Big]_{max}^{q(t)=q} + \text{const.} \tag{4.18}$$

Equation (4.17) is true, since

$$\Big[-\int_{t}^{+\infty} d\tau\; \mathcal{L}(\dot{q},q)\Big]_{max}^{q(t)=q} = \text{const} \tag{4.19}$$

yields just a normalization constant. From eq. (4.19) we derive, for $S(q(t\to +\infty)) = 0$,

$$S(q) = \Big[-\int_{t}^{+\infty} d\tau\, (\phi(\dot{q}-r,\, \dot{q}-r) + \psi(\frac{\partial S}{\partial q}, \frac{\partial S}{\partial q}))\Big]_{max}^{q(t)=q} + \text{const.} \tag{4.20}$$

Adding eqs. (4.18) and (4.20), we obtain

$$S(q) = \Big[-\tfrac{1}{2}\int_{-\infty}^{+\infty} \big(\phi(\dot{q}-r,\, \dot{q}-r) + \psi(\frac{\partial S}{\partial q}, \frac{\partial S}{\partial q})\big)\Big]_{max}^{q(t)=q}, \tag{4.21}$$

a result which also follows directly from eq. (4.16) by doing the integral over $S$. Again it is possible to extend these results to the case of more-general constraints, which have to be compatible with the condition that $S(t\to +\infty) = S(t\to -\infty)$.

$$S(\{C\}) = \Big[-\tfrac{1}{2}\int_{-\infty}^{+\infty} d\tau\, \Big(\phi(\dot{q}-r,\, \dot{q}-r) + \psi(\frac{\partial S}{\partial q}, \frac{\partial S}{\partial q})\Big)\Big]_{max}^{q(t)=q} \tag{4.22}$$

and

$$W(\{q,\, C\}) \sim \exp S(\{q,\, C\}) \quad . \tag{4.23}$$

A general extremum principle for macroscopic systems with small fluctuations in steady states can be read off from the result (4.12), (4.19), namely, the <u>principle of minimum dissipation</u>:

$$\int_{-\infty}^{t} d\tau\;\; \mathcal{L}(\dot{q},q) = \min \tag{4.24}$$

under given constraints. For systems near thermodynamic equilibrium this principle is due to Onsager[7] (cf. Gyarmati[2]).

Equation (4.18) shows the relation of this principle to an extremum property of $S(q)$ if this quantity is considered as a functional of the paths starting at $q = 0$ in the infinite past and reaching the point $q$ at time $t$. Equation (4.20) also shows how the fluctuation

$q(t) = q$ which arises in the steady state with probability density $\exp S(q)$ decays to zero, on the average, as $t \to \infty$. For some time $t' > t$, $q(t)$ will have decayed to some value $q(t')$, on the average. Equation (4.20) then yields

$$S(q(t')) - S(q(t)) = \left[\int_t^{t'} d\tau\left(\phi(\dot{q}-r, \dot{q}-r) + \psi\left(\frac{\partial S}{\partial q}, \frac{\partial S}{\partial q}\right)\right)\right]_{min} \geq 0 \; . \tag{4.25}$$

The right-hand side gives the production of $S$ in the time interval $t' - t$. Thus, the decay of a fluctuation in the steady state is also governed by an extremum principle, the <u>principle of minimum production of S</u>, which follows from the more general variational principle (4.24).

A particularly important class of constraints $C$ is the one where average values of some $\dot{q}^{\nu}$ or $q^{\nu}$ are prescribed. These constraints are realized physically by external "forces" or "fluxes" acting on the system. To obtain these, the equations for the averages are derived from the minimum-dissipation principle by the Lagrange-multiplier method. The Lagrange multipliers which realize the constraints then yield the external forces and fluxes. As long as the constraints are linear in the $q$, $\dot{q}$, which we require because of the linearity assumption already made, the external forces and fluxes enter as inhomogeneities in the equations of motion.

Starting from any steady state, one can obtain a new steady state in its vicinity by fixing the averages of some of the $\dot{q}^{\nu}$ or $q^{\nu}$ to have certain finite (and small) time-independent values $\langle\dot{q}^{\nu}\rangle_0$, $\langle q^{\nu}\rangle_0$. The new steady state is then governed by the same equations as the original one, except that $\dot{q}^{\nu}$ and $q^{\nu}$ in the original equations are replaced by $\dot{q}^{\nu} - \langle\dot{q}^{\nu}\rangle_0$ and $q^{\nu} - \langle q^{\nu}\rangle_0$. This replacement just takes care of the inhomogeneities introduced by the external forces and fluxes. Of particular interest is the class of steady states generated in this way in the vicinity of the state of thermodynamic equilibrium. Here, $S$, $\phi$ and $\psi$ are still given by the same expressions as in thermodynamic equilibrium. In particular, $S$ is then related to the entropy, and the principle of minimum production of $S$ given above reduces to Prigogine's principle of minimum entropy production, which is restricted to steady states near thermodynamic equilibrium[31].

Prigogine and Glansdorff[11] have given a more general theory, which also applies to steady states <u>far</u> from equilibrium, provided one may linearize with respect to the fluctuations from the steady state. This is precisely the assumption that we have also made in this section, and a comparison of our results with those of Glansdorff and Prigogine is therefore possible. In their theory the additional assumption is made that the second-order derivatives of $S(q)$ are still given by the second-order derivatives of the entropy $S_{eq}$. Thus[+)]

$$S(q) = S_{eq}(q) - \lambda_\nu q^\nu ,$$

where the $\lambda_\nu$ are parameters which characterize the steady state. Define $\delta q = q - q_0$ where $q_0$ maximizes $S(q)$. Then

$$\sigma^{-1}_{\nu\mu} = \left.\frac{\partial^2 S_{eq}}{\partial q^\nu \partial q^\mu}\right|_{q_0} = \sigma^{-1}_{eq\,\nu\mu}(q_0) \tag{4.27}$$

$$W_0(q) \sim \exp\left[\tfrac{1}{2} \left.\frac{\partial^2 S_{eq}}{\partial q^\nu \partial q^\mu}\right|_{q_0} \delta q^\nu \delta q^\mu\right] . \tag{4.28}$$

We note that eq. (3.4) with (4.27) becomes the usual fluctuation-dissipation relation, which fixes $Q^{\nu\mu}$ in terms of $\sigma^{-1}_{eq\,\nu\mu}$ .

With the additional assumption (4.27) Prigogine and Glansdorff[11)] proceed to express the exponent in (4.28) as the time integral of a "local potential":

$$\tfrac{1}{2} \left.\frac{\partial^2 S_{eq}}{\partial q^\nu \partial q^\mu}\right|_{q_0} \delta q^\nu \delta q^\mu = - \int_{\substack{t\\ q(t)=q}}^{\infty} d\tau\, \tfrac{1}{4}\, Q^{\nu\mu}(q_0)\, \delta\left[\left(\frac{\partial S_{eq}}{\partial q^\nu} - \lambda_\nu\right)\left(\frac{\partial S_{eq}}{\partial q^\mu} - \lambda_\mu\right)\right] \tag{4.29}$$

where

$$\delta \frac{\partial S_{eq}}{\partial q^\nu} = \sigma^{-1}_{eq\,\nu\mu}\, \delta q^\mu , \qquad \delta \lambda_\nu = 0 \tag{4.30}$$

and the integration on the right-hand side is along the average path of regression of a fluctuation, which is given by the law

$$\dot{q}^\nu = \tfrac{1}{2} Q^{\nu\mu} \left(\frac{\partial S_{eq}}{\partial q^\mu} - \lambda_\mu\right) + r^\nu \tag{4.31}$$

The steady-state drift rate $r^\nu$ drops out if (4.31) is inserted in the right-hand side of (4.29), owing to (3.4), (3.5) and (4.27). The local potential in (4.29) is thus simply

$$\delta\, \tfrac{1}{4} Q^{\nu\mu} \frac{\partial S}{\partial q^\nu}\frac{\partial S}{\partial q^\mu} = \delta \psi\left(\frac{\partial S}{\partial q}, \frac{\partial S}{\partial q}\right) . \tag{4.32}$$

Equations (4.32), (4.29) and (4.28) constitute the basic result of

---

+) In spatially extended systems, eqs. (4.26), (4.27) and (4.28) only need to hold locally, and are then equivalent to the assumption of local equilibrium.

Prigogine and Glansdorff for small fluctuations from the steady state. Their result follows as a special case from eq. (4.20) if we insert there the maximizing path, which is given by the solution of (4.31) with the initial condition $q(t) = q$. Along this path we have

$$\phi(\dot{q}-r,\ \dot{q}-r) = \psi(\partial S/\partial q, \partial S/\partial q) = \tfrac{1}{2}\,\dot{S}(\dot{q},\ \partial S/\partial q) \qquad (4.33)$$

and

$$\delta\psi = 2\psi\,, \text{ since } \quad \delta q^{\nu} = q^{\nu} - q_0 = q^{\nu} \qquad (4.34)$$

according to our conventions about the zero of $q$. Equation (4.20) is more general than (4.28), (4.29), (4.32), and does not require the assumption (4.27). On the other hand, if $\sigma_{\nu\mu}^{-1}$ is not determined by a local-equilibrium assumption, $Q^{\nu\mu}$ can no longer be fixed by a fluctuation-dissipation relation and has to be determined by other means, e.g., by a derivation of the Fokker-Planck equation from the microscopic equations.

With this remark, we want to conclude our consideration of stationary macroscopic systems, linearized with respect to their fluctuations. The main results of this section:

$$W(\{q(\tau)\}) \sim \exp\left[-\int_{-\infty}^{t} \mathcal{L}(\dot{q},q)\,d\tau\right] \qquad (4.35)$$

and

$$\int_{-\infty}^{t} \mathcal{L}(\dot{q},q)\,d\tau = \min \qquad (4.36)$$

considerably generalize well-known theorems of linear irreversible thermodynamics. However, they clearly rest on the Gaussian property of the probability-density functional.

One may wonder to what extent these results are modified in cases where the linearity assumption has to be dropped. The question raised here really has two parts. First, what is the form of $\mathcal{L}(\dot{q},q)$ when the linearity assumption is dropped. Second, how can the evaluation of the path integrals then be related to a variational principle. The answers to these two questions are given in the two following sections.

## 5. PATH-INTEGRAL SOLUTION OF THE FOKKER-PLANCK EQUATION

We now wish to construct a solution of eq. (2.13) in the form

$$P(qt/q_0t_0) = \int_{q_0}^{q} D\mu(\{q\})\exp\left[-\int_{t_0}^{t} \mathcal{L}(\dot{q},q)\,d\tau\right] \qquad (5.1)$$

i.e., we seek expressions for $D\mu(\{q\})$ and $\mathcal{L}(\dot{q},q)$ in (5.1) in terms of the quantities $K^{\nu}(q)$ and $Q^{\nu\mu}(q)$ given in (2.13). In a recent paper[23] the solution to this problem has been given. We will refer to this paper as I.[+] In the following we add some comments on the solution given there.

First, we wish to point out an important connection between the two probability densities

$$\Omega(\xi,q_0;\ t_1,t_0) \quad = \quad P(q_0+\xi t_1;\ q_0 t_0) \tag{5.2}$$

and

$$\bar{\Omega}(\xi,q;\ t_1,t_0) \quad = \quad P(q,t_1\,/\,q-\xi,t_0) \tag{5.3}$$

and the formal solution

$$W(t) \quad = \quad \underline{S}(t,t_0)W(t_0);\qquad \underline{S}(t,t_0) \quad = \quad e^{\underline{L}(t-t_0)} \tag{5.4}$$

of the Fokker-Planck equation (2.13), which is written in the form

$$\dot{W}(t) \quad = \quad \underline{L}\ W(t) \tag{5.5}$$

$$\underline{L} \quad = \quad -\tfrac{1}{2}\,\underline{p}_\nu\,\underline{p}_\mu\,Q^{\nu\mu}(q) - i\,\underline{p}_\nu\,K^{\nu}(q) \tag{5.6}$$

$$\underline{p}_\nu \quad = \quad -i\frac{\partial}{\partial q^\nu} \tag{5.7}$$

It is shown in section 2 of I that $\Omega(\xi,q_0)$ is just the Fourier transform of the c-number function $\tilde{S}(p,q,t,t_0)$ obtained by ordering the operator function $\underline{S}(\underline{p},q,t,t_0)$ in such a way that all $\underline{p}$ appear on the left, i.e., act after the q, and then substituting c-numbers p for the operators $\underline{p}$. $\bar{\Omega}$ is obtained by the same process, but with the p ordered to the right. This observation is useful for the construction of $P(qt+\varepsilon|q_0t)$ for small $\varepsilon$, which is given in section 4 of I.

It is important to realize that some care has to be exercised when constructing this short-time solution, because of the rather pathological analytical properties of the sample paths, which we have discussed in section 3. As we have seen there, these properties arise exclusively from the term quadratic in $\underline{p}$ in eq. (5.6). We can therefore discuss these difficulties by dropping all the other terms in (5.6) for the moment. What we are interested in finally is a solution of

---

+) In I a slightly different notation has been used. We give a short dictionary (present notation → notation in I): $q^\nu \to q_\nu$, $K^\nu \to K_\nu$, $Q^{\nu\mu} \to Q_{\nu\mu}$, $Q_{\nu\mu} \to Q^{-1}{}_{\nu\mu}$.

(5.6) for finite, not necessarily small times, which we can construct from the short-time solution by

$$P(qt/q_0t_0) = \int dq_{N-1} \int dq_{N-2} \cdots \int dq_1 \, P(qt/q_{N-1}t_{N-1}) \cdots P(q_1t_1/q_0t_0) \tag{5.8}$$

with

$$t_j = t_0 + j\varepsilon , \qquad t_N = t .$$

Each of the functions $P(q_{j+1}t_{j+1}|q_jt_j)$ has a typical spread

$$\Delta q_j = |q_{j+1} - q_j| \sim \varepsilon^{\frac{1}{2}} \tag{5.9}$$

which gives rise to the analytical properties discussed in section 3. Applying $\underline{L}$ to each of these functions, therefore, gives a result of the order of

$$|\underline{L}\, P(q_{j+1}\, t_{j+1}/q_jt_j)| \simeq \frac{1}{\varepsilon} |P(q_{j+1}\, t_{j+1}\, /\, q_jt_j)| \tag{5.10}$$

since each application of $\underline{p}$ contributes a factor $\varepsilon^{-\frac{1}{2}}$. This property shows that, at least for some points in the interval $t_0$, t, it would be wrong to use the most straightforward short-time expansion

$$\underline{S}(t_{j+1}, t_j) = 1 + \varepsilon\, \underline{L} + \tfrac{1}{2}\varepsilon^2 L^2 + \cdots \tag{5.11}$$

since the right-hand side of this expansion does not converge if applied to $P(q_jt_j\, /\, q_{j-1}\, t_{j-1})$. The reason is that $\underline{L}^n \sim \underline{p}^{2n}$ contributes a factor of $\varepsilon^{-n}$ and the expansion in $\varepsilon$ gets mixed up. We can avoid this by first introducing the rescaled operator

$$\tilde{\underline{p}}_\nu = \sqrt{\varepsilon}\, \underline{p}_\nu \tag{5.12}$$

and then expanding in powers of $\varepsilon$ , now keeping $\tilde{\underline{p}}_\nu$ fixed. This is done in section 4 of I.

Of course, it is more difficult to put

$$\exp\left[-\tfrac{1}{2}\, \tilde{\underline{p}}_\nu\, \tilde{\underline{p}}_\mu\, Q^{\nu\mu}(q)\right] \tag{5.13}$$

into the desired order with respect to $\tilde{\underline{p}}$, q than it would be just to order

$$1 - \tfrac{1}{2}\, \varepsilon\, \underline{p}_\nu\, \underline{p}_\mu\, Q^{\nu\mu}(q) \tag{5.14}$$

with respect to $\underline{p}$, q. However, the task is simplified by the $\varepsilon$-dependence of the commutation relation

$$[\tilde{\underline{p}}_\nu\, ,\, q^\mu] = -\, i\sqrt{\varepsilon}\, \delta_\nu{}^\mu \tag{5.15}$$

and the fact that it is sufficient to work in second order in $\sqrt{\varepsilon}$ . The ordering of $\exp \varepsilon \underline{L}(\tilde{\underline{p}}/\sqrt{\varepsilon}, q)$ to second order in $\sqrt{\varepsilon}$ is done in the Appendix of I.

The final step to be carried out, then, is to insert the short-time solution in (5.8) and take the limit $\varepsilon \to 0$. This step is discussed in section 3 of I and carried out in section 4 of I.

We want to make here some further comments concerning the limit $\varepsilon \to 0$ — in particular, about the behaviour of the differences $q_j - q_{j-1}$ in the limit $\varepsilon \to 0$. From our experience with linear processes in section 3 we know that for any given point $t$ on the time axis, with probability 1, the time derivative of the process $\dot{q}$ does not exist. Hence the limit $\lim_{\varepsilon \to 0} (q_j - q_{j-1})/\varepsilon$ is not defined as it stands and one has the freedom of requiring certain additional properties of this symbol in order to give it a meaning. It is here that the various different ways in which path integrals can be defined have their origin(32,33)+). Our procedure to fix the meaning of $\lim_{\varepsilon \to 0} (q_j - q_{j-1})/\varepsilon$ is to require that it obeys all the rules of calculus. We therefore also use the same symbol as in ordinary calculus. Thus, if

$$\lim_{\varepsilon \to 0} q_j = \lim_{\varepsilon \to 0} q_{j-1} = q(t) ,$$

we write formally

$$\lim_{\varepsilon \to 0} \frac{q_j - q_{j-1}}{\varepsilon} = \dot{q}(t) . \tag{5.16}$$

For example, if we consider the nonlinear transformation to new coordinates $Q$:

$$q = f(Q) , \tag{5.17}$$

we require that

$$\begin{aligned} \dot{q}(t) = \lim_{\varepsilon \to 0} \frac{q_j - q_{j-1}}{\varepsilon} &= \lim_{\varepsilon \to 0} \frac{f(Q_j) - f(Q_{j-1})}{\varepsilon} \\ &= \frac{\partial f(Q(t))}{\partial Q(t)} \lim_{\varepsilon \to 0} \frac{Q_j - Q_{j-1}}{\varepsilon} \\ &= \frac{\partial f}{\partial Q} \dot{Q}(t) . \end{aligned} \tag{5.18}$$

It is, of course, not trivial that we are allowed to make such a requirement. A necessary condition is its self-consistency, which can be checked by making the alternative assignment

---

+) Cf. also the seminar of Dr. Tirapegui .

$$\dot{q}(t) \quad = \quad \lim_{\quad 0} \frac{q_{j+1} - q_j}{} \tag{5.19}$$

and proving that the limit 0 of (5.8) comes out unchanged. This simple check is sketched in I in a "note added in proof".

The most prominent feature of our way of handling the limit $\varepsilon \to 0$ is the fact that it leads to results whose form is independent of the special coordinates used. Thus, the path integral constructed in this way is manifestly covariant under transformations of the type (5.17). This feature is useful in applications in which the requirement of general covariance carries information about physical properties of the system, as in general relativity[34]. It is also useful in the applications we wish to discuss here, in which $\mathcal{L}(\dot{q},q)$ is used to define dissipative potentials, since one expects that only potentials defined as invariants under general coordinate transformations will have physical meaning. Another equally important consequence of the way we take $\varepsilon \to 0$ is the fact that we may use the rules of calculus to do manipulations under the functional integral sign.

The explicit result obtained in I, section 4, for $\varepsilon \to 0$ is, in our present notation,

$$D\mu(\{q\})dq \quad = \quad \lim_{\varepsilon \to 0} \prod_{j=1}^{N} \frac{dq_j}{\sqrt{(2\pi\varepsilon)^n Q(q_j)}} \tag{5.20}$$

with

$$Q(q) \quad = \quad \mathrm{Det}\, Q^{\nu\mu}(q) \quad = \quad \| Q^{\nu\mu}(q) \| \tag{5.21}$$

and

$$\mathcal{L}(\dot{q},q) \quad = \quad \tfrac{1}{2} Q_{\nu\mu} (\dot{q}^\nu - h^\nu)(\dot{q}^\mu - h^\mu) + \tfrac{1}{2}\sqrt{Q}\frac{\partial}{\partial q^\nu}\frac{h^\nu}{\sqrt{Q}} + \frac{1}{12} R \tag{5.22}$$

with

$$h^\nu(q) \quad = \quad K^\nu(q) - \tfrac{1}{2}\sqrt{Q}\frac{\partial}{\partial q^\mu}\frac{Q^{\nu\mu}}{\sqrt{Q}} \tag{5.23}$$

$$R \quad = \quad Q^{\varkappa\mu} R_{\varkappa\mu} \tag{5.24}$$

$$R_{\varkappa\mu} \quad = \quad \frac{\partial \Gamma^\lambda{}_{\varkappa\lambda}}{\partial q^\mu} - \frac{\partial \Gamma^\lambda{}_{\varkappa\mu}}{\partial q^\lambda} + \Gamma^\lambda{}_{\nu\mu}\Gamma^\nu{}_{\varkappa\lambda} - \Gamma^\lambda{}_{\nu\lambda}\Gamma^\nu{}_{\varkappa\mu} \tag{5.25}$$

$$\Gamma^\lambda{}_{\mu\nu} \quad = \quad \tfrac{1}{2} Q^{\lambda\varkappa}\left(\frac{\partial Q_{\varkappa\nu}}{\partial q^\mu} + \frac{\partial Q_{\varkappa\mu}}{\partial q^\nu} - \frac{\partial Q_{\mu\nu}}{\partial q^\varkappa}\right); \tag{5.26}$$

$R(q)$ is the Riemann curvature scalar if $Q^{\nu\mu}(q)$ is used as a metric tensor to define a Riemann geometry in the space of the $q^{\nu}$. The quantity $h^{\nu}(q)$ defined in (5.23) transforms like a contravariant vector, and $\mathcal{L}(\dot{q},q)$ is a scalar. It is easy to see that the result (3.24), (3.38) for the linear case is contained in (5.22) as a special case.

Before proceeding to a further discussion of the result (5.22) we want to add another remark concerning the limit $\varepsilon \to 0$. We have discussed how $D\mu(\{q\})$ and $\mathcal{L}(\dot{q},q)$ are constructed from the short-time solution in that limit. However, we have not yet discussed here how one does the reversed calculation, i.e., how one constructs $P(q\, t_0 + \varepsilon \,/\, q_0 t_0)$ from the knowledge of $D\mu(\{q\})$ and $\mathcal{L}(\dot{q},q)$. Indeed, this question is nontrivial, since terms like $(q_j - q_{j-1})$ and $(q_j - q_{j-1})^3/\varepsilon$ contribute, with probability 1, with equal order of magnitude in $\sqrt{\varepsilon}$ to $P(q_j t_j | q_{j-1} t_{j-1})$ but do not contribute to $\mathcal{L}(\dot{q},q)$ in the same order in $\sqrt{\varepsilon}$ (the former contribution is $\dot{q}\varepsilon$ while the latter is $\dot{q}^3 \varepsilon^2$). Thus, it might appear at first sight that in this example only the term $\sim (q_j - q_{j-1})$ could be reconstructed from $\mathcal{L}$ and $D\mu$, while the information about $(q_j - q_{j-1})^3/\varepsilon$ is lost in the limit $\varepsilon \to 0$, and this has led some authors to the belief that, e.g., terms of order $(q_j - q_{j-1})^3/\varepsilon$ have been incorrectly omitted when the limit $\varepsilon \to 0$ was taken to obtain $\mathcal{L}(\dot{q},q)$. This, in fact, is not so. As we have shown in a recent paper (contributed to Statphys. 13, Haifa 1977, henceforth quoted as II), one can reconstruct all these terms explicitly from the functional integral. (In addition, an alternative method for constructing $\mathcal{L}$ and $D\mu$ from the Fokker-Planck equation is also given in II.) The way one can construct $P(q\, t_0 + \varepsilon \,/\, q_0 t_0)$ from $\mathcal{L}(\dot{q},q)$ and $D\mu(q)$ to <u>arbitrary</u> order in $\sqrt{\varepsilon}$ is as follows.

We write

$$P(q\, t_0 + \varepsilon \,|\, q_0 t_0) = \frac{\exp\left[-\ \varepsilon \mathcal{L}\left(\frac{q-q_0}{\varepsilon},\ q\right)\right]}{\sqrt{(2\pi\varepsilon)^n Q(q)}}\ Z(\varepsilon, q - q_0) \qquad (5.27)$$

with

$$\begin{aligned} Z(\varepsilon,\eta) &= 1 + C_{\nu\mu}\, \eta^{\nu}\eta^{\mu} \\ &+ \frac{1}{\varepsilon}(D_{\nu\mu\varkappa}\, \eta^{\nu}\eta^{\mu}\eta^{\varkappa} + E_{\mu\nu\varkappa\lambda}\, \eta^{\mu}\eta^{\nu}\eta^{\varkappa}\eta^{\lambda}) \\ &+ \frac{1}{\varepsilon^2} G_{\mu\nu\varkappa\lambda\rho\sigma}\, \eta^{\mu}\eta^{\nu}\eta^{\varkappa}\eta^{\lambda}\eta^{\rho}\eta^{\sigma} \qquad (5.28) \\ \eta^{\nu} &= q^{\nu} - q_0^{\nu} \end{aligned}$$

where $C_{..}$, $D_{...}$, $E_{....}$ and $G_{......}$ are matrices of constants independent of $\varepsilon$ and $\eta$. Several comments are necessary to explain this expansion. First, we see that when we take $\varepsilon \to 0$ and $\eta/\varepsilon \to \dot{q}$ only the first term of the expansion survives (i.e. $Z(\varepsilon, \eta) \to 1$) and contributes to $\mathcal{L}$ and $D\mu$. Thus $P(q\ t_0+\varepsilon | q_0 t_0)$ for arbitrary C, D, E, G leads to the given form of $\mathcal{L}$ and $D\mu$. This condition explains the absence of terms $\sim \eta$ and $\sim \varepsilon$ in (5.28).

The question posed earlier now becomes simply: how can we determine the coefficient matrices C, D, E, G, and why do only these terms appear in the expansion (5.28) and not a term $\sim \frac{1}{\varepsilon^2} F_{\mu\nu\kappa\lambda\rho} \eta^\mu \eta^\nu \eta^\kappa \eta^\rho$ or, generally, terms of the structure $\eta^{2n+1}/\varepsilon^n$ and $\eta^{2n+2}/\varepsilon^n$ for $n \geqslant 3$? The answers follow from two general conditions that have to be satisfied by $P(q\ t_0+\varepsilon / q_0 t_0)$ in the normalization condition

$$\int dq\ P(q\ t_0+\varepsilon | q_0 t_0) = 1 , \tag{5.29}$$

and the Chapman-Kolmogorov equation

$$P(q\ t_0+2\varepsilon | q_0 t_0) = \int dq' P(q\ t_0+\varepsilon | q' t_0) P(q' t_0+\varepsilon | q_0 t_0) . \tag{5.30}$$

If the expansion (5.27), (5.28) is inserted on the right-hand side of (5.30) and the integrals are evaluated to accuracy $\eta^{2n+2}/\varepsilon^n$, an expansion of the same form but with renormalized coefficients results. The requirement that after the substitution $\varepsilon \to \varepsilon/2$ the renormalized expansion be identical to the original one, together with the normalization condition (5.29), then fixes the coefficient matrices C, D, E, G. The absence of terms $\sim F \eta^5/\varepsilon^2$, $\eta^{2n+1}/\varepsilon^2$ and $\eta^{2n+2}/\varepsilon^2$ for $n \geqslant 3$ from the expansion (5.28) can now be explained. Some reflection on the explicit expressions for $\mathcal{L}$ and $D\mu$ shows that terms of this form are not created by the right-hand side of (5.30) if they are absent initially. Therefore, one possible fixed-point value of their coefficients is zero. It is a little laborious but not difficult to check that this fixed point is also unique, by first allowing nonzero values of these coefficients and looking at the way they transform when inserted into the right-hand side of eq. (5.30). The fixed-point values of the matrices C, D, E and G uniquely define the short-time propagator to the desired order (i.e., to order $\varepsilon$).

The procedure we have outlined is independent of the coordinates used and leads to a unique expression for $P(q\ t_0+\varepsilon | q_0 t_0)$ in every order of $\sqrt{\varepsilon}$ for given $\mathcal{L}$ and $D\mu$. From the expression for $P(q\ t_0+\varepsilon | q_0 t_0)$ to order $\varepsilon$ it is straightforward to derive the differential equation satisfied by P, which is, of course, the Fokker-Planck equation. The details of these calculations are a little laborious - they are given in part in II, where $C_{\mu\nu} = -\frac{1}{12} R_{\mu\nu}$ is calculated. More details will be reported elsewhere[36]. These results only concern us here inasmuch as they prove that there is a manifestly covariant procedure by which the equivalence of the Fokker-Planck description and a unique coordinate-independent path-integral description can be established both ways. In addition, we have obtained a definition of what we mean by a path integral for the conditional probability density, which is independent of the coordinates used and independent of the limiting procedure by which the path integral is derived from a finite-dimensional integral.

## 6. THE LAGRANGIAN AND DISSIPATIVE POTENTIALS

Having obtained the path-integral solution of the Fokker-Planck equation (2.13) in the last section, we now want to proceed and see what its relation is to the two central quantities that characterize a nonequilibrium steady state: the potential $S(q)$ and the drift velocity in the steady state $r^{\nu}(q)$, which are defined in (2.22) - (2.24). It is instructive to rewrite (2.24) in terms of the contravariant vector $h^{\nu}(q)$ defined in (5.23):

$$h^{\nu} = \tfrac{1}{2} Q^{\nu\mu} \frac{\partial \tilde{S}}{\partial q^{\mu}} + r^{\nu} , \tag{6.1}$$

where

$$\tilde{S}(q) = S(q) + \tfrac{1}{2} \ln \| Q^{\nu\mu}(q) \| , \tag{6.2}$$

i.e.,

$$W_0(q) = \frac{1}{\sqrt{\| Q^{\nu\mu}(q) \|}} \exp \tilde{S}(q) . \tag{6.3}$$

As is clear from (6.3), the reduced potential $\tilde{S}(q)$ just transforms like a scalar. Therefore, the drift velocity in the steady state $r^{\nu}$ is, in fact, a contravariant vector, i.e., a quantity which is defined independently of the spatial coordinates used, unlike the original drift velocity $K^{\nu}$, which is not a covariant quantity. Since $K^{\nu}(q)$ and $S(q)$ are not covariant quantities it is preferable to work with their covariant counterparts $h^{\nu}(q)$ and $\tilde{S}(q)$ instead[37].

We now define the two dissipative potentials $\phi(\dot{q},\dot{q})$ and $\psi(\frac{\partial \tilde{S}}{\partial q}, \frac{\partial \tilde{S}}{\partial q})$, as in the linear case:

$$\phi(\dot{q},\dot{q}) = Q_{\mu\nu} \dot{q}^{\mu} \dot{q}^{\nu} \tag{6.4}$$

$$\psi\left(\frac{\partial\tilde{S}}{\partial q}, \frac{\partial\tilde{S}}{\partial q}\right) = \tfrac{1}{4} Q^{\mu\nu} \frac{\partial\tilde{S}}{\partial q^{\mu}} \frac{\partial\tilde{S}}{\partial q^{\nu}} \tag{6.5}$$

Both dissipative potentials are scalars under general coordinate transformations. In terms of these, the scalar $\mathcal{L}(\dot{q},q)$ can be rewritten as

$$\mathcal{L} = \tfrac{1}{2}\left[\phi(\dot{q}-r, \dot{q}-r) + \psi\left(\frac{\partial\tilde{S}}{\partial q}, \frac{\partial\tilde{S}}{\partial q}\right) - \dot{\tilde{S}}\left(\dot{q}, \frac{\partial\tilde{S}}{\partial q}\right)\right]$$

$$+ \tfrac{1}{4}\sqrt{Q}\, \frac{\partial}{\partial q^{\nu}}\left(\frac{Q^{\nu\mu}}{\sqrt{Q}} \frac{\partial\tilde{S}}{\partial q^{\mu}}\right) + \frac{1}{12} R \, . \tag{6.6}$$

The steady-state drift velocity $r^{\nu}$ appears only in the dissipative potential $\phi$, since it drops out from $\dot{S}$ and $\tfrac{1}{2}\sqrt{Q}\,\frac{\partial}{\partial q^{\nu}} \frac{h^{\nu}(q)}{\sqrt{Q}}$ owing to the relation

$$\sqrt{Q}\, \frac{\partial}{\partial q^{\nu}} \frac{r^{\nu}(q)}{\sqrt{Q}} + r^{\nu}(q) \frac{\partial\tilde{S}}{\partial q^{\nu}} = 0 \quad , \tag{6.7}$$

which follows from (2.23). The first three terms in (6.6) are therefore the same as in the linear case, eq. (4.1), where we obtained them on the basis of the more special relations (3.6), (3.7), which do not hold in the nonlinear case. The last two terms in (6.6) reduce to a constant in the linear case and could be absorbed there in the measure $D\mu$. Even though these terms are not constants in the general case, we are free to use redefined quantities

$$\tilde{\mathcal{L}}(\dot{q},q) = \tfrac{1}{2}\left[\phi(\dot{q}-r, \dot{q}-r) + \psi\left(\frac{\partial\tilde{S}}{\partial q}, \frac{\partial\tilde{S}}{\partial q}\right) - \dot{\tilde{S}}\left(\dot{q}, \frac{\partial\tilde{S}}{\partial q}\right)\right] \tag{6.8}$$

$$D\tilde{\mu}(\{q\}) = D\mu(\{q\})\exp\left[-\tfrac{1}{4}\int d\tau\left(\sqrt{Q}\, \frac{\partial}{\partial q^{\nu}}\left(\frac{Q^{\nu\mu}}{\sqrt{Q}} \frac{\partial\tilde{S}}{\partial q^{\mu}}\right) + \tfrac{1}{3}R\right)\right]. \tag{6.9}$$

As a result, we see that, apart from the need for a more complicated measure $D\tilde{\mu}(\{q\})$, the relation of the dissipative potentials $\phi$, $\psi$ and the production rate $\dot{\tilde{S}}$ to the probability-density functional is the same as in the Gaussian case. This result is therefore an extensive generalization of the original Onsager-Machlup result. Of course, there is the same drawback as in the linear case, namely, that we do not know, in general, what $r^{\nu}(q)$ is. This drawback is even more serious here than it was in the linear case, since, there at least, a rigorous systematic procedure to determine $r^{\nu}$ was available (cf. eq. (3.5)). Here, no systematic procedure exists, apart from perturbative approaches, which use the Gaussian case as a starting point.

The only exemption from this bleak prospect is the case when $W(\{q(\tau)\})$ has time-reversal symmetry. Let $T$ be the operator of the time-reversal transformation and let

$$\tilde{q}(\tau) = T\, q(\tau) \quad \text{and} \quad \dot{\tilde{q}} = -T\, \dot{q}(\tau) \tag{6.10}$$

be the time-reversed $q(\tau)$ and $\dot{q}(\tau)$. Similarly, if the system is controlled by a set of external parameters $\lambda$, let

$$\tilde{\lambda} = T\lambda \tag{6.11}$$

be the time-reversed version of these parameters. Then $W(\{q(\tau)\},\lambda)$ has time-reversal symmetry if

$$W(\{q(\tau)\},\lambda)\, D\mu(\{q(\tau)\},\lambda) = W(\{\tilde{q}(\tau)\},\tilde{\lambda})D\mu(\{\tilde{q}(\tau)\},\tilde{\lambda}). \tag{6.12}$$

Using

$$W(\{q(\tau)\},\lambda)D\mu(\{q(\tau)\},\lambda) = \exp\left[-\int \mathcal{L}(\dot{q},q,\lambda)d\tau\right]D\mu(\{q(\tau)\},\lambda) \tag{6.13}$$

with $\mathcal{L}(\dot{q},q)$ and $D\mu$ given by (5.22), (5.20), we can compare coefficients on both sides of eq. (6.12). We obtain

$$D\mu(\{q(\tau)\},\lambda) = D\mu(\{\tilde{q}(\tau)\},\tilde{\lambda}) \tag{6.14}$$

and

$$\mathcal{L}(\dot{q},\, q,\lambda) - \frac{dF(q,\lambda)}{d\tau} = \mathcal{L}(-\dot{\tilde{q}},\, \tilde{q},\tilde{\lambda}) + \frac{dF(\tilde{q},\tilde{\lambda})}{d\tau}. \tag{6.15}$$

The total time derivatives on both sides of eq. (6.15) are allowed, since they either drop out from the expression for $W(\{q(\tau)\})$ if $F(\tau\to+\infty) = F(\tau\to-\infty)$, i.e., if the parameters

$$\begin{aligned} F(\tau\to+\infty) &= \lambda_{+\infty} \\ F(\tau\to-\infty) &= \lambda_{-\infty} \end{aligned} \tag{6.16}$$

are considered as external parameters with the time-reversal properties

$$\tilde{\lambda}_{+\infty} = T\lambda_{+\infty} = \lambda_{-\infty},\quad \tilde{\lambda}_{-\infty} = T\lambda_{-\infty} = \lambda_{+\infty}. \tag{6.17}$$

We now use the result for $\mathcal{L}$ (eq. (5.22)) on both sides of eq. (6.15) and compare the coefficients of $\dot{q}^\nu \dot{q}^\mu$, $\dot{q}^\nu$ and the terms without time derivatives, respectively.

Assuming that the variables $q^\nu$ transform as even or odd variables:

$$\tilde{q}^{\nu}(\tau) = \varepsilon^{\nu} q^{\nu}(\tau), \qquad \varepsilon^{\nu} = \pm 1 , \tag{6.18}$$

we obtain the symmetry relations

$$Q^{\nu\mu}(q,\lambda) = \varepsilon^{\nu}\varepsilon^{\mu} Q^{\nu\mu}(\tilde{q},\tilde{\lambda}) . \tag{6.19}$$

These relations automatically ensure that eq. (6.14) is satisfied. Using these relations for the coefficients of $\dot{q}^{\nu}$ we obtain

$$Q_{\nu\mu}(q,\lambda)(h^{\mu}(q,\lambda) + \varepsilon^{\mu} h^{\mu}(\tilde{q},\tilde{\lambda})) = \frac{\partial}{\partial q^{\mu}}(F(\tilde{q},\tilde{\lambda}) + F(q,\lambda)) \tag{6.20}$$

We can use eq. (6.20) to decompose

$$h^{\mu}(q,\lambda) = \tfrac{1}{2} Q^{\mu\nu}\frac{\partial}{\partial q^{\nu}}(F(\tilde{q},\tilde{\lambda}) + F(q,\lambda)) + j^{\mu} \tag{6.21}$$

in which

$$j^{\mu}(q,\lambda) = \tfrac{1}{2}(h^{\mu}(q,\lambda) - \varepsilon^{\mu} h^{\mu}(\tilde{q},\tilde{\lambda})) \tag{6.22}$$

is just the reversible part of $h^{\mu}$. Then the last set of relations obtained from (6.15) takes the concise form

$$j^{\nu}(q,\lambda)\frac{\partial}{\partial q^{\nu}}(F(\tilde{q},\tilde{\lambda}) + F(q,\lambda)) + \sqrt{Q}\,\frac{\partial}{\partial q^{\nu}}\left(\frac{j^{\nu}(q,\lambda)}{\sqrt{Q}}\right) = 0. \tag{6.23}$$

Equations (6.19), (6.20), (6-23) are just the potential conditions for detailed balance[38]. Comparison of eqs. (6.21), (6.23) with (6.1) show that one possible solution to the latter equations is

$$\tilde{S}(q,\lambda) = F(q,\lambda) + F(\tilde{q},\lambda) \tag{6.24}$$

$$r^{\nu}(q,\lambda) = j^{\nu}(q,\lambda), \tag{6.25}$$

from which the symmetry relations

$$\tilde{S}(\tilde{q},\tilde{\lambda}) = \tilde{S}(q,\lambda) \tag{6.26}$$

$$r^{\nu}(\tilde{q},\tilde{\lambda}) = -\varepsilon^{\nu} r^{\nu}(q,\lambda) \tag{6.27}$$

follow. Since the time-independent solution $W_C(q)$ is unique under the assumptions we have made, the solution (6.24), (6.25) is also unique. Hence, the drift velocity in a steady state with time-reversal symmetry is necessarily reversible (if all external parameters are also replaced by their time-reversed versions), i.e., $r^{\nu}$ is then not related to any dissipative processes. In this special case, it is therefore possible to determine $r^{\nu}$ simply from its transformation property under time reversal. In particular, this is always possible if the steady state under consideration is a state of thermodynamic equilibrium or close to such a state in the sense discussed after eq. (4.25). In that case, a time-reversal transformation again leads to a (generally different) state of thermodynamic equilibrium, in which all time-reversed processes occur with the same probability as the original processes in the original equilibrium states. In a nonequilibrium steady state the time-reversal transformation (6.10) is also defined. However, a general nonequilibrium steady state does not transform into another possible nonequilibrium steady state that would be compatible with the time-reversed boundary conditions and constraints. This may seem to be in conflict with the fact that in all cases the underlying microscopic processes are reversible. However, by remembering that the boundary conditions and external constraints of the system also have to be transformed the seeming contradiction is resolved, since for a nonequilibrium steady state the time-reversed state and the time-reversed boundary conditions and constraints do not match in the same way as they do for thermodynamic equilibrium states.

Thus, for nonequilibrium steady states, the only significance of (6.8) is to show how $\tilde{\mathcal{L}}$ is composed of $\phi$, $\psi$ and $\tilde{S}$ in principle. The only expression known explicitly is then eq. (5.22), which can be written in the form

$$\begin{aligned}\mathcal{L} &= \tfrac{1}{2}\,\phi(\dot{q} - h,\ \dot{q} - h) + \tfrac{1}{2}\sqrt{Q}\,\frac{\partial}{\partial q^{\mu}}\,\frac{h^{\mu}}{\sqrt{Q}} + \frac{1}{12}\,R \qquad (6.28)\\ &= \approx{\mathcal{L}} + \tfrac{1}{2}\sqrt{Q}\,\frac{\partial}{\partial q^{\mu}}\,\frac{h^{\mu}}{\sqrt{Q}} + \frac{1}{12}\,R \ .\end{aligned}$$

In either case, the relation of $\mathcal{L}(\dot{q}, q)$ to reduced probability densities is much more complicated than in the linear case, where path integrals could be carried out by extremizing $\int \mathcal{L}(\dot{q}, q)\,d\tau$. The generalization of this simple result to nonlinear processes is considered in the next section.

The relation of the form of the probability-density functional to the maximum-entropy principle is discussed in the Appendix.

## 7. THE ONSAGER-MACHLUP FUNCTION FOR NONLINEAR PROCESSES

We now want to know what is the generalization of the result of the linear theory, that averages of paths under certain constraints C are obtained by the minimization of $\int \mathcal{L}(\dot{q}, q)\,d\tau$ :

$$\int \mathcal{L}(\langle\dot{q}\rangle, \langle q\rangle)d\tau = \text{min.} \tag{7.1}$$

We can obtain averages in the general case from the characteristic functional[25)]

$$Z(\{\eta\}) = \int D\mu(\{q\})\exp\left[-\int_{-\infty}^{+\infty} (\mathcal{L}(\dot{q},q) - \eta_\nu q^\nu)d\tau\right] \tag{7.2}$$

$$Z(\{0\}) = 1. \tag{7.3}$$

For example,

$$\langle q^\nu(\tau)\rangle = \left[\frac{\delta}{\delta\eta_\nu^\nu(\tau)} Z(\{\eta\})\right]_{\eta=0}, \tag{7.4}$$

$$\langle q^\nu(\tau)q^\mu(\tau')\rangle = \left[\frac{\delta^2}{\delta\eta_\nu^\nu(\tau)\,\delta\eta_\mu^\mu(\tau')} Z(\{\eta\})\right]_{\eta=0}. \tag{7.5}$$

Introducing

$$\Psi(\{\eta\}) = \ln Z(\{\eta\}), \quad V(\{\eta\}) = \left.\frac{\delta\Psi}{\delta\eta}\right|_{\eta=0}, \tag{7.6}$$

we have

$$\Psi(\{0\}) = 0, \quad v^\nu(\{0\},\tau) = \langle q^\nu(\tau)\rangle. \tag{7.7}$$

The Legendre transform

$$\Gamma(\{v\}) = -\Psi(\{\eta\}) + \int v^\nu(\tau)\,\eta_\nu(\tau)d\tau \tag{7.8}$$

then has the properties

$$\frac{\delta\Gamma}{\delta v^\nu(\tau)} = \eta_\nu(\tau), \tag{7.9}$$

$$\Gamma(\{\langle q\rangle\}) = 0; \qquad \left.\frac{\delta\Gamma}{\delta v^\nu(\tau)}\right|_{v=\langle q\rangle} = 0. \tag{7.10}$$

Thus, $\langle q(\tau)\rangle$ is obtained from the stationarity principle

$$\Gamma(\{\langle q\rangle\}) = \text{stationary.} \tag{7.11}$$

Whether $\Gamma(\{v\})$ attains a maximum or minimum for $v = \langle q\rangle$ can be learned from the second-order derivative

$$C^{-1}_{\nu\mu}(\tau,\tau') = \left.\frac{\delta^2\Gamma}{\delta v^\nu(\tau)\,\delta v^\mu(\tau')}\right|_{v=\langle q\rangle}, \tag{7.12}$$

which is obtained from eq. (7.9) as

$$C^{-1}_{\nu\mu}(\tau,\tau') = \left.\frac{\delta\eta_\nu(\tau)}{\delta v^\mu(\tau')}\right|_{v=\langle q\rangle} \tag{7.13}$$

i.e.,

$$\int d\tau'\, C^{-1}_{\nu\mu}(\tau,\tau')C^{\mu\lambda}(\tau',\tau'') = \delta_\nu{}^\lambda\,\delta(\tau-\tau'') \tag{7.14}$$

with

$$C^{\nu\mu}(\tau,\tau') = \left.\frac{\delta v^\nu(\tau)}{\delta\eta_\mu(\tau')}\right|_{\eta=0} = \left.\frac{\delta^2\psi}{\delta\eta_\nu(\tau)\,\delta\eta_\mu(\tau')}\right|_{\eta=0}$$

$$= \langle\!\langle (q^\nu(\tau) - \langle q^\nu(\tau)\rangle)(q^\mu(\tau') - \langle q^\mu(\tau')\rangle)\rangle\!\rangle \tag{7.15}$$

According to eq. (7.15) the kernel $C^{\nu\mu}(\tau,\tau')$ is positive-definite and hence the same must be true of its inverse (7.12). Therefore, $\Gamma(\{v\})$ attains a minimum for $v(\tau) = \langle q(\tau)\rangle$. In summary, we see that $\Gamma(\{v\})$ is a potential which yields the time-dependent average path upon minimization.

We now consider the minimization of $\Gamma(\{v\})$ under various constraints. First we take the case

$$\Gamma(\{v\}) = \min, \qquad v(\tau) = v \quad \text{for} \quad \tau = t\,. \tag{7.16}$$

With a Lagrange multiplier $\eta$ we obtain

$$\frac{\delta\Gamma}{\delta v(\tau)} = \eta\,\delta(\tau-t),$$

i.e., from (7.9),

$$\eta_\nu(\tau) = \eta_\nu\,\delta(\tau-t),$$

$$Z(\{\bar\eta\}) = \langle e^{\eta_\nu q^\nu(t)}\rangle \equiv \psi_1(\eta)\,, \tag{7.17}$$

$$\Gamma_{\min}(\{v\}) = -\psi_1(\eta) + v^\nu\eta_\nu \equiv \Gamma_1(v)\,.$$

From $\Gamma_1(v)$ the average of $q(t)$ can still be obtained by minimization:

$$\left.\frac{\partial\Gamma}{\partial v}\right|_{v=\langle q\rangle} = 0, \quad \left.\frac{\partial^2\Gamma_1}{\partial v^\nu\,\partial v^\mu}\right|_{v=\langle q\rangle} = C^{-1}_{\nu\mu} \tag{7.18}$$

$$\Gamma(\langle q\rangle) = 0\,,$$

i.e., $\Gamma_1(v)$ is the potential for the case when all constraints on $q(\tau)$ are relaxed for $\tau \neq t$ and all $q(\tau)$ for $\tau \neq t$ are integrated out. The integration over $q(\tau)$ for $\tau \neq t$ corresponds to the minimization of $\Gamma(\{q(\tau)\})$. It is straightforward to generalize this result to the constraint

$$C_\ell : v(\tau_1) = v_1, \ldots, v(\tau_\ell) = v_\ell$$

to obtain

$$\Gamma_\ell(v_\ell \tau_\ell ; \ldots ; v_1 \tau_1) = \left[\Gamma(\{v(\tau)\})\right]^{C_\ell}_{\min} \tag{7.19}$$

from which correlation functions involving $v(\tau_1), \ldots, v(\tau_\ell)$ can be obtained by constructing

$$\psi_\ell = -\Gamma_\ell + \sum_{i=1}^{\ell} v_i^\nu \eta_{i\nu} ; \quad \eta_{i\nu} = \frac{\partial \Gamma_\ell}{\partial v_i^\nu} \tag{7.20}$$

$$Z_\ell = \exp \psi_\ell$$

and taking derivatives with respect to $\eta_i$. Thus $\Gamma(\{v(\tau)\})$ replaces $\int d\tau \, \mathcal{L}(\dot{q}, q)$ of the linear theory in all respects concerned with evaluating the path integral.

We now want to find out the physical meaning of $\Gamma(\{v(\tau)\})$. To this end we consider the quantity

$$K(\{P(\{q\})\}) = \int D_\mu(\{q\}) P(\{q\}) \ln \frac{P(\{q\})}{W(\{q\})} , \tag{7.21}$$

which is a functional of an arbitrary probability-density functional $P(\{q\})$ analogous to $K$ in eq. (2.15). The functional $W(\{q\})$ in eq. (7.21) is just the probability-density functional of the system under consideration. The quantity $K(\{P\})$ has a clear physical meaning for any macroscopic system. Its thermodynamic and statistical significance+) has been studied by Schlögl in a series of papers[13-15]. In information theory, $K(\{P\})$ is a measure of the "gain of information" obtained by an observer who observes the probability-density functional $P(\{q\})$ in a system whose steady-state probability-density functional is given by $W(\{q\})$.

The thermodynamic significance of $K(\{P\})$ derives from the fact that it is closely related to the entropy difference in the two states described by $P(\{q\})$ and $W(\{q\})$, respectively, if we define entropy in the usual way (cf. also the Appendix):

$$S(\{P(\{q\})\}) = -k \int D_\mu(\{q\}) P(\{q\}) \ln \frac{P(\{q\})}{N(\{q\})} . \tag{7.22}$$

Here $N(\{q\})$ is the density of microstates in function space $\{q(\tau)\}$ for fixed $q(\tau)$ (see the Appendix):

+) for ordinary probability densities, not for functionals.

$$N(\{q\})D\mu(\{q\}) \quad = \quad \text{number of microstates in } D\mu(\{q\}) \,.$$

It is then easy to show[14] that the entropy difference for the functionals $P(\{q\})$ and $W(\{q\})$ can be written as

$$\delta S = \delta_{\ell} S + \delta_{n\ell} S \tag{7.23}$$

where $\delta_{\ell} S$ is that part of $\delta S$ which is linear+) in $P(\{q\}) - W(\{q\})$ (or linear in the difference of the averages $\langle \ell n \frac{W(\{q\})}{N(\{q\})} \rangle$ in the two states):

$$\delta_{\ell} S = - \int D\mu(\{q\})(P(\{q\}) - W(\{q\}))\, \ell n\, \frac{W(\{q\})}{N(\{q\})} \,, \tag{7.24}$$

and the remainder $\delta_{n\ell} S$ is nonlinear in $P(\{q\}) - W(\{q\})$ and given by

$$\delta_{n\ell} S(\{P\}) = - K(\{P\}) \,. \tag{7.25}$$

Let us now minimize $K(\{P\})$ with respect to $P$ under the constraint that only test probability density functionals with an average path

$$\langle q^{\nu}(\tau)\rangle = v^{\nu}(\tau) \qquad (\text{all } \tau) \tag{7.26}$$

are allowed, where $v^{\nu}(\tau)$ is prescribed in an arbitrary way. The result, obtained in a straightforward fashion, is

$$P(\{q,v\}) = Z^{-1}(\{\eta\})\, W(\{q\}) \exp \int d\tau\, \eta_{\nu}(\tau) q^{\nu}(\tau) \,, \tag{7.27}$$

with $Z(\{\eta\})$ given by eq. (7.2). The Lagrange parameter $\eta(\tau)$ is fixed by the constraint (7.26), which yields

$$\frac{\delta Z(\{\eta\})}{\delta \eta_{\nu}(\tau)} = v^{\nu}(\tau)$$

from which $\eta_{\nu} = \eta_{\nu}(\{v\})$ follows. Inserting the minimizing functional in (7.21) we obtain

$$K_{min}(\{P\}) = K(\{v\}) = - \ell n Z(\{\eta\}) + \int d\tau\, \eta_{\nu}(\tau) v^{\nu}(\tau) = \Gamma(\{v\}). \tag{7.28}$$

Thus $\Gamma(\{v\})$ is the minimum of the functional $K(\{P\})$ under the constraint that $\langle q(\tau)\rangle = v(\tau)$. It is therefore the "gain of information" obtained by observing an average path $\langle q(\tau)\rangle = v(\tau)$.

---

+) In the case where $P(\{q\})$ <u>also</u> satisfies the constraint under which $W(\{q\})$ minimizes $S$, $\delta_{\ell} S = 0$ (see the Appendix).

At the same time it is that part of the entropy decrease which is non-linear in $\langle \ln \frac{W}{N} \rangle$ and accompanies the observation of $\langle q(\tau)\rangle = v(\tau)$. If $\langle q(\tau)\rangle$ is observed to coincide with the average path in the steady state, both the "gain of information" and the entropy decrease are, of course, zero.

Similarly, the reduced potential $\Gamma_1(v)$ given by eq. (7.17) is obtained as the minimum of the functional $K(\{W(q)\})$, defined in eq. (2.15), under the constraint that $\langle q^\nu(\tau)\rangle = v^\nu(\tau)$. Its statistical and thermodynamic meaning is therefore obvious.

For macroscopic systems one usually wants to fix attention on the long-time behaviour of the system. In the limit of long time scales a useful representation of the functional $\Gamma(\{v(\tau)\})$ is given by an expansion of $\Gamma(\{v(\tau)\})$ in terms of time derivatives of increasing order:

$$\Gamma(\{v(\tau)\}) = \int d\tau\, \gamma(v(\tau), \dot{v}(\tau), \ddot{v}(\tau) \dots) \tag{7.29}$$

with+)

$$\gamma = V(v) + A_{\nu\mu}(v)A^\mu(v)\dot{v}^\nu + \tfrac{1}{2} A_{\nu\mu}(v)\dot{v}^\nu\dot{v}^\mu + B_\nu(v)\ddot{v}^\nu + \dots \; . \tag{7.30}$$

The minimum condition for the average path $\langle q(\tau)\rangle$ starting from a given initial value $\langle q(0)\rangle$ becomes

$$\left.\frac{\partial\gamma}{\partial v^\nu}\right|_{v=\langle q\rangle} = 0\,, \qquad \left.\frac{\partial\gamma}{\partial \dot{v}^\nu}\right|_{v=\langle q\rangle} = 0\,,$$
$$\left.\frac{\partial\gamma}{\partial \ddot{v}^\nu}\right|_{v=\langle q\rangle} = 0\,, \quad \text{etc.,} \tag{7.31}$$

which yields the equations (with $\langle q\rangle = q$)

$$0 = \frac{\partial V}{\partial q^\nu} + \frac{\partial A_{\kappa\mu}A^\mu}{\partial q^\nu}\dot{q}^\kappa + \tfrac{1}{2}\frac{\partial A_{\kappa\lambda}}{\partial q^\nu}\dot{q}^\kappa\dot{q}^\lambda + \frac{\partial B_\kappa}{\partial q^\nu}\ddot{q}^\kappa + \dots,$$
$$0 = A_{\nu\mu}A^\mu + A_{\nu\mu}\dot{q}^\mu + \dots, \tag{7.32}$$
$$0 = B_\nu + \dots \; .$$

Retaining only terms including second-order time derivatives we obtain for $v = \langle q\rangle \equiv q$

$$B_\nu = 0, \qquad \dot{q}^\nu = -\Lambda^\nu(q)$$
$$\frac{\partial V}{\partial q^\nu} = \tfrac{1}{2}\frac{\partial}{\partial q^\nu}\left[A_{\mu\kappa}\Lambda^\mu A^\kappa\right]. \tag{7.33}$$

---

+) We do not prove that such an expansion exists - we simply assume it. At least in higher orders it does not exist if there are "long time tails" (cf. Dr. Alder's lectures).

From the latter equation we obtain

$$V(q) = \tfrac{1}{2} A_{\mu\nu} A^{\mu} A^{\nu} + \text{const} \tag{7.34}$$

and, therefore,

$$\gamma = \gamma(\dot{q},q) = \tfrac{1}{2} A_{\nu\mu}(q)(\dot{q}^{\nu} + A^{\nu}(q))(\dot{q}^{\mu} + A^{\mu}(q)) + \text{const} . \tag{7.35}$$

Since $\gamma(\dot{q},q) = 0$, we have to take const = 0 in view of eq. (7.33). As a result, the long-time limit of the functional $\Gamma(\{v\})$ has the same form as the Lagrangian $\tilde{\mathcal{L}}$ as given by eq. (6.28), provided that the original quantities $h^{\nu}(q)$, $Q_{\nu\mu}(q)$ are replaced by renormalized quantities:

$$\begin{aligned} h^{\nu}(q) &= - A^{\nu}(q) \\ Q_{\nu\mu}(q) &= A_{\nu\mu}(q) \quad . \end{aligned} \tag{7.36}$$

The function $\gamma(\dot{v},v)$, which allows us to represent the functional $\Gamma(\{v\})$ by a mere function of $\dot{v}$ and $v$ in the long-time limit, is now the complete analogue of the Onsager-Machlup function for linear Gaussian processes. In fact, for Gaussian processes $\gamma(\dot{v}, v)$ is easily calculated and found to be

$$\gamma(\dot{v},v) = \mathcal{L}(\dot{v},v) + \text{const}, \tag{7.37}$$

i.e.,

$$\begin{aligned} Q_{\nu\mu} &= A_{\nu\mu} , \\ h^{\nu}(v) &= B^{\nu}{}_{\mu} v^{\mu} = - A^{\nu}(v) , \end{aligned} \tag{7.38}$$

i.e., $\gamma(\dot{v},v)$ reduces exactly to the Onsager-Machlup function in that case. For non-Gaussian processes $Q_{\nu\mu}$ and $h^{\nu}$ get renormalized into $A_{\nu\mu}$ and $A^{\nu}$. If the coefficients $Q_{\nu\mu}$ and $h^{\nu}$ satisfy the symmetry relations implied by time-reversal symmetry, a set of corresponding symmetry relations is inherited by the renormalized quantities and $A_{\nu\mu}$ and $A^{\nu}$. In order to obtain these relations it is only necessary to repeat the arguments given in eqs. (6.15) - (6.23) for $\mathcal{L}(\dot{q},q)$ using the renormalized function $\gamma(\dot{v},v)$. If under time reversal the $v^{\nu}$ transform as even or odd variables:

$$\tilde{v}^{\nu}(\tau) = \varepsilon^{\nu} v^{\nu}(\tau), \quad \dot{\tilde{v}}^{\nu}(\tau) = - \varepsilon^{\nu} \dot{v}^{\nu}(\tau), \quad \varepsilon^{\nu} = \pm 1 ,$$

we obtain the time-reversal symmetry relations

$$A_{\nu\mu}(v,\lambda) = \varepsilon^{\nu}\varepsilon^{\mu} A_{\nu\mu}(\tilde{v},\tilde{\lambda}) , \tag{7.39}$$

$$\tfrac{1}{2} A_{\nu\mu}(A^{\mu}(v,\lambda) + \varepsilon^{\mu} A^{\mu}(\tilde{v},\tilde{\lambda})) = \tfrac{1}{2} \frac{\partial \Gamma(v,\lambda)}{\partial v^{\nu}} , \tag{7.40}$$

where

$$\Gamma(v,\lambda) \quad = \quad \Gamma(\tilde{v},\tilde{\lambda}), \tag{7.41}$$

and

$$\tfrac{1}{2}(A^{\nu}(v,\lambda) \; - \; \varepsilon^{\nu} A^{\nu}(\tilde{v},\tilde{\lambda})) \; = \; - \bar{r}^{\nu}(v,\lambda) \; , \tag{7.42}$$

where

$$\bar{r}^{\nu}(v,\lambda) \quad = \quad - \varepsilon^{\nu} \, \bar{r}(\tilde{v},\tilde{\lambda}), \tag{7.43}$$

and, finally,

$$\bar{r}^{\mu}(v,\lambda) \frac{\partial \Gamma(v,\lambda)}{\partial v^{\mu}} \quad = \quad 0 \; . \tag{7.44}$$

The function $\Gamma(v,\lambda)$ appearing in eqs. (7.40), (7.41) and (7.44) is identical to $\Gamma_1(v)$, i.e.,

$$\Gamma(v,\lambda) \; = \left[\int_{-\infty}^{+\infty} d\tau \;\; \gamma(\dot{v},v,\lambda)\right]_{\min}^{v(t)=v} \; = \; \Gamma_1(v) \, , \tag{7.45}$$

or

$$\Gamma(v,\lambda) \; = \left[\int_{-\infty}^{t} d\tau \;\; \gamma(\dot{v},v,\lambda)\right]_{\min}^{v(t)=v} \; = \; \Gamma_1(v) \, , \tag{7.46}$$

since

$$\left[\int_{t}^{\infty} d\tau \;\; \gamma(\dot{v},v,\lambda)\right]_{\min}^{v(t)=v} \quad = \quad 0 \; . \tag{7.47}$$

We can prove the equivalent eqs. (7.45), (7.46) solely on the basis of the relations

$$A^{\mu}(v) \quad = \quad \tfrac{1}{2}\left[A^{-1}\right]^{\mu\nu} \frac{\partial \Gamma(v,\lambda)}{\partial v^{\nu}} \; - \bar{r}^{\nu} \; , \tag{7.48}$$

$$0 \quad = \quad \bar{r}^{\mu} \; \frac{\partial \Gamma(v,\lambda)}{\partial v^{\mu}} \; , \tag{7.49}$$

which define $\Gamma(v)$ and $\bar{r}^{\mu}$ in terms of $A^{\mu}$ <u>independently</u> of the presence of time-reversal symmetry.

In fact, taking eqs. (7.48) and (7.49) together, we see that $\Gamma(v)$ is obtained as a solution of the "Hamilton-Jacobi" equation

$$\tfrac{1}{2}\left[A^{-1}\right]^{\nu\mu} \; \frac{\partial \Gamma}{\partial v^{\nu}} \frac{\partial \Gamma}{\partial v^{\mu}} \quad = \quad A^{\mu} \frac{\partial \Gamma}{\partial v^{\mu}} \qquad . \tag{7.50}$$

The same equation is satisfied by the minimum "action" on the right-hand side of eq. (7.46), as is well known from Hamilton's principle in mechanics. Indeed, if we interpret $\gamma(\dot{v},v)$ given by eq. (7.35) as a Lagrangian, the "canonically conjugate momenta" are defined by

$$p_{\nu} \quad = \quad \frac{\partial \gamma(\dot{v},v)}{\partial \dot{v}^{\nu}} \quad = \quad A_{\nu\mu}(v)(\dot{v}^{\mu} + A^{\mu}(v)) \, , \tag{7.51}$$

the "Hamiltonian" is defined by

$$H(p,v) = p_\nu \dot{v}^\nu - \gamma = \tfrac{1}{2} (A^{-1})^{\nu\mu} p_\nu p_\mu - p_\nu A^\nu , \qquad (7.52)$$

and the "Hamilton-Jacobi" equation reads

$$\tfrac{1}{2}(A^{-1})^{\nu\mu} \frac{\partial \Gamma_1}{\partial v^\nu} \frac{\partial \Gamma_1}{\partial v^\mu} - \frac{\partial \Gamma_1}{\partial v^\nu} A^\nu = \frac{\partial \Gamma_1}{\partial t} , \qquad (7.53)$$

where $\Gamma_1$ is the minimum action, considered as a function of the initial and final point:

$$\Gamma_1(vt;v_0t_0) = \int_{t_0,v(t_0)=v_0}^{t,v(t)=v} d\tau \, \gamma(\dot{v},v) \quad . \qquad (7.54)$$

If the initial point is taken towards $t_0 \to -\infty$, $\Gamma_1$ cannot depend on $t$ and the right-hand side of eq. (7.53) vanishes. Thus eq. (7.50) is obtained again, and thus has the solution $\Gamma(v,\lambda) = \Gamma_1(v)$.

Since we have been able to identify $\Gamma(v,\lambda)$ with $\Gamma_1(v)$ we are again in a position to write $\gamma(\dot{v},v)$ in terms of dissipative potentials and the production rate of $\Gamma_1(v)$. With the definition of the renormalized dissipative potentials

$$\begin{aligned} \phi_r(\dot{v},\dot{v}) &= A_{\nu\mu} \dot{v}^\mu \dot{v}^\nu , \\ \psi_r\left(\frac{\partial \Gamma_1}{\partial v}, \frac{\partial \Gamma_1}{\partial v}\right) &= \tfrac{1}{4} \left[A^{-1}\right]^{\nu\mu} \frac{\partial \Gamma_1}{\partial v^\nu} \frac{\partial \Gamma_1}{\partial v^\mu} , \\ \dot{\Gamma}_1\left(\dot{v}, \frac{\partial \Gamma_1}{\partial v}\right) &= \dot{v}^\nu \frac{\partial \Gamma_1}{\partial v^\nu} , \end{aligned} \qquad (7.55)$$

we obtain

$$\gamma(\dot{v},v) = \tfrac{1}{2}(\phi_r(\dot{v}-\bar{r}, \dot{v}-\bar{r}) + \psi_r\left(\frac{\partial \Gamma_1}{\partial v}, \frac{\partial \Gamma_1}{\partial v}\right) + \dot{\Gamma}_1(\dot{v}, \frac{\partial \Gamma_1}{\partial v} ). \qquad (7.56)$$

We may therefore summarize the results of this section in the following way: The nonlinear part of the excess entropy

$$\delta_{nl} S(\{v\}) = - \Gamma(\{v\})$$

associated with the average path $v(\tau)$ generalizes all the properties of the Onsager-Machlup function $\int \mathcal{L}(\dot{q},q)d\tau$ for linear Gaussian processes to the nonlinear domain. In the long-time limit, $\Gamma(\{v\})$ may be written in terms of a function of $v$ and $\dot{v}$ alone; this is the generalized Onsager-Machlup function, which has all the formal properties of the original Onsager-Machlup function except that it is no longer a quadratic form in the $\dot{v}$ and $v$.

## 8. RELATION OF THE ONSAGER-MACHLUP FUNCTION AND THE PATH-INTEGRAL SOLUTION

We now want to study more closely the relation between $\Gamma(\{v\})$ and the function $\mathcal{L}(\dot{q},q)$. This relation is completely specified by eqs. (7.2) - (7.10). Only in the case where $\mathcal{L}(\dot{q},q)$ is a quadratic form in $\dot{q}$ and $q$ is one able to relate $\Gamma(\{v\})$ to $\mathcal{L}(\dot{q},q)$ in an explicit and rigorous way. One has, simply,

$$\Gamma(\{v\}) = \int d\tau \ \mathcal{L}(\dot{v},v) \ . \tag{8.1}$$

In all other cases it is necessary to resort to approximations. We shall consider an approximation scheme based on the assumption that $Q^{\nu\mu}$ is proportional to a small parameter — an assumption that is usually satisfied in macroscopic systems. If we replace $Q^{\nu\mu}$ by $\nu^2 Q^{\nu\mu}$ everywhere, where $\nu^2$ is a small parameter, we obtain for $\mathcal{L}(\dot{q},q)$:

$$\mathcal{L}(\dot{q},q) = \mathcal{L}_0 + \mathcal{L}_1 + \mathcal{L}_2 = \frac{1}{2\nu^2} Q_{\nu\mu} (\dot{q}^\nu - h^\nu)(\dot{q}^\mu - h^\mu)$$

$$+ \tfrac{1}{2}\sqrt{Q} \frac{\partial}{\partial q^\nu} \frac{h^\nu}{\sqrt{Q}} + \frac{\nu^2}{12} R \ . \tag{8.2}$$

Furthermore, in the following we shall choose coordinates for which

$$\sqrt{Q} \frac{\partial}{\partial q^\mu} \frac{Q^{\nu\mu}}{\sqrt{Q}} = 0 \ . \tag{8.3}$$

The choice (8.3) is always possible[37] and has the advantage that we need not distinguish $h^\nu$ and $K^\nu$ (cf. eq. (5,23)), so that $h^\nu = K^\nu$ can be assumed to be independent of $\nu^2$.

The dependence of $\mathcal{L}(\dot{q},q)$ on $\nu^2$ suggests a saddle-point approximation for the integral (7.2). If we introduce the three action integrals associated with the $\mathcal{L}_i$ (i = 1, 2, 3):

$$A_i = \int \mathcal{L}_i \, d\tau \tag{8.4}$$

we can write the path integral for $\Psi(\{\eta\})$:

$$\exp \Psi(\{\eta\}) = \exp\left[-A_0(\{q_0\}) + \int \eta_\nu q_0^\nu \, d\tau\right] \times$$

$$\times \int D\mu(\{q\})\exp\Big[-A_1(\{q_0\}) - \tfrac{1}{2}\int d\tau d\tau' \frac{\delta^2 A_0}{\delta q_0^\nu(\tau)\delta q_0^\nu(\nu')}(q^\nu(\tau)-q_0^\nu(\tau)(q^\nu(\tau')-q_0^\nu(\tau') + \ldots\Big] . \qquad (8.5)$$

Here $q_0(\tau)$ is the maximizing path, which satisfies

$$\frac{\delta A_0(\{q_0\})}{\delta q_0^\nu(\tau)} = \eta_\nu(\tau) . \qquad (8.6)$$

Higher-order terms, including terms of order $\nu^2$, have not been written down. By combining the various orders in $\nu^2$ we have assumed that $q^\nu(\tau) - q_0^\nu(\tau)$ contributes a factor $\nu$. This need not always be the case, particularly near instabilities, where a modification of this expansion may become necessary.

The zeroth-order approximation to $\psi(\{\eta\})$ is obtained by replacing the functional integral in (8.5) by 1. We obtain

$$\psi_0(\{\eta\}) = -A_0(\{q_0(\{\eta\})\}) + \int \eta_\nu q_0^\nu(\{\eta\}) d\tau . \qquad (8.7)$$

With eqs. (8.6) and (7.6) we obtain, in this order,

$$v^\mu = q_0^\mu$$

and hence, by eq. (7.8),

$$\Gamma_0(\{v\}) = A_0(\{v\}); \quad \gamma_0(\dot{v},v) = \mathcal{L}_0(\dot{v},v) . \qquad (8.8)$$

We remark that $\psi_0(\{\eta\})$ and $\Gamma_0(\{v\})$ are both correctly normalized. This would not have been automatically the case if we had applied the saddle-point approximation independently of the smallness $\nu^2 Q^{\nu\mu}$. The fluctuation corrections to $\gamma(\dot{v},v)$ are contained in the higher-order terms in (8.5). Unfortunately, these are somewhat difficult to compute, in general. The general formula obtained from (8.5) for the first order is[24)]

$$\psi_1(\{\eta\}) = \psi_0(\{\eta\}) - A_1(\{q_0(\{\eta\})\}) + \tfrac{1}{2}\int d\tau \; C^\nu{}_\nu(\tau,\tau) \qquad (8.9)$$

where the kernel $C^\nu{}_\mu(\tau,\tau') = \underline{C}$ is given by

$$\underline{C} = \ln \underline{\tilde{C}}$$

$$(\tilde{C}^{-1})^\nu{}_\mu(\tau,\tau') = Q^{\nu\lambda}(q_0(\tau)) \frac{\delta^2 A_0(\{q_0\})}{\delta q_0^\lambda(\tau)\,\delta q_0^\mu(\tau')} \qquad (8.10)$$

and multiplication of the kernels implies integration over time and matrix multiplication. We note that the kernel $\tilde{C}^{\nu}_{\mu}(\tau, \tau')$ is essentially the second-order correlation function in the lowest-order approximation. In the same order we obtain

$$\Gamma_1(\{v\}) = A_0(\{v\}) + A_1(\{v\}) + \tfrac{1}{2}\int d\tau\, C^{\nu}_{\nu}(\tau,\tau,\{v\}) \, . \qquad (8.11)$$

Further evaluation of these formulae and of higher-order corrections can be given if the system is specified further. The necessary techniques have already been developed in the quantum-theory and field-theory literature, to which we refer the reader who is interested in solving a specific problem(24-28).

## 9. APPLICATION TO THE BÉNARD PROBLEM

As a simple illustration of the general theory we want to consider the instability of a liquid layer heated from below. We are interested in the behaviour of the liquid near the point of instability. This problem has recently been studied theoretically and experimentally in great detail. In essence we shall follow closely the treatment given in ref. 39. We shall not be concerned, therefore, with the algebraic details of the problem, which can be found in ref. 39. Rather we shall try to work out the relation to our present theoretical framework.

In the Boussinesq approximation the hydrodynamic equations in appropriate dimensionless units are given by

$$\begin{aligned} \nabla_i v_i &= 0 \\ \dot{v}_i + v_j \nabla_j v_i &= -\nabla_i p + \nabla^2 v_i + \sqrt{R}\, T\, \delta_{i3} + \nabla_j s_{ij} \qquad (9.1) \\ P(\dot{T} + v_j \nabla_j T) &= \nabla^2 T + \sqrt{R}\, v_3 - \nabla_j q_j \, . \end{aligned}$$

Here $\vec{v}$, $T$ and $p$ are the deviations (in dimensionless units) of the streaming velocity, temperature and pressure from their values in the non-convecting, purely heat-conducting state. $R$ is the Rayleigh number, which is proportional to the temperature difference between the bottom and the top of the layer. $P$ is the Prandtl number, i.e., the ratio of viscosity and heat conductivity. The quantities $s_{ij}$ and $q_j$ are Gaussian random quantities with zero average. We shall take their strength to be constant phenomenological parameters describing the noise in the experiment. We introduce a five-component vector

$$u = (\vec{v}, T, p) = (u_i) \qquad (9.2)$$

and a scalar product

$$(u^{(1)}, u^{(2)}) = \sum_i \int \frac{d^3x}{F} (u_i u_i), \qquad (9.3)$$

where $F$ is the surface of the fluid layer. Then the Boussinesq equations (9.1) are written as

$$\underline{L}(u) = \underline{I} \qquad (9.4)$$

where

$$\underline{I} = (\nabla_j s_{ij}, -\nabla_j q_j, 0) \qquad (i = 1, 2, 3) \qquad (9.5)$$

Let

$$Q^{-1}_{ij} (x,x')\delta(t-t') = \langle I_i(x,t) I_j(x't')\rangle. \qquad (9.6)$$

If the experimental noise is very small we can write

$$\Gamma(\{v\}) = \int d\tau\, \gamma(\dot{v},v) \approx \int d\tau\, \gamma_0(\dot{v},v) \qquad (9.7)$$

immediately. In our notation $\gamma_0(\dot{v},v)$ is given by

$$\gamma_0 = \tfrac{1}{2} \int dt(\underline{L}(u), \underline{Q}\, \underline{L}(u)), \qquad (9.8)$$

where $v^\nu \rightarrow u_i(x)$. In our present example the curvature scalar $R$ is zero. Furthermore, it can be checked also that $\frac{\partial K^\nu}{\partial q^\nu} = \text{const}$, in spite of the nonlinearity of the Boussinesq equations. Thus $\gamma_0$ as given by eq. (9.9) also gives the function $\mathcal{L}(\dot{q},q)$, where $q^\nu \rightarrow u_i(x)$, and can be used to compute fluctuation corrections to $\gamma_0$.

We now want to study eqs. (9.1), (9.4) in the vicinity of the point of instability

$$R = R_c \quad (= 27\pi^4/4 \text{ for free-free boundary conditions, assumed in the following for simplicity}) \qquad (9.9)$$

of the heat-conducting state. For simplicity we restrict attention to a layer that is long in the 1-direction and short in the 2-direction, so that we may put all derivatives $\nabla_2$ equal to zero. We now introduce a small parameter $\varepsilon$, to be identified later:

$$R = R_c(1 + r^{(1)}\varepsilon + r^{(2)}\varepsilon^2 + \ldots) \qquad (9.10)$$

and apply a multiple time-and-length-scale analysis[40,41] by putting

$$\nabla_1 \to \nabla_1 + \varepsilon \frac{\partial}{\partial \xi} + \dots,$$

$$\frac{\partial}{\partial t} \to \frac{\partial}{\partial t} + \varepsilon \frac{\partial}{\partial \tau_1} + \varepsilon^2 \frac{\partial}{\partial t^2} + \dots, \tag{9.11}$$

$$u = \varepsilon (u^{(0)} + \varepsilon u^{(1)} + \varepsilon^2 u^{(2)} + \dots ) .$$

We insert all these expansions in the expression for $\underline{L}(u)$, eq. (9.1). The result is an expansion of $\underline{L}(u)$ in powers of $\varepsilon$ :

$$\begin{aligned} \underline{L}(u) = {} & \varepsilon \underline{L}_0(u^{(0)}) + \varepsilon^2(\underline{L}_0(u^{(1)}) + \underline{L}_1(u^{(0)}) \\ & + \varepsilon^3(\underline{L}_0(u^{(2)}) + \underline{L}_1(u^{(1)}) + \underline{L}_2(u^{(0)})) + \dots \end{aligned} \tag{9.12}$$

where each $\underline{L}_i$ is a linear matrix differential operator. We do not write down the explicit expressions for the $L_i$, which can be found in Ref. 39 (for the case $\frac{\partial}{\partial t} = 0$, $\frac{\partial}{\partial \tau_1} = 0$, $r^{(1)} = 0$ which is the only case that will interest us in dètail, cf. below).

In our multiple time-and-length scale analysis, the time and space integrals in eq. (9.8) factorize into integrals over the short times and lengths $t$, $\vec{x}$, and integrals over the long times $\tau_1, \tau_2$ and lengths $\xi$:

$$\begin{aligned} \int dt & \to \int d\tau_2 \int d\tau_1 \int dt \; \delta(\tau_1 - \tfrac{t}{\varepsilon})\, \delta(\tau_2 - \tfrac{t}{\varepsilon^2}) \\ \int d^3x & \to L_2 \int dx_1\, dx_3 \int d\xi \; \delta(\xi - \tfrac{x_1}{\varepsilon}) \end{aligned} \tag{9.13}$$

where $L_2$ is the length of the layer in 2-direction. Henceforth we put $L_2 = 1$. In the following we shall define the scalar product (9.3) to imply integration over $\vec{x}$ alone and shall write all other integrals explicitly.

We can now insert the expansion (9.12) into the expression for $\gamma_0$, eq. (9.8), and analyse each order in $\varepsilon$ separately. In the lowest order we have

$$\begin{aligned} \gamma_0^{(0)} = \frac{\varepsilon^2}{2} \int dt \int d\tau_1 d\tau_2 d\xi \, (\underline{L}_0(u^{(0)}), \delta(\xi - \tfrac{x_1}{\varepsilon}) \underline{Q}\, \underline{L}_0(u^{(0)})) \\ \delta(\tau_1 - \tfrac{t}{\varepsilon})\, \delta(\tau_2 - \tfrac{t}{\varepsilon^2}) . \end{aligned} \tag{9.14}$$

This expression describes the most dominant contribution to the regression of a fluctuation at the point of instability. Minimizing it we obtain the fast part of the regression law of a fluctuation:

$$\underline{L}_0(u^{(0)}) = 0 . \tag{9.15}$$

We now want to restrict attention to a long time scale, where the fast part of the fluctuation has already gone by. In this case we are only interested in solutions of (9.15) with $\partial/\partial t = 0$. With free-free boundary conditions $u^{(0)}$ is obtained in the form

$$u^{(0)} = W(\xi, \tau_1, \tau_2)\, \psi_0(x_1, x_3) + \text{c.c.} \tag{9.16}$$

where the complex amplitude $W$ is arbitrary, while $\psi_0(x_1, x_3)$ is fixed:

$$\psi_0 = e^{i \frac{\pi}{\sqrt{2}} x_1} (i\sqrt{2}\cos\pi x_3,\ 0,\ \sin\pi x_3,\ \sqrt{3}\sin\pi x_3,\ -3\pi\cos\pi x_3). \tag{9.17}$$

Inserting (9.16) into (9.14) we obtain $\gamma_0^{(0)} = 0$. We see from (9.16), (9.17) that after the decay of the fast part of the fluctuation, described by the Onsager-Machlup function $\gamma_0^{(0)}$ (eq. (9.14)), the <u>relative</u> amplitudes of the velocity, temperature and pressure fluctuations are completely fixed, while the overall size and spatial location of the fluctuation, described by the function $W(\xi, \tau_1, \tau_2)$, remain arbitrary. It is remarkable that, after this first stage, the relative amplitudes are completely independent of the overall size and shape of the fluctuation.

The next stage of the decay of a fluctuation is governed by

$$\gamma_0^{(1)} = \frac{\varepsilon^4}{2}\int d\tau_1 d\tau_2 d\xi\, \delta(\tau_2 - \frac{\tau_1}{\varepsilon})\Big((\underline{L}_0(u^{(1)}) + \underline{L}_1(u^{(0)})), \delta(\xi - \frac{x_1}{\varepsilon}) \times \times \underline{Q}(\underline{L}_0(u^{(1)}) + \underline{L}_1(u^{(0)}))\Big) . \tag{9.18}$$

Minimizing this part of the Onsager-Machlup function we obtain

$$\underline{L}_0(u^{(1)}) = -\underline{L}_1(u^{(0)}) . \tag{9.19}$$

The operator $\underline{L}_0$ is singular in view of eq. (9.15). Since $\underline{L}_0$ also turns out to be self-adjoint, it has the left eigenvectors $\psi_0$ and $\psi_0^*$ with eigenvalue 0. Hence, the solubility condition of (9.19) is

$$(\psi_0^*, L_1(u^{(0)})) = 0 = (\psi_0, L_1(u(0))) \tag{9.20}$$

which requires

$$r^{(1)} = 0 . \tag{9.21}$$

Let us now again restrict attention to long times, at which the second stage of the fluctuation has also had enough time to pass by. We are then interested in solutions of (9.19) with

$$\frac{\partial}{\partial \tau_1} = 0 \ . \tag{9.22}$$

They are given by

$$u^{(1)} = u^{(1)\ell} + u^{(1)n\ell}$$

$$u^{(1)\ell} = \exp[i\frac{\pi}{\sqrt{2}}x_1]\Big(-\frac{2}{\pi}\frac{\partial W}{\partial \xi}\cos\pi x_3,\ 0,\ 0,\ \frac{2\sqrt{2}i}{\sqrt{3}\pi}\frac{\partial W}{\partial \xi}\sin\pi x_3,$$
$$-4\sqrt{2}\ i\frac{\partial W}{\partial \xi}\cos\pi x_3\Big) + \text{c.c.}$$

$$u^{(1)n\ell} = \Big(0,\ 0,\ 0,\ -\frac{\sqrt{3}P}{4\pi}|W|^2\sin 2\pi x_3,\ (1+\frac{9}{8}P)\Big[\frac{|W|^2}{2}\cos 2\pi x_3 +$$
$$+ W^2 e^{i(\pi\sqrt{2})x_1}\Big]\Big) + \text{c.c.} \tag{9.23}$$

$u^{(1)\ell}$, $u^{(1)n\ell}$ describe the readjustments of the relative amplitudes of the streaming velocity, temperature, and pressure that have occurred after the second stage. Taken separately, $u^{(1)\ell}$ and $u^{(1)n\ell}$ give the readjustments that depend on the shape and size, respectively, of the fluctuation. These effects, and even their time dependence, are given by the minimum condition for $\gamma_0^{(1)}$. The size of the fluctuation still remains unaffected.

The long-time behaviour of a fluctuation, after these two stages have passed by, is governed by

$$\gamma_0^{(2)} = \frac{\varepsilon^6}{4}\int d\tau_2\, d\xi\, \Big((\underline{L}_0(u^{(2)}) + \underline{L}_1(u^{(1)}) + \underline{L}_2(u^{(0)}),$$
$$\delta(\xi - \frac{x_1}{\varepsilon})\underline{Q}(\underline{L}_0(u^{(2)}) + \underline{L}_1(u^{(1)}) + \underline{L}_2(u^{0}))\Big). \tag{9.24}$$

Minimizing this part we obtain

$$\underline{L}_0(u^{(2)}) = -(\underline{L}_1(u^{(1)}) + \underline{L}_2(u^{(0)}) \ . \tag{9.25}$$

The solubility condition

$$(\psi_0^*, \underline{L}_0(u^{(2)})) = (-\psi_0^*, (\underline{L}_1(u^{(1)}) + \underline{L}_2(u^{(0)}))) = 0 \tag{9.26}$$

yields the condition[39, 40]

$$\Big(\psi_0^*, (\underline{L}_1(u^{(1)}) + \underline{L}_2(u^{(0)}))\Big) =$$

$$\frac{3}{2}(1+P)\frac{\partial W}{\partial \tau_2} - \Big\{\frac{3}{2}\pi^2 r^{(2)} - \tfrac{1}{2}P^2|W|^2\Big\}W - 4\frac{\partial^2 W}{\partial \xi^2} = 0 \ . \tag{9.27}$$

Since we stop the procedure at this point, we may put $\varepsilon = 1$ and thus from (9.10),

$$r^{(2)} = \frac{R - R_c}{R_c} \tag{9.28}$$

is the true small parameter of the expansion.

Solving eqs. (9.25) under this condition, we first obtain a further shape- and size-dependent readjustment of the relative amplitudes of the velocity, temperature and pressure. In addition, eq. (9.27) describes the stage when the actual size of the fluctuation changes in time. This change is so slow for small $r^{(2)}$ that it was not visible during the last two stages. The validity of this equation has been verified experimentally in great detail[42].

Let us now try to separate the Onsager-Machlup function $\gamma_0$ into two parts which govern by a minimum principle the readjustment of the relative amplitudes and the change of the overall size of the fluctuation. This is easily done by expanding the kernel $\underline{Q}_{ij}(\vec{x}, \vec{x}')$ as

$$\underline{Q}_{ij}(\vec{x},\vec{x}') = \psi_{0_i}(x_1,x_3)(\psi_0^*, Q\psi_0)\, \psi_{0_j}^* (x_1{}'x_3{}') + \text{c.c} + \underline{Q}_{r_{ij}}(\vec{x},\vec{x}') \tag{9.29}$$

with

$$(\psi_0^*, \underline{Q}_r \psi_0) = 0 \; . \tag{9.30}$$

We assume now that $Q(\vec{x}, \vec{x}')$ does not correlate different plane-wave modes of the system. Then

$$\underline{Q}_r \psi_0 = \underline{Q}_r \psi_0^* = 0 \; . \tag{9.31}$$

As a result of eqs. (9.29) - (9.31), $\gamma_0$ neatly splits into the two desired parts:

$$\gamma_0 = \gamma_{fl} + \gamma_r \tag{9.32}$$

where $\gamma_r$ describes the adjustment of the relative amplitudes, which, as we have seen, occurs mostly in short time intervals:

$$\gamma_r = \tfrac{1}{2} \int dt\, (\underline{L}(u), \underline{Q}_r\, \underline{L}(u)) \tag{9.33}$$

and $\gamma_{fl}$ governs the dynamics of the overall size of the fluctuation, which occurs on long time scales

$$\gamma_{fl} = \tfrac{1}{2}\int dt \left[(L(u), \psi_0)(\psi_0^*, Q\,\psi_0)(\psi_0^*, \underline{L}(u) + \text{c.c.}\right] . \tag{9.34}$$

Thus, on short time scales, $\gamma_0 \approx \gamma_r$, while on long time scales, $\gamma_0 \approx \gamma_{fl}$. Notice that on short time scales the Onsager-Machlup function is practically indistinguishable from the one in equilibrium, while on long time scales it differs radically from the equilibrium function.

We are now in a position to write down an explicit expression for $\gamma_{f\ell}$ in the order we have considered above. We only have to make use of the solubility conditions in the various orders to obtain

$$\gamma_{f\ell} = \frac{9}{8}\int d\xi \, d\tau_2 \, Q_0 \left|(1+P)\frac{\partial W}{\partial \tau_2} - \left\{\frac{3\pi^2}{2}\frac{R-R_c}{R_c} - \tfrac{1}{2}P^2|W|^2\right\}W - 4\frac{\partial^2 W}{\partial \xi^2}\right|^2$$

where

$$Q_0 = 2\,\mathrm{Re}\,(\psi_0^*, \underline{Q}\,\psi_0) \quad . \tag{9.35}$$

The Onsager-Machlup function $\gamma_{f\ell}$ can now be brought into the form (7.56) with

$$\Gamma_1 = \frac{9}{4}Q_0(1+P)\int d\xi \left[-\frac{3\pi^2}{2}\frac{R-R_c}{R_c}|W|^2 + \frac{P^2}{4}|W|^4 + 4\left|\frac{\partial W}{\partial \xi}\right|^2\right], \tag{9.36}$$

$$\phi_r = \int d\xi \; Q_0\left(\frac{3}{2}(1+P)\right)^2 \left|\frac{\partial W}{\partial \tau}\right|^2 \quad , \tag{9.37}$$

$$\psi_r = \frac{1}{4Q_0\left(\frac{3}{2}(1+P)\right)^2}\int d\xi \left|\frac{\delta \Gamma_1}{\delta W(\xi)}\right|^2 \quad , \tag{9.38}$$

$$\dot{\Gamma}_1 = \int d\xi \left(\dot{W}(\xi)\,\frac{\delta \Gamma_1}{\delta W(\xi)} \quad + \quad \mathrm{c.c.}\right). \tag{9.39}$$

$\Gamma_1$ is the "generalized thermodynamic potential" of the Ginsburg-Landau form, which has been determined in Ref. 38. It has been determined experimentally in Ref. 41. Our derivation here clarifies its significance in the context of nonequilibrium thermodynamics. Indeed, as we have generally shown in section 7, $\Gamma_1(\{W\})$ is the "gain of information" obtained by observing the fluctuation of size W in the steady state.

It should be remembered that we have determined $\Gamma_1(\{W\})$ here only without fluctuation corrections. It is therefore equal to $-\tilde{S}$, which determines the fluctuations:

$$\tilde{S}(\{W\}) = -\Gamma_1(\{W\}) \quad .$$

Indeed one may use $\tilde{S}(\{W\})$ or $\mathcal{L}(\dot{W},W) = \gamma_{f\ell}(\dot{W},W)$ to calculate fluctuation corrections[39]. Experimentally, however, fluctuation corrections have turned out to be unobservably small.

## APPENDIX: DERIVATION OF THE PROBABILITY-DENSITY FUNCTIONAL FROM THE MAXIMUM-ENTROPY PRINCIPLE

Let us define the entropy, which is associated with a probability density functional $W(\{q(\tau)\})$ by

$$S(\{W(\{q\})\}) = -k \int D\mu(\{q\})\, W(\{q\}) \ln \frac{W(\{q\})}{N(\{q\})}, \tag{A.1}$$

where

$$N(\{q\}) D\mu(\{q\}) = \text{number of trajectories of the micro-variables in } D\mu(\{q\}) . \tag{A.2}$$

Equation (A.1) is a straightforward generalization of well-known expressions.

The quantity $N(\{q\})$ can be determined if we consider the Langevin equations associated with the Fokker-Planck eqs. (2.13). Interpreted as stochastic differential equations in the sense of Stratonovich, they read

$$\begin{aligned} \dot{q}^{\nu} &= K^{\nu}(q) - \tfrac{1}{2} \frac{\partial g^{\nu}{}_{i}(q)}{\partial q^{\lambda}} g^{\lambda}_{j}(q)\, \delta^{ij} + g^{\nu}{}_{i}(q)\, \xi^{i}(t) \\ &= f^{\nu}(q) + g^{\nu}{}_{i}\, \xi^{i}(t), \end{aligned} \tag{A.3}$$

with

$$g^{\nu}{}_{i}(q)\, g^{\mu}{}_{k}(q)\, \delta^{ik} = Q^{\nu\mu}(q), \tag{A.4}$$

$$\langle \xi^{i}(t) \rangle = 0 , \tag{A.5}$$

$$\langle \xi^{i}(t)\, \xi^{k}(t') \rangle = \delta^{ik}\, \delta(t - t') . \tag{A.6}$$

The quantity $N(\{q\})\, D\mu(\{q\})$ is now taken as the number of trajectories of the microvariables in $[\prod_j dq_j]$ for $\varepsilon \to 0$ that satisfy eq. (A.3). This number is given by

$$N(\{q\}) D\mu(\{q\}) = \int D\xi\, \delta(q - q(\{\xi\}))\, Dq \tag{A.7}$$

where the $q^{\nu}$ are determined as functionals of the $\xi^{i}$ by eq. (A.3).

If we carry out the integral in (A.7) we obtain

$$N(\{q\})\, D\mu(\{q\}) = \lim_{\varepsilon \to 0} \left[ \left\| \frac{\partial \xi^{i}}{\partial q^{\nu}} \right\| dq(t_j) \right], \tag{A.8}$$

i.e. $N(\{q\})$ is the Jacobian of the transformation $\xi \to q$ defined by eq. (A.3). The same Jacobian for the <u>linear</u> equation (A.3) has already appeared in eq. (3.16) $(J = N)$. There we have seen that the evaluation of the path-integral representation of the Fokker-Planck equation is equivalent to an evaluation of N. Since we know the path-integral solution for the general case we also know $N(\{q\})$.

We now maximize the entropy (A.1) under the constraint (A.3), (A.6), and obtain

$$W(\{q\})D\mu(\{q\}) \sim \exp\left[-\tfrac{1}{2}Q_{\mu\nu}(\dot{q}^{\mu}-f^{\mu})(\dot{q}^{\nu}-f^{\nu})\right]N(\{q\})D\mu(\{q\}) .$$

The same result is obtained if the Gaussian of the $\xi(t)$ is transformed via (A.3) into the distribution $W(\{q\})$, which is again equivalent to the path-integral solution of the Fokker-Planck equation. Notice that we do not need the Gaussian assumption for the $\xi(t)$ for the present derivation; rather it comes out as a result.

REFERENCES


1) S.R. DE GROOT and P. MAZUR, Nonequilibrium Thermodynamics, North Holland, Amsterdam, 1962.

2) I. GYARMATI, Nonequilibrium Thermodynamics, Springer, Berlin, 1975.

3) P. GLANSDORFF and I. PRIGOGINE, Thermodynamic Theory of Structure, Stability and Fluctuations, Wiley Interscience, New York, 1971.

4) L.D. LANDAU and E.M. LIFSHITZ, Statistical Physics, Pergamon Press, Oxford, 1969.

5) R. KUBO, J. Phys. Soc. Jap. 12, 570 (1957).

6) D.N. ZUBAREV, Nonequilibrium Statistical Thermodynamics, Plenum, New York, 1974 (translated from the Russian).

7) L. ONSAGER, Phys. Rev. 37, 405 (1931); 38, 2265 (1931).

8) L. ONSAGER and S. MACHLUP, Phys. Rev. 91, 1505 (1953); 91, 1512 (1953).

9) P. GLANSDORFF and I. PRIGOGINE, Physica 20, 773 (1954).

10) P. GLANSDORFF and I. PRIGOGINE, Physica 30, 351 (1964).

11) P. PRIGOGINE and P. GLANSDORFF, Physica 31, 1242 (1965).

12) P. GLANSDORFF and I. PRIGOGINE, Physica 46, 344 (1970).

13) F. SCHLÖGL, Ann. Phys. 45, 155 (1967).

14) F. SCHLÖGL, Z. Physik 243, 309 (1971).

15) F. SCHLÖGL, Z. Physik 244, 199 (1971).

16) R. GRAHAM, Coherence in Quantum Optics, eds. L. Mandel and E. Wolf, Plenum, New York, 1972, p. 851.

17) R. GRAHAM, Springer Tracts in Modern Physics, vol. 66, Heidelberg, 1973.

18) R. GRAHAM in: Fluctuations, Instabilities and Phase Transitions ed. T. Riste, Plenum Press, New York, 1975.

19) M.S. GREEN, J. Chem. Phys. 20, 1281 (1952).

20) R. ZWANZIG, Phys. Rev. 124, 983 (1961)

21) R.P. FEYNMAN and A.R. HIBBS, Quantum Mechanics and Path Integrals, Gordon and Breach, New York, 1965.

22) N, WIENER, Acta Math. 55, 117 (1930).

23) R. GRAHAM, Z. Physik B26, 281 (1977).

24) E.S. ABERS and B.W. LEE, Phys. Reports 9C, no. 1, 1 (1973).

25) E. BRÉZIN, J.C. LE GUILLOU and J. ZINN-JUSTIN, Phys. Rev. D15, 1544 (1977); D15, 1558 (1977).

26) S. COLEMAN and E. WEINBERG, Phys. Rev. D7, 1888 (1973).

27) R. RAJARAMAN, Phys. Reports 21C, 5, 227 (1975).

28) F.R. DASHEN, B. HASSLACHER and A. NEVEU, Phys. Rev. D10, 4114 (1974); D11, 3424 (1975).

29) J.L. LEBOWITZ and P.G. BERGMANN, Ann. Phys. 1, 1 (1957).

30) H.K. JANSSEN, Z. Physik B23, 377 (1976).

31) I. PRIGOGINE, Ac. Roy. Belg. Bull. Sc. 31, 600 (1945).

32) H. LESCHKE and M. SCHMUTZ, Z. Physik B27, 85 (1977).

33) H. HAKEN, Z. Physik B24, 321 (1976).

34) R. GRAHAM, Phys. Rev. Lett. 38, 51 (1977).

35) R. GRAHAM, Short-Time Propagators in Riemann Geometries, Statphys. 13, Haifa 1977, to appear in the Proceedings (available on request).

36) U. DEININGHAUS and R. GRAHAM, to be published.

37) R. GRAHAM, Z. Physik B26, 397 (1977).

38) R. GRAHAM and H. HAKEN, Z. Physik 243, 289 (1971); 245, 141 (1971)

39) R. GRAHAM, Phys. Rev. Lett. 31, 1479 (1973); Phys. Rev. A10, 1762 (1974).

40) A. NEWELL and J. WHITEHEAD, J. Fluid Mech. 38, 279 (1969).

41) L.A. SEGEL, J. Fluid Mech. 38, 203 (1969).

42) J. WEISFREID, Y. POMEAU, M. DUBOIS, C. NORMAND and P. BERGÉ, J. de Physique, to be published.

# SYNERGETICS - A FIELD BEYOND IRREVERSIBLE THERMODYNAMICS

H. Haken
Institut für theoretische Physik
der Universität Stuttgart

## 1. INTRODUCTION

Over recent years I have given numerous lectures on synergetics. A number of these lectures have been published in conference or Summer School proceedings. Furthermore, I have written a textbook on this subject [1].

These things have brought a certain dilemma for me. On the one hand it is rather simple for me to give lectures on this field based on material already in existence. On the other hand there is a certain danger of duplicating publications, which might be boring for readers following up this item. To escape this dilemma, at least to some extent, I have organized these lecture notes in the following way. First of all I shall try to give an introduction to this field not yet published elsewhere, and I shall also communicate some new results. On the other hand I shall present topics which are necessary for an audience to follow a lecture, but which I have published elsewhere only in a sketchy way. The reader is advised to fill in the gaps with aid of my book on synergetics.

## 2. THERMODYNAMICS, IRREVERSIBLE THERMODYNAMICS, SYNERGETICS

When we ask physicists or chemists what they think is the most universal field in physics or chemistry many of them will most probably answer "thermodynamics". Indeed, the most striking feature of thermodynamics is its seemingly universal validity. Thermodynamics applies, for instance, to entirely different substances, and it needs only rather few assumptions. Of course, there is a price to be paid for this universality. We must assume that matter is in thermal equilibrium. Furthermore, the processes considered must proceed infinitely slowly. Among the most striking phenomena in the realm of thermodynamics are certainly phase transitions, i.e., abrupt changes of properties of matter. Such changes occur in particular when the temperature is changed. The most prominent concept of thermodynamics is that of entropy. We know that closed systems approach a unique final state, namely, that of thermodynamic equilibrium characterized by maximum entropy. Among further concepts which are used to construct the beautiful building of thermodynamics are free energy, or, more generally, thermodynamic potentials, and equations of state.

In spite of this beautiful building we know that there are other phenomena in nature which do not completely fit into this building. These phenomena are connected with dissipative processes, such as the electric current, heat conduction, or diffusion. In these processes, typically, certain forces, such as an electric potential, a temperature or a concentration gradient, cause corresponding fluxes, i.e., an electrical current, a heat flux or diffusion (concentration flux). If the systems are not too far away from thermal equilibrium, concepts such as entropy are still valid, at least in a certain sense. For instance, one may speak of local entropy or local temperature. The relations between fluxes and forces are governed by quite general laws. As long as the law of detailed balance is valid the famous Onsager relations may be derived. Further concepts used in this field are entropy production and excess entropy production. The formulation of a stability criterion for thermodynamic states using the excess entropy production has led irreversible thermodynamics to a definite climax[2].

Table 1

| Field | Phenomena | Concepts and methods |
|---|---|---|
| Thermodynamics | Thermal equilibrium, infinitely slow changes, phase transitions | Entropy, temperature, free energy, etc., thermodynamic potentials, equations of state |
| Irreversible thermodynamics | Relaxation phenomena, electric current, heat conduction, concentration fluxes, diffusion | Onsager relations, fluxes, forces, Glansdorff-Prigogine stability criterion |

However, there is a further step which is still more crucial than that between thermodynamics and irreversible thermodynamics. In Nature there are processes far from thermal equilibrium which do not lead to the usual relaxation phenomena. Quite on the contrary, these processes create structures out of chaotic states. I shall give a number of explicit examples below and show that very often the transition from chaos to order strongly resembles phase transitions. The phenomena occurring here are closely related to problems of biology, for instance, the question of self-organization, and, quite generally, to the birth of new qualities. In the following I want to discuss what methods and concepts are required to cope with these phenomena. While the methods of thermodynamics and of irreversible thermodynamics have been very successful in the realm of their fields, it will turn out that now new ideas are needed here. To discuss the kind of problems we are confronted with, let me start with an example lying seemingly entirely outside physics, namely, language. The sentence "this is a man" represents a system composed of subsystems, namely, the words. These again are composed of a further subsystem, the letters. While the letters by themselves do not make sense, the words and, still more fully, the sentence, make sense.

Table 2

| | hierarchy |
|---|---|
| this is a man | system |
| this is a man | subsystem I |
| t h i s i s a m a n | subsystem II |

Table 3

| | |
|---|---|
| system ↑ | sentence makes sense: new quality |
| subsystem | |

Making sense means that a new quality has arisen. Evidently, by a proper putting together of subsystems or, more generally speaking, by a proper cooperation of subsystems, the total system exhibits an entirely new quality which cannot be attached to the individual subsystems. There are further characteristic connections between systems and subsystems. For instance, when we mutilate the sentence "this is a man" by taking away a letter, for instance, "thi is a man", still the sentence makes sense. But the sense tells us that a certain letter is missing and must be inserted. This example shows that the new quality (the sense of the sentence) determines the behavior of the subsystems. This reflects a general feature, namely, that systems "slave" the subsystems. On the other hand the subsystems bring the systems into existence in a well-regulated way. When we arrange the subsystems differently, for instance,

a a h i i m n s s t

the sense is lost; the total system has changed. Thus the way the subsystems cooperate is essential for the evolving structure. Such relations between subsystems and systems can be found in many scientific disciplines. The essential problems are now to define the new qualities of the total system and to show how the subsystems cooperate so as to produce that new quality. The field dealing with this kind of problems is called "Synergetics" which is a word taken from Greek, meaning "science of cooperation".

Table 4

| Field | Phenomena | Concepts and methods |
|---|---|---|
| Synergetics | Ordered states far from thermal equilibrium, nonequilibrium phase transitions, creation of new qualities, self-organization | Order parameters, slaving principle, cooperation or competition |

In this article I want to show mainly by means of examples that so far few new principles suffice to characterize a good deal of the features of such new kinds of systems. Essential concepts will be the order parameter which describes the new quality, the slaving principle, and the competition and cooperation of order parameters. Among the mathematical tools are stochastic differential equations of different kinds such as Langevin equations, Fokker-Planck equations and master equations, and newly-developed methods of bifurcation theory.

## 3. EXAMPLES

Fig. 1 represents several examples in which a disordered state goes over into an ordered state.

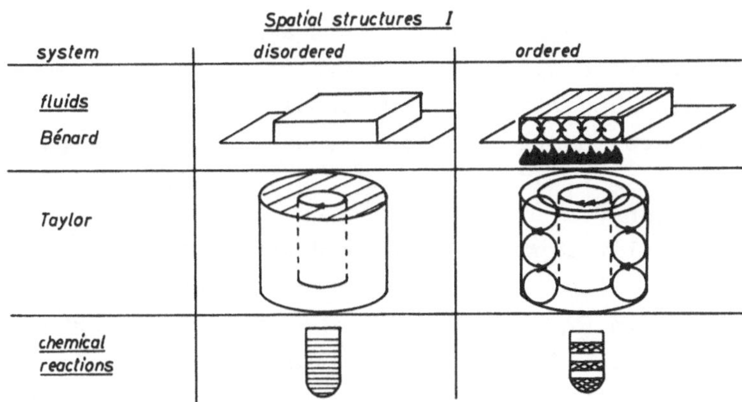


Fig. 1: Compare text

A famous example is the so-called convection instability. A special form of it, namely, the formation of hexagons, was discovered about 1900 by Bénard. Let us consider a fluid layer which is confined by a vessel and which is heated from below. Since the temperature at the upper surface is maintained at a fixed value, a temperature gradient is built up in the liquid layer. If the temperature gradient is small, heat is transferred from the lower to the upper surface by conduction. This means that there is an increased microscopic but no visible macroscopic motion of the liquid. Beyond a certain critical temperature gradient an entirely new phenomenon sets in, or, in other words, a new quality occurs. Parts of the heated liquid expand, move to the upper surface, cool down and fall down again. Quite amazingly, this motion occurs not at random but in a well-defined pattern, namely, of rolls. Evidently, a new structure has arisen. A similar feature is observed in the so-called Taylor instability. A liquid is confined by two coaxial cylinders, the inner one rotating. At small rotation speeds the liquid also rotates, in a laminar flow. However, beyond a certain critical velocity, suddenly a macroscopic motion in the form of closed rolls appears.

The Bénard instability need not always show rolls; it can also show a hexagonal pattern (when we look at the surface from above). At higher temperature gradients more and more complicated patterns may occur. Quite generally we may state that by change of an external parameter, in our case the temperature, abruptly a number of qualitatively entirely different states may be created. More exactly speaking, these states are created by the system itself. An entirely different field also showing pattern formation is chemistry. Here it has been known since the last century that certain reactions may proceed in the so-called Liesegang rings. A very popular demonstration nowadays is the Belusov-Zhabotinskiĭ reaction. It can lead to the patterns given schematically in fig. 1, or it can lead to concentric rings, or spirals, or oscillations.

Bénard instabilities can be observed under various circumstances, namely, not only in liquids but also in the formation of cloud streets in the atmosphere, in crystal growth out of the melt, granulae of the sun, etc. A wide field of pattern formation is known in biology. The solution of such problems leads into fundamental questions of morphogenesis (see below).

The next group of phenomena refers to oscillations which occur on a macroscopic scale in a self-organized manner (fig. 2).

*temporal structures: oscillations*

| *system* | *disordered* | *ordered* |
|---|---|---|
| *laser* | $E$ … $t$ | $E$ … $t$ |
| *chemical reactions*<br>*biological clocks* | *no oscillation noise* | |

Fig. 2: Compare text

A well-known example nowadays is the laser. Here a rod of laser-active material with two mirrors at its end-faces is pumped energetically from the outside and the atoms emit light. When the energy flux is small the electric-field strength of the emitted light consists of random wave tracks. Beyond a critical value of the pump power the emitted light is nearly perfectly sinusoidal, i.e., the whole set of the laser atoms oscillates in a macroscópic manner and so does the emitted light. A number of chemical reactions show oscillating patterns in which the colour changes, for instance, from red to blue.

A further group of phenomena is related to the formation of pulses (fig. 3), which occur in quite different disciplines.

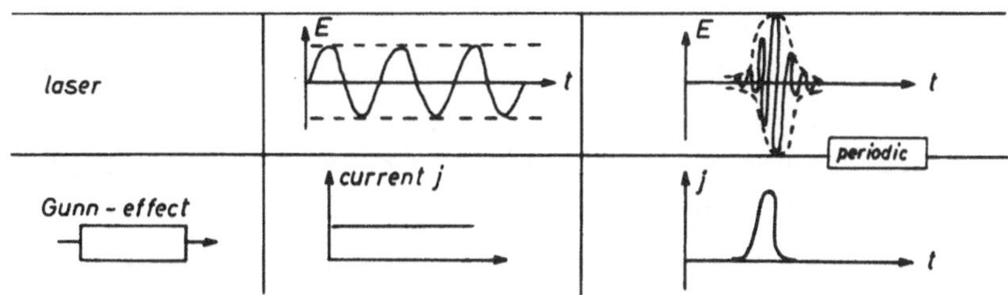


Fig. 3: Compare text

When the laser is pumped still more it ceases emitting a constant wave and starts emitting ultrashort pulses. A similar phenomenon occurs in the Gunn effect, where in certain semiconductors the current $j$ suddenly breaks into pulses. During recent years a new kind of behavior has attracted the attention of mathematicians, physicists and biologists, namely, completely random motion caused by deterministic equations (fig. 4). Again it appears that such behavior is found in different disciplines, e.g., in models of turbulence of fluids, in the laser, in models of the earth's magnetic field, in chemical-reaction models, and in population dynamics.

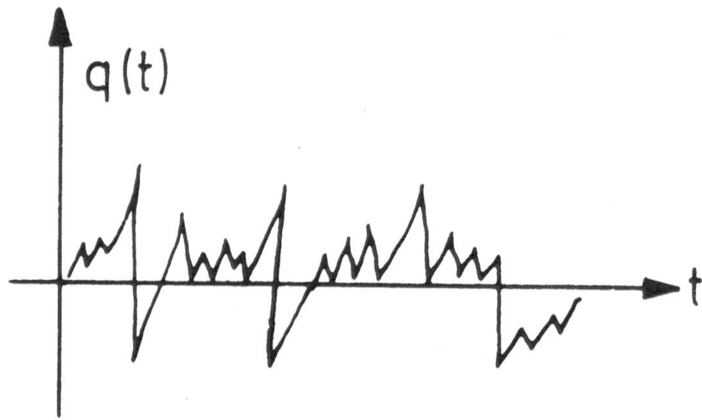


Fig. 4: Compare text

So far I have been talking about phenomena which can be found in quite different disciplines, mainly in the natural sciences. Processes leading to structures occur also in other domains: in ecology, society, development of cities, industrial companies, languages, etc. When we consider the above examples (and many others) we discover a number of common features. All of the systems are composed of many subsystems (the laser of atoms, the fluid of molecules, etc.). When an external parameter (energy flux, temperature gradient, etc.) is changed, suddenly a new ordered state occurs. Below I want to show that these phenomena can be treated from a unified point of view. Such an approach has to deal with a number of problems.

<u>Problems</u> :

1) Under which conditions is a structure replaced by a new one ("stability and instability")?
2) What is the new structure? How can we describe it adequately?
3) Are there several new structures possible? Can they be put into classes?
4) How is the new structure selected out of all the possible ones?
5) How is it built up? Smoothly? Abruptly?
6) What is the role of fluctuations?
7) Is the new structure stable?
8) How does it relax towards its stable state?
9) What hierarchy of new structures is possible?

Now I shall give a sketch of general concepts and methods of synergetics.

## Self-organization and slaving

### Nonlinear equations and local stability

I describe a system by a set of variables $q_j$ which, in general, change in the course of time. This change is determined by equations of the form

$$\dot{q}_j = N_j(q_1,\ldots,q_n;\sigma) + F_j(t), \tag{3.1}$$

where $N_j$ are nonlinear functions of the variables and $\sigma$ symbolizes external "control" parameters. Because in many cases the variables $q$ depend on space, $N_j$ may also contain derivatives with respect to the spatial coordinates $\underset{\sim}{x}$ .

It is important to note that systems are usually subject to fluctuations of the surroundings or to internal fluctuations. These are taken care of by time-dependent fluctuating forces, $F_j(t)$ . Let me first neglect these fluctuations and let me consider a rather simple situation,which, however, reveals a good deal of the essential facts. I assume that I have found a stationary solution $q_j^o$ . The solution may, for instance, represent a fluid at rest or a stationary state of a chemical reaction. Since under certain circumstances new patterns may occur, we check the stability of a given configuration. Therefore I consider small deviations $u_j = q_j - q_j^o$ , which will depend on space and time. I then, as a first step, follow up the usual stability analysis by inserting $q_j = q_j^o + u_j$ into the nonlinear equations and linearizing them with respect to the $u'$s . This leaves me with equations of the type

$$\dot{u}_j = \sum_{j'} L_{jj'}u_{j'} . \tag{3.2}$$

In their simplest form the equations read $\dot{u} = \lambda u$ , where a negative eigenvalue $\lambda$ indicates a stable solution and a positive eigenvalue an unstable solution. Of course, in general I shall find a whole set of such solutions which can represent, for instance, waves or dynamic configurations of the system.

### Slaving

To determine the variables $q_j$ without restriction to a linear approach I represent $q_j$ as a superposition of the modes $u_j^{(\lambda)}$ with still unknown amplitudes $\xi_\lambda$ . Again I confine my analysis to a typical and most simple case, taking into account only two modes, one stable, $u_j^{(s)}(\underset{\sim}{x})$ , the other one unstable, $u_j^{(u)}(\underset{\sim}{x})$ .

$$q_j(\underset{\sim}{x},t) = q_j^o + \xi_u(t)u_j^{(u)}(\underset{\sim}{x}) + \xi_s(t)u_j^{(s)}(\underset{\sim}{x}). \quad (3.3)$$

Skipping all the details, I eventually find two equations, one for the amplitude $\xi_u$ of the unstable mode and another one for $\xi_s$, the stable mode. A typical example is

$$\dot{\xi}_u = \lambda_u \xi_u - \xi_u \xi_s, \quad (3.4)$$

$$\dot{\xi}_s = -|\lambda_s|\xi_s + \xi_u^2. \quad (3.5)$$

Now a very simple but equally important point follows. When I change the external parameters $\sigma$ in such a way that the first mode with the index $u$ passes from the stable to the unstable regime, $\lambda_u$, or at least its real part, must pass through zero. That means, however, that this mode varies very slowly. Now let me assume that the behavior of the total system is eventually governed by $\xi_u$. According to eq.(3.5) this means that $\xi_s$ also changes very slowly. Therefore I can neglect the temporal derivative, $\dot{\xi}_s$, compared to $|\lambda_s|\xi_s$. This allows me to solve this equation immediately and to express $\xi_s$ by $\xi_u$:

$$\xi_s(t) = |\lambda_s|^{-1} \xi_u^2(t). \quad (3.6)$$

I shall say in the following that $\xi_s$ is <u>slaved</u> by $\xi_u$, i.e., $\xi_s$ adapts itself immediately. Such phenomena are not limited to physics or chemistry but are found in quite different disciplines. Consider, for instance, humans. After their birth they quickly learn the language so that they are slaved by the language. We see immediately that these slaved entities then carry on the macroscopic features. Inserting (3.6) into (3.4) I obtain the equation

$$\dot{\xi} = \lambda\xi - |\lambda_s|^{-1}\xi^3 + F(t), \quad (3.7)$$

where I have dropped the index $u$. I have now again included the fluctuating force, $F(t)$. In the general case with many modes, all stable modes can be eliminated. A systematic elimination technique including all higher orders of $\xi_u$, complex $\xi_u$'s, "small-band" excitations and fluctuating forces has been given in my previous papers and is described in my 1977 textbook on synergetics (see references). In practice it often turns out that only a few modes are unstable so that an enormous reduction of the numbers of degrees of freedom results. This procedure allows us to study most of the phenomena I have described in the first part of my lecture in an adequate way, if a number of refinements are taken into account.

Order parameters

The slaving principle has a very important consequence: once $\xi_u$ is known, all the $\xi_s$'s are also known. Therefore the $\xi_u$'s determine the macroscopic behavior or, in other words, the order. I shall call them order parameters. They play a similar role, both in physical systems far from thermal equilibrium and in nonphysical systems, to that of the order parameters of conventional phase-transition theory of systems in thermal equilibrium.

To interpret eq. (3.7) I add an acceleration term $m\ddot{\xi}$ and interpret $\xi$ for the time being as the coordinate of a particle under the action of a deterministic force $K(\xi) = \lambda\xi - \xi^3$ and a fluctuating force $F$. Both in physics and in mathematics it is known that we can write the right-hand side as the derivative of a potential function $V(\xi)$, $K(\xi) = -\partial V/\partial\xi$. Depending on the size of the eigenvalue $\lambda$ we find two configurations, indicated by a dashed and a solid line in fig. 5. This allows me to establish links with phase-transition theory in statistical physics. For a detailed discussion of these phase-transition analogies, see, for instance, my 1977 textbook on synergetics. When I plot the equilibrium position as a function of the parameter $\lambda$ I find the following situation (fig. 5, r.h.s.).

$\xi_u$ order parameters
one degree of freedom:

$(m\ddot{\xi}+) \quad \dot{\xi} = \lambda\xi - b\xi^3 + F(t)$

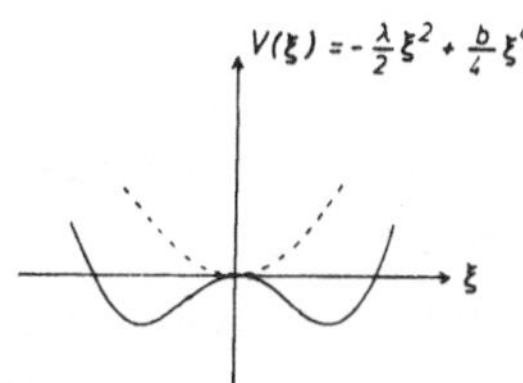


critical slowing down, soft mode
critical fluctuations
broken symmetry

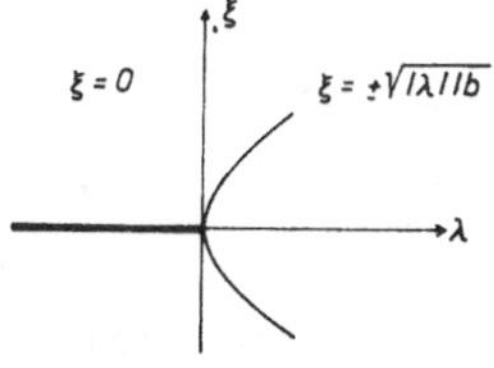


bifurcation

l.h.s. Potential as a function of coordinate $\xi$
dashed line $\lambda < 0$
solid line $\lambda > 0$

r.h.s. Equilibrium coordinate $\xi$ as function of control parameter $\lambda$

Fig. 5: Compare text

As long as $\lambda$ is negative the equilibrium position is $\xi = 0$. However, for positive values of $\lambda$ I obtain two stable configurations. Since the resulting curve has the form of a fork this phenomenon is called bifurcation. I have presented here the simplest case one can presumably think of, which already represents, however, some typical features of certain classes of nonlinear differential equations dealt with by a branch of mathematics called bifurcation theory.

## 4. SELF-ORGANIZATION IN CONTINUOUSLY EXTENDED MEDIA. AN OUTLINE OF THE MATHEMATICAL APPROACH [+]

In this and the following sections we deal with equations of motion of continuously extended systems containing fluctuations. We first assume external parameters permitting only stable solutions and then linearize the equations, which define a set of modes. When the external parameters are changed the modes becoming unstable are taken as the order parameters. Since the relaxation time of the order parameters tends to infinity the damped modes can be eliminated adiabatically, leaving us with a set of nonlinear coupled order-parameter equations. In two or three dimensions they allow, for example, for hexagonal spatial structures. Our procedure has numerous practical applications.

To explain our procedure we look at the general form of the equations which are used in hydrodynamics, lasers, nonlinear optics, chemical-reaction models and related problems. To be concrete we consider macroscopic variables, although in many cases our procedure is also applicable to microscopic quantities. We denote the physical quantities by $\underset{\sim}{U} = (U_1, U_2, \ldots)$, mentioning the following examples. In lasers, $\underset{\sim}{U}$ stands for the electric-field strength, the polarization of the medium, and the inversion density of laser-active atoms. In nonlinear optics, $\underset{\sim}{U}$ stands for the field strengths of several interacting modes. In hydrodynamics, $\underset{\sim}{U}$ stands, e.g., for the components of the velocity field, the density and the temperature. In chemical reactions, $\underset{\sim}{U}$ stands for the numbers (or densities) of molecules participating in the chemical reaction. In all these cases, $\underset{\sim}{U}$ obeys equations of the following type

$$\frac{\partial}{\partial t} U_\mu = G_\mu(\nabla, \underset{\sim}{U}) + D_\mu \nabla^2 U + F_\mu(t); \; \mu = 1,2,\ldots,n. \quad (4.1)$$

[+] For references see Ref. 1

In it, $G_\mu$ are nonlinear functions of $\underset{\sim}{U}$ and perhaps of a gradient. In most applications, like lasers or hydrodynamics, $G_\mu$ is a linear or bilinear function of $\underset{\sim}{U}$, though in certain cases (especially in chemical-reaction models), a cubic coupling term may equally well occur. The next term in (4.1) describes diffusion (D real) or wave-type propagation (D imaginary). In this latter case the second-order time derivative in the wave equation has been replaced by the first derivative by use of the "slowly-varying amplitude approximation". The $F_\mu(t)$'s are fluctuating forces which are caused by external reservoirs and internal dissipation and which are connected with the damping terms occurring in (4.1).

We shall not be concerned with the derivation of equations (4.1). Rather, our goal will be to derive from (4.1) equations for those undamped modes which acquire a macroscopic size and determine the dynamics of the system in the vicinity of the instability point. These modes form a mode skeleton which grows out from fluctuations above the instability and thus describes the "embryonic" state of the evolving spatio-temporal structure.

## 5. GENERALIZED GINZBURG-LANDAU EQUATIONS FOR NONEQUILIBRIUM PHASE TRANSITIONS

We now start with a treatment of (4.1). We assume that the functions $G_\mu$ in (4.1) depend on external parameters $\sigma$ (e.g. energy pumped into the system). First we consider values of $\sigma$ such that $\underset{\sim}{U} = \underset{\sim}{U}_0$ (independent of space and time) is a stable solution of (4.1). We decompose $\underset{\sim}{U}$

$$\underset{\sim}{U} = \underset{\sim}{U}_0 + \underset{\sim}{q} \tag{5.1}$$

with

$$\underset{\sim}{q} = \begin{pmatrix} q_1(\underset{\sim}{x},t) \\ \cdot \\ \cdot \\ \cdot \\ q_n(\underset{\sim}{x},t) \end{pmatrix}. \tag{5.2}$$

Splitting the r.h.s. of (4.1) into a linear part $K\underset{\sim}{q}$ and a nonlinear part $\underset{\sim}{g}$, we obtain (4.1) in the form

$$\left( \frac{\partial}{\partial t} - K(\nabla^2) \right) \underset{\sim}{q} = \underset{\sim}{g}(\underset{\sim}{q}) + \underset{\sim}{F}(t). \tag{5.3}$$

In it the matrix

$$K = (\hat{K}_{\mu\nu}) \tag{5.4}$$

has the form

$$\hat{K}_{\mu\nu} = K_{\mu\nu} + \delta_{\mu\nu} D_\mu \nabla^2 \,, \text{ where } K_{\mu\nu} = \frac{\partial G_\mu}{\partial U_\nu}\bigg|_{U_{\nu,0}} . \tag{5.5}$$

Our whole procedure applies, however, to a matrix $K$ which depends in a general way on $\nabla$. $\underset{\sim}{g}$ is assumed in the form

$$g_i(\underset{\sim}{q}) = \sum_{\mu\nu} q_\mu g^{(2)}_{i\mu\nu}(\nabla) q_\nu + \sum_{\mu\nu\kappa} g^{(3)}_{i\mu\nu\kappa} q_\mu q_\nu q_\kappa ; \tag{5.6}$$

$g^{(2)}$ may or may not depend on $\nabla$, and equally $g^{(3)}$ may or may not depend on $\nabla$ (if $g^{(3)}$ depends on $\nabla$, the sequence of $g$ and $q$ must be chosen properly). We first deal with $K$ and introduce operators

$$\underset{\sim}{O}^{(j)} = \underset{\sim}{O}^{(j)}(\nabla) \tag{5.7}$$

which still depend on $\nabla$ and which are defined as eigenvectors satisfying the equation

$$K(\nabla)\underset{\sim}{O}^{(j)} = \lambda_j(\nabla)\underset{\sim}{O}^{(j)} . \tag{5.8}$$

When $\nabla$ is replaced by $i\underset{\sim}{k}$ (5.8) becomes a linear algebraic equation so that (5.8) can easily be solved without resorting to any operator techniques. We furthermore introduce eigenfunctions of the wave equation

$$\nabla^2 \chi_{\underset{\sim}{k}}(x) = -k^2 \chi_{\underset{\sim}{k}}(\underset{\sim}{x}) , \tag{5.9}$$

provided $K$ depends on $\nabla^2$. $\chi_{\underset{\sim}{k}}$ can be chosen later on adequately. We shall use a notation which suggests that the $\chi_k$'s are plane-wave solutions, but it may be advantageous to use other representations as well, e.g., Bessel functions or spherical wave functions, depending on the problem. If $K$ depends on $\nabla$ in odd powers (and in even powers), the $\chi_k$'s are taken as complex plane wave solutions of (5.9). We represent $\underset{\sim}{q}(\underset{\sim}{x})$ as the superposition

$$\underset{\sim}{q}(\underset{\sim}{x},t) = \Sigma_{\underset{\sim}{k},j}\, \underset{\sim}{O}^{(j)} \xi_{\underset{\sim}{k},j} \chi_{\underset{\sim}{k}}(\underset{\sim}{x}) . \tag{5.10}$$

Now we take a decisive step for what follows. It can be shown by means of explicit examples like laser or hydrodynamic instabilities that there exist small-band excitations of unstable modes. This suggests that we construct wave packets (as in quantum mechanics), so that we take sums over small regions of $\underset{\sim}{k}$ together. We thus obtain carrier modes with discrete wave numbers $\underset{\sim}{k}$ and slowly-varying amplitudes $\xi_{\underset{\sim}{k},j}(\underset{\sim}{x})$. Our first goal is to derive a general set of equations for the mode amplitudes $\xi$. To do so we insert (5.10) into (5.3), multiply from the left-hand side by $\chi^{*}_{\underset{\sim}{k}'}(\underset{\sim}{x})\underset{\sim}{\bar{O}}^{(j')}$ and integrate over a region which contains many oscillations of $\chi_{\underset{\sim}{k}}$ but in which $\xi$ changes very little. After some analysis the basic set of equations has the following structure:

$$\dot{\xi}_{\underset{\sim}{k},j} - \hat{\lambda}_j(\nabla,\underset{\sim}{k})\,\xi_{\underset{\sim}{k},j} = H_{\underset{\sim}{k},j}(\{\xi(\underset{\sim}{x})\}) + \hat{\underset{\sim}{F}}_{\underset{\sim}{k},j} \tag{5.11}$$

where

$$H_{\underset{\sim}{k},j}(\{\xi(\underset{\sim}{x})\}) \quad \sum_{\substack{\underset{\sim}{k}'\underset{\sim}{k}''\\ j'j''}} a_{\underset{\sim}{k}\underset{\sim}{k}'\underset{\sim}{k}'',\,j,j'j''}\, I_{\underset{\sim}{k}\underset{\sim}{k}'\underset{\sim}{k}''}\,\xi_{\underset{\sim}{k}'j'}\,\xi_{\underset{\sim}{k}''j''} + \sum_{\substack{\underset{\sim}{k}'\underset{\sim}{k}''\underset{\sim}{k}'''\\ j'j''j'''}} b_{\underset{\sim}{k}\underset{\sim}{k}'\underset{\sim}{k}''\underset{\sim}{k}''',j,j'j''j'''}\, J_{\underset{\sim}{k}\underset{\sim}{k}'\underset{\sim}{k}''\underset{\sim}{k}'''}\,\xi_{\underset{\sim}{k}'j'} \times \times\ \xi_{\underset{\sim}{k}''j''}\,\xi_{k'''j'''} \quad ; \tag{5.12}$$

a, b, I and J are certain constants which we define explicitly elsewhere[1]. $\hat{\lambda}$ is essentially an eigenvalue. So far practically no approximations have been made, but to cast (5.11) into a practicable form we have to <u>eliminate</u> the unwanted or uninteresting modes, which are the <u>damped modes</u>. Accordingly, we put

$$j = u(\text{unstable}) \quad \text{if} \quad \mathrm{Re}\,\hat{\lambda}_u(0,\underset{\sim}{k}) \gtrsim 0 \tag{5.13}$$

and

$$j = s(\text{stable}) \quad \text{if} \quad \mathrm{Re}\,\hat{\lambda}_s(0,\underset{\sim}{k}) < 0 \;. \tag{5.14}$$

An important point should be observed at this stage: Although $\xi$ bears two indices, $\underset{\sim}{k}$ and $u$, these indices are not independent of each other. Indeed the instability usually occurs only in a small region of $k = k_c$ (compare fig. 6). Thus we must carefully distinguish between the $\underset{\sim}{k}$ values at which (5.19) and (5.20) (see below) are evaluated. Because of the "wave packets" introduced earlier, $\underset{\sim}{k}$ runs over a set of discrete values with $|\underset{\sim}{k}| = k_c$. The prime on the sums in the following formulas indicates this restriction of the summation over $\underset{\sim}{k}$.

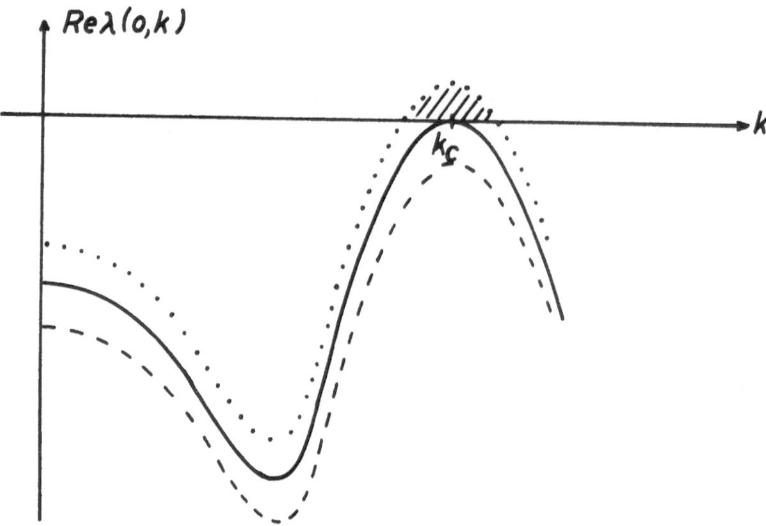


Fig. 6 : The real part of the eigenvalue $\hat{\lambda}$ plotted as function of the k number of the modes (an example).
Dashed line: below the instability all modes are damped
Solid line : marginal situation, one mode at $k = k_s$ is about to become unstable.
Dotted line: shaded area, a whole range of modes has become unstable, giving rise to small-band excitations.
This situation is taken care of by the formation of wave packets and slowly varying amplitudes.

Fig. 6 provides us with an example of a situation where branch j of $\xi_{\underset{\sim}{k},j}$ can become unstable at $k = k_c$ . The basic idea of our further procedure is this. Because the undamped modes may grow without limit provided the nonlinear terms are neglected, we expect that the amplitudes of the undamped modes are considerably bigger than those of the damped modes. Since, on the other hand, close to the "phase-transition" point the relaxation time of the undamped modes tends to infinity, i.e., the real part of $\hat{\lambda}$ tends to zero, the damped modes must adiabatically follow the undamped modes. Although the amplitudes of the damped modes are small, they must not be neglected completely. This neglect would lead to a catastrophe if in (5.12) the cubic terms were lacking. As one convinces oneself very quickly, quadratic terms cannot lead to a globally stable situation. Thus cubic terms are necessary for stabilization. Such cubic terms are introduced, even in the absence of those in the original equations, by the elimination of the damped modes. To exhibit the main features of our elimination procedure more clearly, we put, for the moment,

$$\xi_{\underset{\sim}{k},j} \rightarrow (\underset{\sim}{k},j) \tag{5.15}$$

and drop all coefficients in (5.11). We assume $|\xi_s| \ll |\xi_u|$ and, in a selfconsistent way, $\xi_s \propto \xi_u^2$ . Keeping in (5.11) only terms up to third order in $\xi_u$ , we obtain

$$\left(\frac{d}{dt} - \hat{\lambda}_u\right)(\underset{\sim}{k},u) = \sum_{\underset{\sim}{k}'\underset{\sim}{k}''u's}(\underset{\sim}{k}',u')\cdot(\underset{\sim}{k}'',s) + \sum_{\substack{\underset{\sim}{k}'\underset{\sim}{k}''\\ \underset{\sim}{u}'\underset{\sim}{u}''}}(\underset{\sim}{k}',u')(\underset{\sim}{k}'',u'')$$
$$+ \sum_{\substack{\underset{\sim}{k}'\underset{\sim}{k}''\underset{\sim}{k}'''\\ \underset{\sim}{u}'\underset{\sim}{u}''\underset{\sim}{u}'''}}(\underset{\sim}{k}',u')(\underset{\sim}{k}'',u'')(\underset{\sim}{k}''',u''') + \hat{F}_{\underset{\sim}{k},u'} \, . \tag{5.16}$$

Consider now the corresponding equation for $j=s$ . In it we keep only the terms necessary to obtain an equation for the unstable modes up to third order:

$$\left(\frac{d}{dt} - \hat{\lambda}_s\right)(\underset{\sim}{k},s) = \sum_{\substack{\underset{\sim}{k}'\underset{\sim}{k}''\\ \underset{\sim}{u}'\underset{\sim}{u}''}}(\underset{\sim}{k}',u')(\underset{\sim}{k}'',u'') + \ldots \quad . \tag{5.17}$$

If we adopt an iteration scheme using the inequality $|\xi_s| \ll |\xi_u|$ we readily convince ourselves that $\xi_s$ is at least proportional to $\xi_u^2$ , so that the only relevant terms in (5.17) are those exhibited explicitly. We now use our second hypothesis, namely, that the stable modes are damped much more quickly than the unstable ones, which is well fulfilled for the soft-mode instability. In the case of a hard mode we must be careful to remove the oscillatory part of $(\underset{\sim}{k}',u')(\underset{\sim}{k}'',u'')$ in (5.17). This is achieved by keeping the time derivative in (5.17). We therefore write the solution of (5.17) in the form

$$(\underset{\sim}{k},s) = \left(\frac{d}{dt} - \hat{\lambda}_s\right)^{-1} \sum_{\substack{\underset{\sim}{k}'\underset{\sim}{k}''\\ \underset{\sim}{u}\,\underset{\sim}{u}}}(\underset{\sim}{k}',u')(\underset{\sim}{k}'',u'') \, . \tag{5.18}$$

The prescription to evaluate $d/dt$ is this: in the soft-mode case it can be neglected, whereas in the case $(\underset{\sim}{k},u) \propto \exp(i\omega_{\underset{\sim}{k}}t)$ , $d/dt$ is to be replaced by $i(\omega_{\underset{\sim}{k}'} + \omega_{\underset{\sim}{k}''})$ . The equation of type (5.17) can now be readily solved. If time and space derivatives are neglected the solution of (5.17) is a purely algebraic problem, so that one could also keep higher-order terms in (5.17) without difficulties, at least in principle. Inserting the result into (5.16) we obtain the fundamental set of equations for the order parameters:

$$\left(\frac{d}{dt} - \hat{\lambda}_u(\nabla,\underset{\sim}{k})\right)\xi_{\underset{\sim}{k},u} = \sum_{\substack{\underset{\sim}{k}'\underset{\sim}{k}''\\ u'u''}} a_{\underset{\sim}{k}\underset{\sim}{k}'\underset{\sim}{k}'';u u' u''} I_{\underset{\sim}{k}\underset{\sim}{k}'\underset{\sim}{k}''}\xi_{\underset{\sim}{k}'u'}\xi_{\underset{\sim}{k}''u''}$$

$$+\sum_{\substack{\underset{\sim}{k}'\underset{\sim}{k}''\underset{\sim}{k}'''\\ u'u''u'''}} C_{\underset{\sim}{k}\underset{\sim}{k}'\underset{\sim}{k}''\underset{\sim}{k}''',u u' u'' u'''}\xi_{\underset{\sim}{k}'u'}\xi_{\underset{\sim}{k}''u''}\xi_{\underset{\sim}{k}'''u'''}$$

$$+ \tilde{F}_{\underset{\sim}{k},u}$$

$$\equiv H^{(r)}_{\underset{\sim}{k},u}(\{\xi_u(x)\}) + \tilde{F}_{\underset{\sim}{k},u}\,, \tag{5.19}$$

where we have used the abbreviations

$$C_{\underset{\sim}{k}\underset{\sim}{k}'\underset{\sim}{k}''\underset{\sim}{k}''',\,u u' u'' u'''} = b_{\underset{\sim}{k}\underset{\sim}{k}'\underset{\sim}{k}''\underset{\sim}{k}''',\,u u' u'' u'''} J_{\underset{\sim}{k}\underset{\sim}{k}'\underset{\sim}{k}''\underset{\sim}{k}'''}$$

$$+ 2{\sum_{\hat{\underset{\sim}{k}} s}}' a_{\underset{\sim}{k}\underset{\sim}{k}'\hat{\underset{\sim}{k}},u u' s}\, I_{\underset{\sim}{k}\underset{\sim}{k}'\hat{\underset{\sim}{k}}}\left\{\frac{d}{dt} - \lambda_s(0,\hat{\underset{\sim}{k}})\right\}^{-1} a_{\hat{\underset{\sim}{k}}\underset{\sim}{k}''\underset{\sim}{k}''';\,s u'' u'''}\,\times$$

$$\times\, I_{\hat{\underset{\sim}{k}}\underset{\sim}{k}''\underset{\sim}{k}'''}\;; \tag{5.20}$$

$\tilde{F}_{\underset{\sim}{k},u}$ is defined by

$$\tilde{F}_{\underset{\sim}{k},u}(\underset{\sim}{x},t) = \hat{F}_{\underset{\sim}{k},u}(\underset{\sim}{x},t) + 2\sum_{\substack{\underset{\sim}{k}'u'\\ \hat{\underset{\sim}{k}}\, s}} a_{\underset{\sim}{k}\underset{\sim}{k}'\hat{\underset{\sim}{k}},u u' s} I_{\underset{\sim}{k}\underset{\sim}{k}'\hat{\underset{\sim}{k}}}\,\xi_{\underset{\sim}{k}'u'}(\underset{\sim}{x})\,\times$$

$$\times\left\{\frac{d}{dt} - \hat{\lambda}_s(0,\hat{\underset{\sim}{k}})\right\}^{-1}\hat{F}_{\hat{\underset{\sim}{k}},s}(\underset{\sim}{x},t)\,. \tag{5.21}$$

The reason I have called the equations (5.19) generalized Ginzburg-Landau equations becomes clear when I assume that there exists only one parameter. In some typical cases of practical interest $\hat{\lambda}_u$ can be written in the form

$$\hat{\lambda}(\nabla,\underset{\sim}{k}) = \lambda(0) + \gamma\nabla^2\;;\quad \lambda(0),\gamma \text{ real}\,. \tag{5.22}$$

Eqs. (5.19) then acquire the form

$$\frac{d\xi}{dt} = (\lambda(0) + \gamma\nabla^2)\xi + aI\xi^2 + C\xi^3 + F(t)\,. \tag{5.23}$$

Furthermore, in a number of systems, $a = 0$ for symmetry reasons. Eq. (5.23) then reduces to (with $b = -C$)

$$\frac{d\xi}{dt} = (\lambda(0) + \gamma\nabla^2) - b\xi^3 + F(t) \tag{5.24}$$

for real $\xi$, and to

$$\frac{d\xi}{dt} = (\lambda(0) + \gamma\nabla^2)\xi - b|\xi|^2 + F(t) \tag{5.25}$$

for complex $\xi$.

Eq. (5.25) is the famous Ginzburg-Landau equation of a superconductor, describing its phase transition, but now it applies to non-equilibrium phase transitions.

In most cases of practical interest the original equations (4.1) cannot be written in the form

$$\dot{\underset{\sim}{q}} = -\mathrm{grad}_{\underset{\sim}{q}}\, V(\underset{\sim}{q}), \tag{5.26}$$

where $V$ plays the role of a potential. Quite surprisingly, however, in quite a number of explicit examples the order-parameter equations (5.19) have the form

$$\dot{\xi}_u = -\frac{\partial V}{\partial \xi_u}, \tag{5.27}$$

or, for complex variables,

$$\dot{\xi}_u = -\frac{\partial V}{\partial \xi_u^*}. \tag{5.28}$$

One can gain in all these cases an overall picture of the possible configurations. To this end we consider the order-parameter equations (5.27) including fluctuating forces. Since systems relax slowly at critical points, in many cases we may assume that the F's are $\delta$-correlated in time. Furthermore, very often they can be assumed to be Gaussian-distributed. We can write down the Fokker-Planck equation belonging to the eqs. (5.27) when these are interpreted as Langevin equations. When the fluctuating forces obey the relation

$$\langle F_\mu(t)\, F_\nu(t')\rangle = Q\, \delta_{\mu\nu}\, \delta(t-t'), \tag{5.29}$$

the time-independent Fokker-Planck equation can be solved explicitly. Its solution, which describes the stationary probability distribution, reads

$$f(\{\xi_u\}) = N \exp\left(-2V(\{\xi_u\})/Q\right), \tag{5.30}$$

where $N$ is a normalization factor and $V$ is identical with the $V$ occurring in (5.27). By determining the global or local minima of $V(\{\xi_u\})$ we find the globally or locally stable configurations.

## 6. AN APPLICATION OF GENERALIZED GINZBURG-LANDAU EQUATIONS TO MORPHOGENESIS IN BIOLOGY[3)]

In the development of organisms one observes that totipotent cells can undergo differentiations, giving rise to spatial patterns. To explain this phenomenon the following mechanism has been proposed: the cells may produce two kinds of substances, an activator and an inhibitor. Both kinds of chemicals can diffuse in the body. At a high enough concentration of the activator, genes are switched on, causing cell differentiation. This process is counteracted by the inhibitor molecules. The equations for activator concentration $a$ and inhibitor concentration $h$ read (Gierer-Meinhardt model):

$$\dot{a} = \rho + \frac{a^2}{h} - \mu a + \nabla^2 a \ , \tag{6.1}$$

$$\dot{h} = a^2 - h + D\nabla^2 h \ , \tag{6.2}$$

where we have used dimensionless variables. The stationary homogeneous solution of (6.1) and (6.2) reads

$$a_0 = \frac{1}{\mu}(\rho + 1) \ , \tag{6.3}$$

$$h_0 = a_0^2 \ . \tag{6.4}$$

The usual linear stability analysis yields the eigenvalue equation

$$\lambda^{\pm}(k) = \frac{\alpha(k)}{2} \pm \sqrt{\frac{\alpha^2(k)}{4} - \beta(k)} \ , \tag{6.5}$$

$$\alpha(k) = -(D + 1)k^2 + \frac{2\mu}{\rho+1} - \mu - 1 \ , \tag{6.6}$$

$$\beta(k) = (k^2 + \mu)(1 + Dk^2) - \frac{2\mu Dk^2}{\rho+1} \ ; \tag{6.7}$$

$k$ is the wave number of the test mode. The condition for the soft-mode instability, i.e., nonoccurrence of oscillations, is

$$\mathrm{Re}\,\lambda^+(k) \geq 0 \ , \quad \mathrm{Im}\,\lambda^j(k) = 0 \ . \tag{6.8}$$

A brief analysis of (6.5) reveals that (6.8) is fulfilled if

$$(1) \qquad \alpha < 0 \ , \qquad (2) \qquad \beta \leq 0 \tag{6.9}$$

From condition (1) we obtain

$$\rho > \frac{2\mu}{(\mu+1) + (D+1)k^2} - 1 , \tag{6.10}$$

whereas condition (2) yields

$$\rho \leqslant \frac{2\mu Dk^2}{(1+Dk^2)(\mu+k^2)} - 1 . \tag{6.11}$$

For a critical $\rho_c$ the instability condition $\beta = 0$ can first be met for two critical values of k, namely, $k = + k_c$ and $k = - k_c$ . For $\rho = \rho_{max}$ the condition $\alpha = 0$ , which indicates the onset of a hard-mode instability, can be fulfilled.

In our following analysis we will focus our attention on the soft-mode case. $k_c$ , $\rho_c$ and $\rho_{max}$ are given by

$$k_c = \sqrt[4]{\frac{\mu}{D}} , \tag{6.12}$$

$$\rho_c = \frac{2\sqrt{\mu D}}{2 + \sqrt{\mu D} + \frac{1}{\sqrt{\mu D}}} - 1 , \tag{6.13}$$

$$\rho_{max} = \frac{\mu - 1}{\mu + 1} . \tag{6.14}$$

In order that the soft-mode instability occur first we have to require $\rho_c > \rho_{max}$ , from which it follows that

$$D > 2\mu + 1 + 2\sqrt{\mu + 1} . \tag{6.15}$$

a) Solution of nonlinear equations

We assume a two-dimensional layer with side lengths $L_1$ and $L_2$ and $\rho \approx \rho_c$ . We first adopt periodic boundary conditions. We make the hypothesis (5.10):

$$\underset{\sim}{q} = \sum_j \underset{\sim}{O}^{(j)}(\nabla^2) \sum_{\underset{\sim}{k}} \xi^j_{\underset{\sim}{k}}(t) e^{i\underset{\sim}{k}\cdot\underset{\sim}{r}} \tag{6.16}$$

The wave vector $\underset{\sim}{k}$ is assumed in the form

$$\underset{\sim}{k} = 2\pi \begin{pmatrix} \frac{n}{L_1} \\ \frac{m}{L_2} \end{pmatrix} , \qquad n,m = 0, \pm 1, \pm 2 , \ldots \tag{6.17}$$

Since the solution must be real we have to require

$$\xi^j_{\underset{\sim}{k}} = \xi^{j*}_{-\underset{\sim}{k}} , \quad \xi^+_0 = \xi^{-*}_0 . \tag{6.18}$$

We now use the order-parameter equations (5.19). We introduce a new notation by which we replace the vector $\underset{\sim}{k}$ by its modulus $|\underset{\sim}{k}|$ and by the angle $\varphi$ which this vector forms with a fixed axis. We let $\varphi$ run from 0 to $\pi$. The resulting equations read

$$\begin{aligned} \dot{\xi}^{+}_{k_c,\varphi} &= \lambda \xi^{+}_{k_c,\varphi} + c\xi^{+}_{k_c,\varphi+\frac{\pi}{3}}\, \xi_{k_c,\varphi-\frac{\pi}{3}} \\ &+ \xi^{+}_{k_c,\varphi} \sum_{\varphi'} d(|\varphi-\varphi'|)\, |\xi_{k_c,\varphi'}|^2 , \quad (0 \leqslant \varphi, \varphi' \leqslant \pi) \end{aligned} \tag{6.19}$$

$\lambda$ and c are constants. The constants $d(|\varphi-\varphi'|)$ have been evaluated on a computer and are exhibited in fig. 7. Eqs. (6.19) represent a set of coupled equations for the time-dependent functions $\xi^{+}_{k_c,\varphi}(t)$.

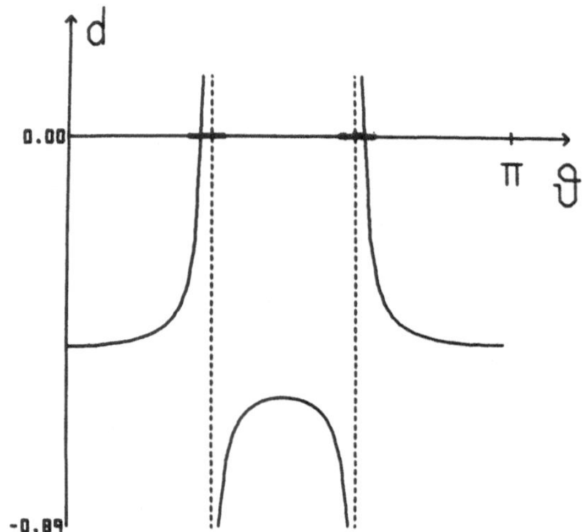


Fig. 7 : $d(\vartheta)$ is plotted as a function of $\vartheta$. In practical calculations the region of divergence is cut out, as indicated by the bars. This procedure can be justified by the construction of wave packets.
(After Haken and Olbrich, to be published)

These equations can be written in the form of potential equations

$$\dot{\xi}_{\varphi} = - \frac{\partial V}{\partial \xi^{*}_{\varphi}} , \tag{6.20}$$

where the potential function is given by

$$V = -\sum_{\varphi=0}^{\pi} \left[ \lambda |\xi_\varphi|^2 + \frac{c}{3} \left( \xi_\varphi \xi_{\varphi+\frac{\pi}{3}} \xi_{\varphi-\frac{\pi}{3}} + \text{c.c.} \right) \right.$$

$$\left. + \frac{1}{2} |\xi_\varphi|^2 \sum_{\varphi'=0}^{\pi} d(|\varphi-\varphi'|) |\xi_{\varphi'}|^2 \right. . \tag{6.21}$$

We assume that the pattern eventually resulting is determined by a configuration of $\xi$'s at which the potential $V$ acquires a local minimum. Thus, we have to seek such $\xi$'s for which

$$\frac{\partial V}{\partial \xi_\varphi^*} = 0 \tag{6.22}$$

and

$$\sum_{\varphi,\varphi'} \frac{\partial V}{\partial \xi_\varphi \partial \xi_{\varphi'}^*} \delta \xi_\varphi \delta \xi_\varphi > 0 \tag{6.23}$$

hold. The system is globally stable if

$$d(|\varphi - \varphi'|) < 0 \tag{6.24}$$

holds or if the matrix $-d(|\varphi-\varphi'|)$ has only positive eigenvalues. We discuss three typical examples.

1) The entirely homogeneous state for which all $\xi_\varphi$ are equal 0 is stable for $\lambda < 0$.
2) We obtain a "roll" pattern for which

$$\xi_{\varphi_1} = x_1 , \tag{6.25}$$

$$\xi_\varphi = 0 \quad \text{for} \quad \varphi \neq \varphi_1 \tag{6.26}$$

where

$$x_1^2 = -\frac{\lambda}{d(\pi)} . \tag{6.27}$$

$\varphi_1$ is a fixed angle which can be chosen arbitrarily. This arbitrary choice is typical of a symmetry-breaking instability, well-known from phase-transition theory.

This configuration is locally stable for

$$\lambda = 0 , \quad d(\vartheta \neq \pi) < d(\pi) < 0 . \tag{6.28}$$

The resulting spatial pattern can be obtained by inserting (6.25), (6.26) into (6.16). The corresponding pattern is exhibited in fig. 8.

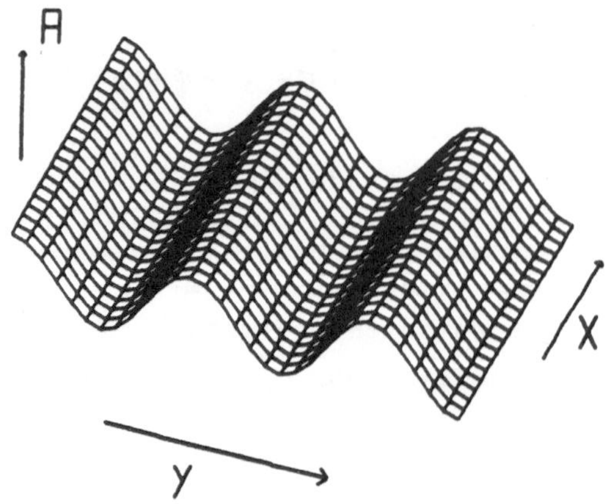


Fig. 8 : A roll-type pattern (after Haken and Olbrich, to be published)

3) Another interesting pattern, in which we find hexagons, is realized when the minimum of $V$ is attained for

$$\xi_{\varphi_1} = \xi_{\varphi_1+\frac{\pi}{3}} = \xi_{\varphi_1-\frac{\pi}{3}} = x_1 , \tag{6.29}$$

$$\xi_{\varphi} = 0 \quad \text{otherwise} ; \tag{6.30}$$

$x_1$ is given by

$$x_1 = \frac{1}{2F}\left(-c \pm \sqrt{c^2-4F\lambda}\right), \quad F = d(\pi) + 2d(\pi/3). \tag{6.31}$$

Again, $\varphi_1$ can be arbitrarily chosen. This configuration is locally stable for

$$F_{\varphi} < F < 0 \text{ , } cx_1 > \frac{c^2}{-F} \text{ , } F_{\varphi} = d(|\varphi - \varphi_1|) + d(|\varphi - \varphi_1 + \frac{\pi}{3}|) + d(|\varphi - \varphi_1 - \frac{\pi}{3}|). \tag{6.32}$$

b) Cylindrical non-flux boundary conditions

In this case we introduce polar coordinates $r$ and $\varphi$ and replace the formerly-used plane waves of the expansion (6.16) by cylinder functions of the form $e^{im\varphi} J_m(kr)$ . The non-flux boundary condition requires that

$$\frac{\partial}{\partial r} J_m(kr)\bigg|_{r=R} = 0 . \tag{6.33}$$

This equation fixes a series of k-values for which (6.33) is fulfilled.

The expansion of $\underset{\sim}{q}$ now reads

$$\underset{\sim}{q} = \sum_j \sum_k \underset{\sim}{O}^{(j)}(k) \sum_m \xi^j_{k,m}(t) e^{im\varphi} J_m(kr). \tag{6.34}$$

Inserting (6.34) into the original equation (6.1, 6.2) leads eventually to equations for the $\xi$'s . The slaving principle allows us to eliminate the stable modes and we obtain equations for the order parameters alone. Since we have a discrete sequence of k-values, for symmetry reasons we may assume that only one mode becomes unstable first. Some typical patterns are exhibited in figs. 10, 11 and 12 .

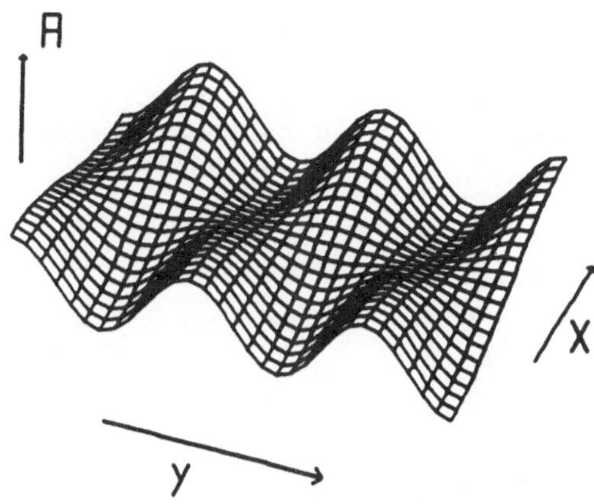


Fig. 9 : The activator concentration belonging to the mode with $k_x = \frac{\pi}{L_1}$ and $k_y = \frac{5\pi}{L_2}$ (zero-flux boundary conditions)

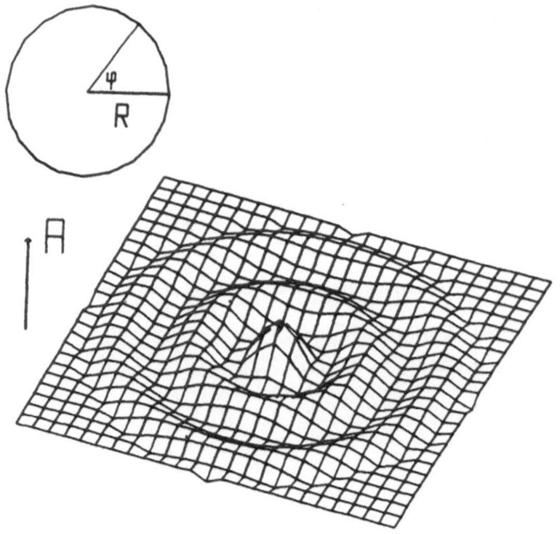


Fig. 10: The activator concentration with cylindrical non-flux boundary conditions described by a rotationally symmetric Bessel function with m = 0 . (After Haken and Olbrich, to be published)

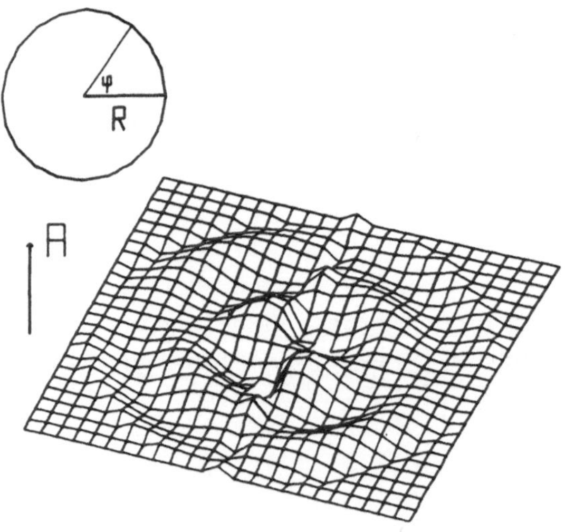


Fig. 11: Ditto with m = 2.(After Haken and Olbrich, to be published)

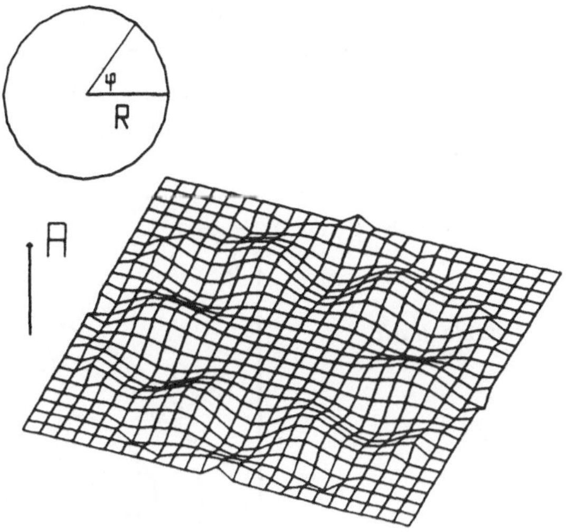


Fig. 12: Ditto with m = 3. (After Haken and Olbrich, to be published)

## 7. NONEQUILIBRIUM PHASE TRANSITIONS AND BIFURCATION OF LIMIT CYCLES

In the preceding chapters we started from a spatially homogeneous and quiescent state. We then derived equations which are capable of quantitatively describing the new structure. This was made possible by the slaving principle, which allows us to do away with an enormous number of degrees of freedom. We know from explicit examples (lasers, fluids, chemical reactions, etc.) that the newly-formed structures can again become unstable when external parameters are further changed. This leads to the interesting and important question as to whether the order-parameter idea and the slaving principle can be extended to situations where the new structure evolves from another one which is spatially inhomogeneous or is oscillating (or both). I have shown quite recently that this extension is indeed possible in a quite remarkable way. Since the corresponding mathematics is rather involved, I shall only sketch the basic ideas.

We again start from nonlinear equations of the form

$$\dot{q}_j = N_j(q_1,\ldots,q_n) + F_j(t), \tag{7.1}$$

where the $N_j$ are nonlinear functions of $q_1, \ldots, q_n$. In spatially-extended media the $q$'s are functions of the space coordinates. In this case the $N_j$ may contain derivatives with respect to the space coordinates and may depend on $x$ itself. Furthermore, the $N_j$ depend on external control parameters. The $F_j$ are fluctuating forces. We assume that for a certain set of control parameters there exists a stable limit-cycle solution (i.e., a periodic solution) or a motion on a torus (quasi-periodic motion). We then assume that, by change of a control parameter, that motion becomes unstable.

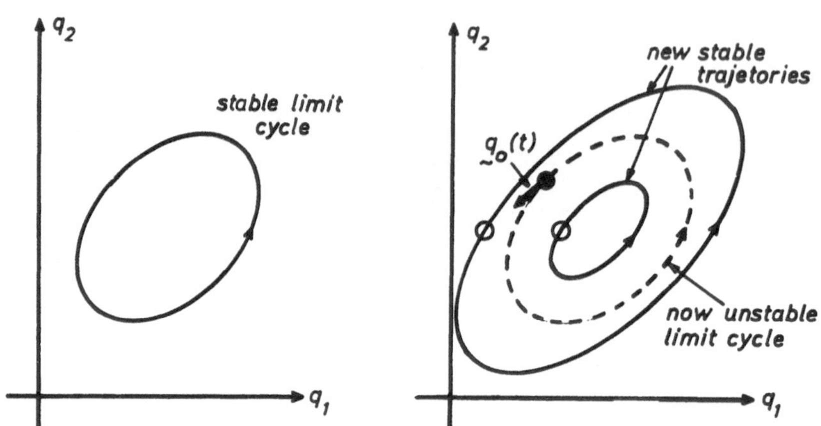


Fig. 13: Example of a limit cycle (left) which bifurcates into two new stable limit cycles (right)

We now perform a stability analysis. Owing to the Floquet theorem or its generalization to quasi-periodic motion we can again distinguish between stable and unstable modes. But in addition to our previous work we must include phase shifts, along the limit cycle or along the torus, by phase angles that are introduced as additional variables serving as part of the order parameters. It is then possible to eliminate the stable modes by a procedure which converges rapidly provided the order parameters are still small and the damping constant of the damped modes sufficiently large. The resulting order-parameter equations that include the phases are capable of describing new large classes of higher bifurcation schemes, i.e., new classes of nonequilibrium phase transitions.

## 8. CONCLUDING REMARKS

Synergetics is a very young field of interdisciplinary research. I hope that the reader has got at least a feeling of what the typical problems in this field are and how one can cope with quite a number of them at present. Surely many interesting developments are still ahead of us, and probably every participant at the Summer School will come across these fascinating problems sooner or later.

REFERENCES

1) H. HAKEN, Synergetics, An Introduction, Nonequilibrium Phase Transitions and Self- Organization in Physics, Chemistry and Biology, Springer-Verlag, Berlin, Heidelberg, New York, 1977. This book contains numerous references.

2) P. GLANSDORFF and I. PRIGOGINE, Thermodynamic Theory of Structure, Stability and Fluctuations, Wiley, New York, 1971.

3) H. HAKEN and H. OLBRICH, to be published.

# COMPUTER RESULTS ON TRANSPORT PROPERTIES

B.J. Alder
University of California, Livermore, California 94550, U.S.A.

Following a broad discussion of generalized hydrodynamics, three examples are given to illustrate how useful this approach is in extending hydrodynamics to nearly the scale of molecular dimensions and the time between collisions, principally by including viscoelastic effects. The three examples concern the behaviour of the velocity autocorrelation function, the decay of fluctuations in a resonating system, and the calculation of the dynamic structure factor obtained from neutron scattering. In the latter case the molecular-dynamics results are also compared with the predictions of generalized kinetic theory. Finally it is shown how to implement generalized hydrodynamics, both on a microscopic and on a macroscopic level.

Hydrodynamics is unable to account for the long-time tails in the velocity autocorrelation functions and the divergent Burnett coefficients observed for the Lorentz gas. Instead, the long-time behaviour of the Burnett coefficient and the distribution of displacements (the self-part of the dynamic structure factor) can be accounted for by a random walk with a waiting-time distribution which is chosen to give the correct velocity autocorrelation function. This random walk predicts, in agreement with the observations, that this displacement distribution is Gaussian at long times for the Lorentz gas, while for hard discs it has been found not to be so.

## 1. GENERAL REMARKS

Before we can speak of generalized hydrodynamics, we must first discuss hydrodynamics. By hydrodynamics we mean the temporal behaviour of fluids from the continuum point of view, as expressed through the Navier-Stokes equations. These equations consist of the conservation laws supplemented by some empirical subsidiary relations, usually called constitutive laws, which relate the fluxes in the system to the applied gradients. We shall be concerned with two such laws, namely Fick's law of diffusion, where the diffusion coefficient, $D_0$, is defined as the proportionality constant between the current and the density gradient $\partial n/\partial x$, or

$$\frac{\partial n}{\partial t} = D_0 \frac{\partial^2 n}{\partial x^2} ,$$

and the proportionality of the stress $S$ to the rate of strain (velocity gradient) $\varepsilon$, which defines the viscosity $\eta$ through

$$S = \eta \varepsilon .$$

If these laws are valid at all, and we shall see instances where they are not, they are valid only in the limit that the gradients are small in amplitude and are applied over a large distance and a long period of time.

It is these restrictions that generalized hydrodynamics tries to remove by generalizing these constitutive relations in such a way that hydrodynamics can be applied with greater accuracy to phenomena on the molecular scale, where gradients may be large but certainly the distances over which they occur range down to molecular diameters and the relevant times are of the order of collision times. The generalization consists in replacing the transport coefficients by transport functions that are functions of space and time. The alternative approach to deal with phenomena at the molecular scale is to use kinetic theory, which in practice means solutions of the Boltzmann equation or of its extension to higher densities - for hard spheres, the Enskog equation. These equations are valid only at the opposite end of the time scale from hydrodynamics, namely, very short times. This is because only in that limit is the molecular-chaos or independent-binary-collision approximation rigorously justifiable, since for short times particles do not have a chance either to recollide or have collisions with a common partner. If one wants to deal by the kinetic theory with phenomena depending on some distance scale, where in the limit of small distances the molecular-chaos approximation is again rigorously correct, one must introduce the generalized Enskog equation to which we shall refer later.

Hence, generalized hydrodynamics tries to bridge the gap in the description of phenomena on a time-and-distance scale between those over which kinetic theory and hydrodynamics are valid. Not so long ago it was thought that this gap was enormous, but the central result of the computer simulation studies[1)] has been that these two stages in the time evolution of a system are intimately coupled. The realization that hydrodynamics is a reasonable description for liquids at the molecular scale gives rise to the hope that generalized hydrodynamics is nearly quantitative. One of the many examples of the accuracy of hydrodynamics is the applicability of the Stokes-Einstein relation to a particle of atomic size diffusing in solvent particles of comparable dimensions.[2)] On the other hand, kinetic theory, which is basically a low-density gas theory, through neglect of all many-body effects, is a pretty inaccurate description at liquid density except at short times. Improvements in the kinetic or generalized kinetic theory are difficult, involving summation of a class of diagrams called ring diagrams in the graph-theoretical development. These graphs represent, only to lowest order, the effect of correlated collisions, and many more types of graphs must be included before a dense fluid can be dealt with effectively.

Thus, the more practical procedure appears to be to introduce the generalized transport coefficients and then to solve all sorts of hydrodynamic problems of interest. An immediately practical application would be to shock problems, in which a steep and large gradient is present for short periods. Next to be discussed is the randomization of the translational velocity of a particle in a fluid, from which the velocity autocorrelation function and the diffusion constant can be deduced. Although hydrodynamics gives the correct long-time behaviour of the decay of the velocity,[1] the correct shorter-time behaviour requires the introduction of a generalized viscosity by means of which, for example, viscoelastic effects can be represented. These are required to reproduce the observed reversal of the velocity at liquid densities, called back-scattering. It is in the detailed description of many such relaxation processes at a molecular level, another example being the rotational relaxation of a molecule in a fluid, that generalized hydrodynamics, and, in particular, the generalized viscosity, will be most useful. The generalized viscosity is equivalent to what is often called the transfer function, the response function, or the memory function. In its present context, it has the virtue of being theoretically calculable and experimentally measurable, although neither of these aims are as yet fully realized.

## 2. THE VELOCITY AUTOCORRELATION FUNCTION

The molecular-dynamics observation that the decay of the velocity autocorrelation function at intermediate densities is nonexponential at long times led to the realization that hydrodynamic modes could be observed at molecular time scales. The fact that a weak positive correlation between the initial velocity of a particle and its velocity some thirty collisions later could be observed meant that some highly collective mode was involved which could only be described by hydrodynamics. An elementary calculation of the decay of the velocity $v$ of a sphere, moving in a continuum characterized by a kinematic viscosity $\nu = \eta/n$ and a compressibility, can be obtained from momentum conservation. The momentum of the sphere is either dissipated by sound waves spreading with distance as $ct$, where $c$ is the sound speed derived from the compressibility and $t$ is time, or as a vortex which spreads as $\sqrt{\nu t}$. This is because the vortex obeys a diffusion type of equation, or, more elementarily, because of dimensional considerations. Thus at long times the vortex mode will dominate, and momentum conservation

$$m \int v \, d\tau \quad = \quad \text{constant}$$

then leads one to conclude that $v$ is proportional to $(\nu t)^{-d/2}$ where $d$ is the dimensionality of the system, since the volume element $d\tau$ spreads as $(\nu t)^{d/2}$. The proportionality constant is also easily determined from the fraction of the momentum that is dissipated in the sound wave relative to that in the vortex mode, namely, 1/2 in each, in two dimensions.

This hydrodynamic result has been known for many years. What is new is that it is quantitatively applicable at intermediate densities at the molecular scale, beyond about 20 collision times. This time, though long from the kinetic point of view, is very short from the hydrodynamic point of view, namely, about $10^{-13}$ sec. After this time has elapsed the vortex has spread only about 3 molecular diameters, as was observed[1)] from a detailed molecular-dynamics calculation of the velocity field, namely,

$$\left\langle v_{\ell}(0) \sum_{j=2}^{\ell} v_j(t) \right\rangle \quad ,$$

where $j$ are the neighbouring particles to the centrally located particle $\ell$, whose velocity decay is followed. The observed velocity field after about 20 collisions quantitatively agreed with that deduced from the hydrodynamic calculation, thus establishing the quantitative applicability of hydrodynamics at a remarkably short distance-and-time scale. At shorter times and distances it is first of all no longer possible to obtain the solution to the hydrodynamic problem analytically, because the sound-wave and vortex mode are not sufficiently separated. A numerical solution of the hydrodynamic problem, furthermore, failed to be quantitative at earlier times.

The greatest failing of the hydrodynamic picture, however, lies in its inability to reverse the velocity of the moving particle. This reversal occurs at higher densities at short times in the velocity autocorrelation function, before the vortex-decay mode is dominant. The physical origin of this reversal is clearly the response of the medium to a nearly instantaneous impulse of the moving particle, and this response cannot be characterized by a viscosity that measures the long-time response of the medium to a shear wave. If the medium cannot readjust sufficiently fast, it acts elastically or viscoelastically, that is, as a solid where the particles can only adjust their location in a limited way to any perturbation. In order to remedy the hydrodynamic picture a time-dependent viscosity must be introduced, since a constant viscosity is only able to slow the particle down. This can easily be done by integrating the stress autocorrelation function, $\rho_\eta(t)$, only up to a definite time, the viscosity being the long-time limit of this integral, i.e., $\eta(t)$ is proportional to $\int_0^t \rho_\eta(s)ds$. Indeed, this time-dependent viscosity obtained from molecular dynamics led to a negative velocity autocorrelation function at short times at high density when introduced into the Navier-Stokes equation, but only semiquantitative results were obtained.[3] The problem is that this viscosity is still a local quantity and it is hoped that, by generalizing it further by including spatial dependences as well, quantitative agreement will be obtained between the velocity autocorrelation function and the hydrodynamic model down to very short times. Encouraging results have been obtained by these means with estimates for the generalized viscosity[4]. The fact that the time-dependent viscosity already introduces the qualitative features of viscoelastic behaviour can be seen from the initial value of $\rho_\eta(0)$, which can be identified as the bulk elastic modulus.

## 3. FLUCTUATION-DRIVEN RESONANCE

This example gives more-direct evidence of the distance scale down to which hydrodynamics is applicable. This is because the observations concern the fluctuations of definite wavelengths that are trapped between two parallel walls. The distance between the walls, which can be made to be only a few molecular diameters in a computer experiment, determines that, of all the possible wavelengths that arise naturally in the fluid placed between the walls, only waves of wavelengths equal to an odd-integer fraction of the distance resonate. This condition arises because of the boundary conditions imposed on the system and is entirely analogous to the problem of repeated beating of a drum. The decay of these resonating fluctuations can be analysed by the standard linear-fluctuation theory of the Navier-Stokes equation of Landau and Placzek[5], and compared with the results of molecular dynamics.

The molecular-dynamics calculation for a hard-sphere system placed between two parallel walls a distance $L$ apart, with periodic boundary conditions in the other two directions parallel to the wall, evaluates the drift or centre-of-mass velocity. This drift velocity or current

$\dot{X}_D = \sum_{i=1}^{N} \dot{X}_i$ is the sum over all particles $i$ of the velocity $\dot{X}_i$ in the $X$ direction, perpendicular to the wall. The coherence in the drift velocity is then measured by its autocorrelation function

$$D(t) = \langle \dot{X}_D(0)\dot{X}_D(t) \rangle ,$$

where, as usual, the brackets indicate statistical-mechanical averaging over many initial conditions of the drift. Since only the resonating modes lead to coherence, the drift autocorrelation function can alternatively be written as

$$D(t) = \left(\frac{4}{\pi}\right)^2 \sum_{n=1,3,5}^{\infty} \frac{1}{n^2} \sum_{i,j}^{N} \left\langle \dot{X}_i(t)\dot{X}_j(0) \sin\left[\frac{n\pi}{L} X_i(t)\right] \sin\left[\frac{n\pi}{L} X_j(0)\right]\right\rangle$$

consistent with the boundary conditions. The factor $n^2$ arises because of the amount of momentum carried in each mode and the factor $(4/\pi)^2$ leads to proper normalization. This drift autocorrelation function can furthermore be rewritten as a mode summation over the intermediate longitudinal-current autocorrelation function $C_\ell(k,t)$ at wave vector $k$:

$$C_\ell(k,t) = \sum_{i,j}^{N} \langle k\dot{X}_i(t)\; k\dot{X}_j(0)\exp[ik(X_i(t) - X_j(0)]\rangle .$$

Thus,

$$D(t) = \frac{1}{2}\left(\frac{4}{\pi}\right)^2 \left(\frac{L}{\pi}\right)^2 \sum_{n=1,3,5}^{\infty} \frac{1}{n^4} C_\ell\left(\frac{n\pi}{L}, t\right) .$$

The current autocorrelation function is usually calculated for an infinite system and, as we shall see, the difference in boundary conditions is not significant. The temporal Fourier transform of $C_\ell(k,t)$ is simply related to the familiar van Hove neutron scattering function $S(k,\omega)$:

$$-\omega^2 S(k,\omega) = C_\ell(k,\omega)$$

although, in the next section, for technical reasons we shall deal primarily with the temporal Fourier transform of $S(k,\omega)$, called the intermediate scattering function $F(k,t)$.

The comparison given in fig. 1 between the molecular-dynamics results and hydrodynamics for the drift autocorrelation function uses, in the hydrodynamic theory, values of the thermodynamic and transport properties appropriate to the pressure perpendicular to the wall. This,

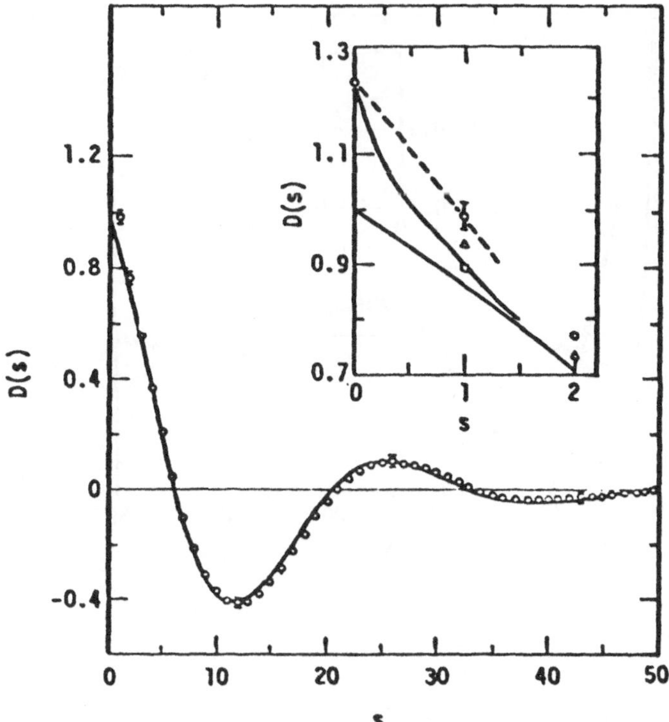


Fig. 1 : The drift autocorrelation function D(s) , represented by circles, as a function of mean collision time s at an effective $v/v_0$ of 1.94 . The solid curve is the hydrodynamic theory for the lowest mode. The insert shows for short times the difference between the contribution of the lowest mode (solid line intersecting at 1) and that of all modes (solid line intersecting at $\pi^2/8 = 1.23$) , calculated hydrodynamically, as well as between the lowest mode (square) and the sum of the first two modes (triangle), calculated by molecular dynamics by means of the longitudinal-current autocorrelation function. Also shown are the initial slope predicted for a system with a wall by kinetic theory (dashed line) and the results of molecular dynamics (circles) for the drift autocorrelation function. The vertical bars indicate the statistical uncertainty of the results.

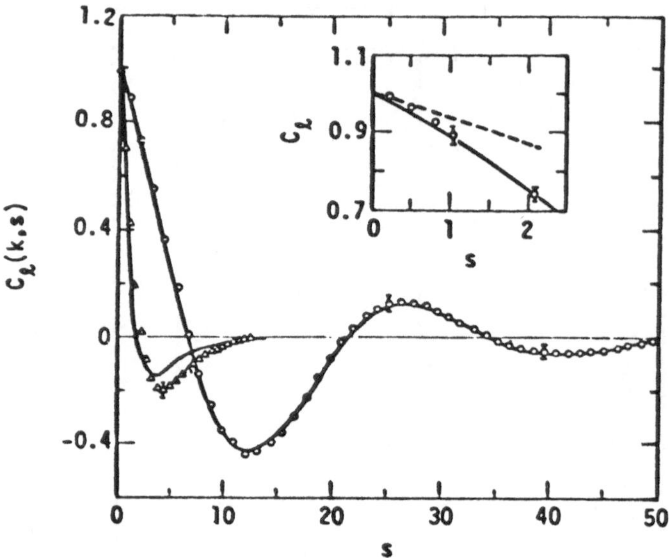


Fig. 2 : The normalized intermediate longitudinal current autocorrelation function, $C_\ell(k,s)$ for a periodic 864 hard-sphere system at $v/v_0 = 1.94$ and at two wave numbers, $k\sigma = 0.594$ (circles) and $k\sigma = 1.78$ (triangles), as a function of mean collision time s . The solid curves represent the hydrodynamic theory. The insert shows in greater detail the early-time behaviour for $k\sigma = 0.594$ as well as the kinetic-theory prediction (dashed line). The vertical bars indicate the statistical uncertainty of the results.

to first order, takes care of the wall-boundary effects. Even though the walls are only about six hard-sphere diameters apart, the agreement is remarkable. Only at very short times, less than two mean collision times (shown in the inset), is any discrepancy apparent. The hydrodynamic calculations are made only for the lowest mode and, if higher modes are included, they improve the agreement even at earlier times. Beyond a few collision times the higher modes make a negligible contribution, partly because they decay much faster and partly because the weight of their contribution is greatly reduced, namely, by the factor $1/n^2$. This is more clearly demonstrated in fig. 2, where the first two modes are separately given for the intermediate longitudinal-current autocorrelation function. It is seen there that the hydrodynamic theory fails to be quantitative for the second-lowest mode, which corresponds to about two molecular diameters. This demonstration of the remarkable validity of hydrodynamics at a distance scale of a few molecular diameters and down to a time scale of a few mean collision times can perhaps be rationalized by the following considerations. First of all, we are dealing with a function which seems remarkably insensitive to the details of the molecular motion, particularly at this intermediate density, which is about 30% lower than the solidification density. This will be shown more clearly in the next section, but suffice it to say for the moment that the velocity autocorrelation function at this density has, as yet, no negative region, and the transport coefficients are nearly quantitatively given by the simple uncorrelated-binary-collision theory of Enskog. Furthermore, a more relevant parameter by which to measure the validity of hydrodynamics than the distance scale given by the molecular diameter is the mean free path, and at these densities the mean free path is a small fraction of a molecular diameter.

## 4. THE NEUTRON SCATTERING FUNCTION

The intermediate scattering function $F(k,t)$ for hard spheres has been calculated in a straightforward manner from the density autocorrelation function at a couple of densities and a series of $k$ values. At the lower density, even somewhat lower than in the previous section, it is not surprising that hydrodynamics works well at the lowest $k$ value that could be studied in the molecular-dynamics system consisting of 500 particles with periodic boundary conditions. This is shown in fig. 3. Perhaps more surprising is the good

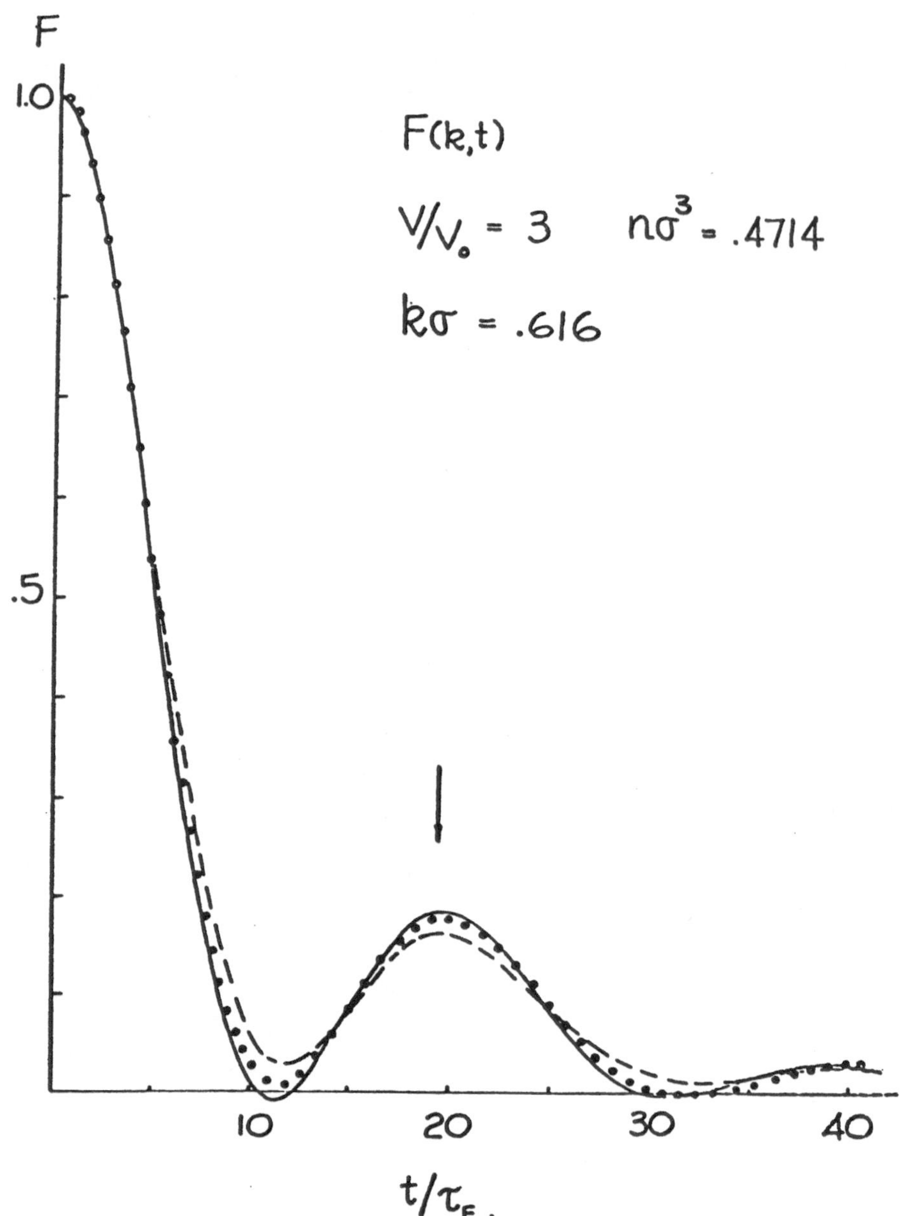


Fig. 3 : The intermediate scattering function, F(k,t) of a hard-sphere gas. Time, t, is given in units of Enskog mean collision time $\tau_E = [4\sqrt{\pi} n v_0 \sigma^2 g(\sigma)]^{-1}$ . Molecular-dynamics results for 500 particles are denoted by open circles, the generalized Enskog result by the solid curve, and the hydrodynamic result by the dashed curve. The arrow indicates the points beyond which the periodic boundary conditions affect the molecular-dynamics results.

agreement, also shown in fig.3, with the generalized kinetic theory at this relatively high density.

Enskog, in order to make the kinetic theory of Boltzmann applicable for hard spheres at higher density, introduced in front of the collision operator a function $\chi$ that takes into account that at higher density the particle size is no longer negligible compared with the distance particles travel between collisions. The factor $\chi$ can be determined in many ways, but one way is such that the rigorously known initial slope (short-time behaviour) of any autocorrelation function is correctly given. For the transport coefficient $\chi$ can then easily be shown to be the radial distribution function at contact, which measures rigorously the enhanced collision rate at higher density. For the space- or k-dependent transport function a completely analogous procedure can be followed, except now $\chi$ is a function of space. Again, the initial slope of these transport functions is rigorously known, and is often expressed in terms of the so-called sum rules.[6] Thus, the generalized kinetic equations that have to be solved are well established and an accurate solution of them will soon be published. These generalized kinetic equations at long times go over to the hydrodynamic model but involve the transport properties as given by the Enskog theory. Since, as pointed out earlier, the Enskog theory at this density gives these transport coefficients well, and the known thermodynamic properties and equilibrium structure factors are introduced into the kinetic theory, it is not surprising that good agreement is obtained.

At a density near solidification the generalized kinetic theory shows significant deviations at two different k values, as shown in figs. 4 and 5. The kinetic theory is, of course, always rigorous at short times, and would also be at sufficiently high k (short wavelength), but evidently the higher k value shown, typical of neutron experiments, is not high enough. The ordinary hydrodynamic theory is also not quantitative but can be made quantitative by using a crude version of generalized hydrodynamics. If we introduce into the hydrodynamic theory a crude version of a time-dependent viscosity, namely, at short times the Enskog viscosity, and at longer times the molecular dynamically-determined viscosity, which is about twice as large, the hydrodynamic theory gives vastly improved results, as shown in fig. 4. For neutron-diffraction experimentalists it should be pointed out, however, that the deviations of these two extreme theories (namely, the kinetic and hydrodynamic theories) from the correct result are quite small, of the order of a few per cent, so that these experiments

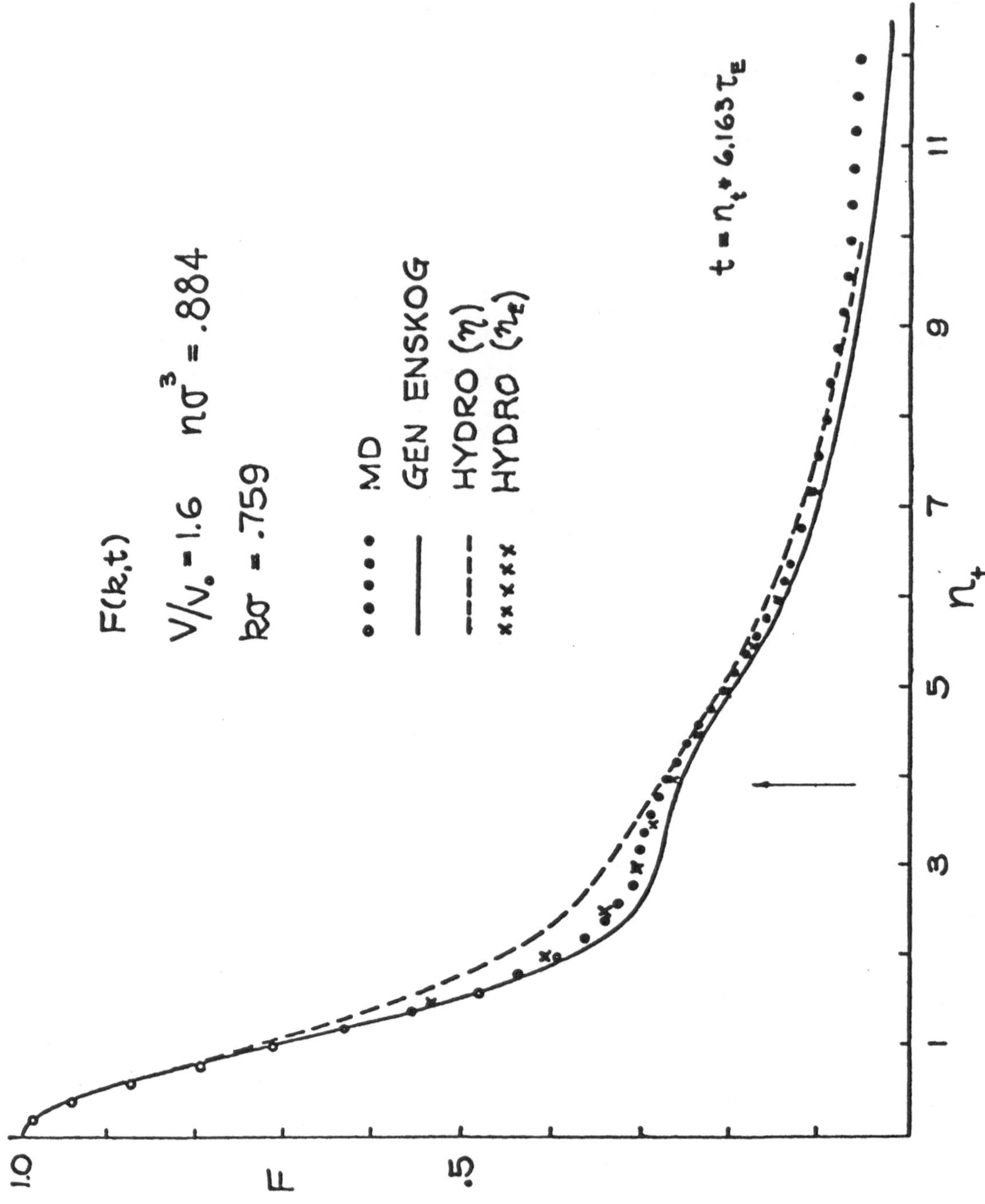


Fig. 4 : As Fig. 3 except for different conditions.

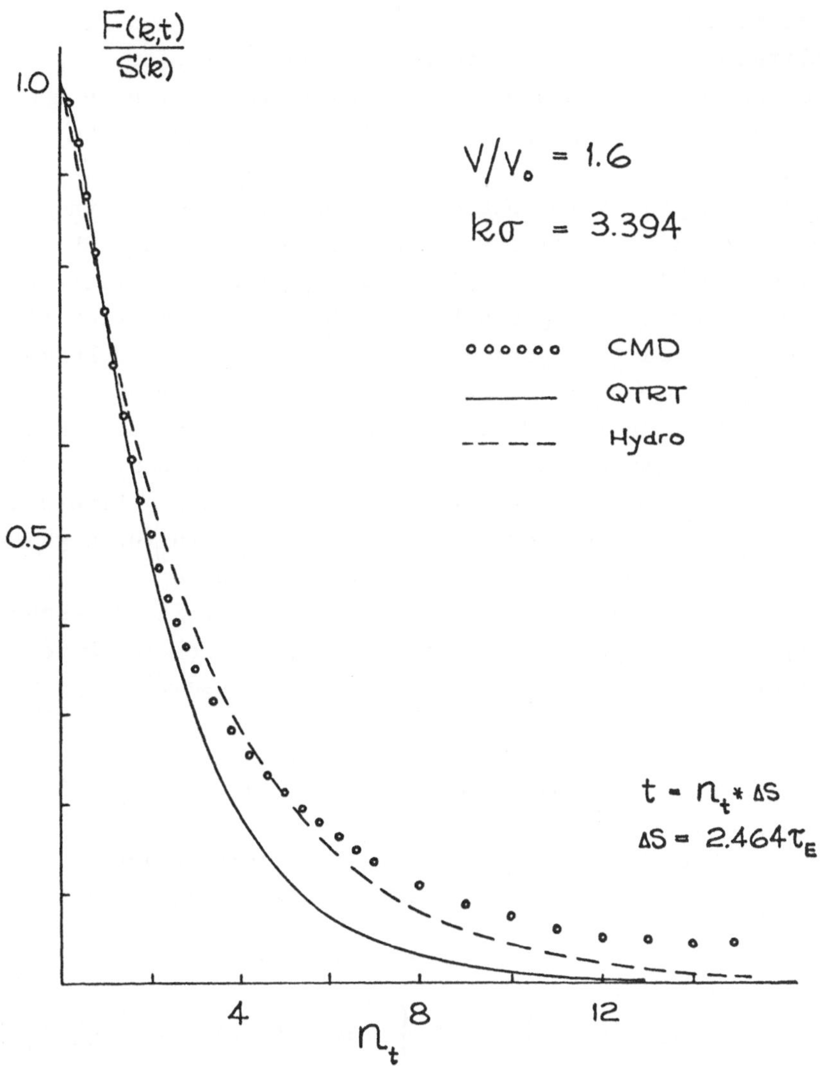


Fig. 5 : As Fig. 3 except for different conditions.

have to be more precise than hitherto if they are to give details of correlated collisions not included in these theories.

Other types of autocorrelation functions that have been calculated at this higher density lead to similar conclusions. The longitudinal-current correlation function, here called $J_\ell(k,t)$ and previously $C_\ell(k,t)$, disagrees with both kinetic and hydrodynamic theory at the lowest possible $k$ value, although the agreement with kinetic theory improves at higher $k$ values(see inset, fig. 6). Again, agreement with hydrodynamic theory can be extended to much shorter times if instead of the full transport coefficients one uses the Enskog (short-time) transport coefficients. The transverse-current autocorrelation function, shown in fig. 7, gives clear evidence at this high density of the need to introduce frequency-dependent viscosities. Neither kinetic nor hydrodynamic theory can predict the negative region of this transverse-current autocorrelation, equivalent to the negative feature in the velocity autocorrelation function discussed earlier. Incidentally, this evidence for viscoelastic behaviour in hard-sphere systems at sufficiently high density and large $k$ values leads to the conclusion that even in hard-sphere fluids shear waves can be propagated.

## 5. GENERALIZED HYDRODYNAMICS

The first objective is to calculate microscopically $\eta(k,\omega)$ or $\eta(r,t)$, so as to introduce it into the generalized relationship

$$S(k,\omega) = \eta(k,\omega)\,\varepsilon$$

or

$$S(r,t) = \int_{-\infty}^{t} dt' \int_V dr'\, \eta(r-r', t-t')\,\varepsilon(r',t') .$$

From this representation it can be seen that the generalized viscosity is the response of the system to a perturbation at a point $r'$ at a time $t'$, as felt at position $r$ at some later time $t$. The straightforward calculation starts with the consideration of the ordinary viscosity which can be written,[2)] except for constants, in either the Einstein form, as $\eta = \langle \Delta G^2 \rangle / t$, or as the time integral of the stress autocorrelation function: $\eta = \int_0^t dt \langle S_{xy}(0) S_{xy}(t) \rangle$ . Here $G$ is the appropriate dynamical variable $\sum_{\ell=1}^{N} \dot{X}_\ell Y_\ell$ , and the time derivative $\dot{G}_{xy} = S_{xy}$ is the corresponding current. The symbol $S$ is not to be confused with the neutron scattering function introduced earlier. The generalization starts with the Fourier transform of the momentum density

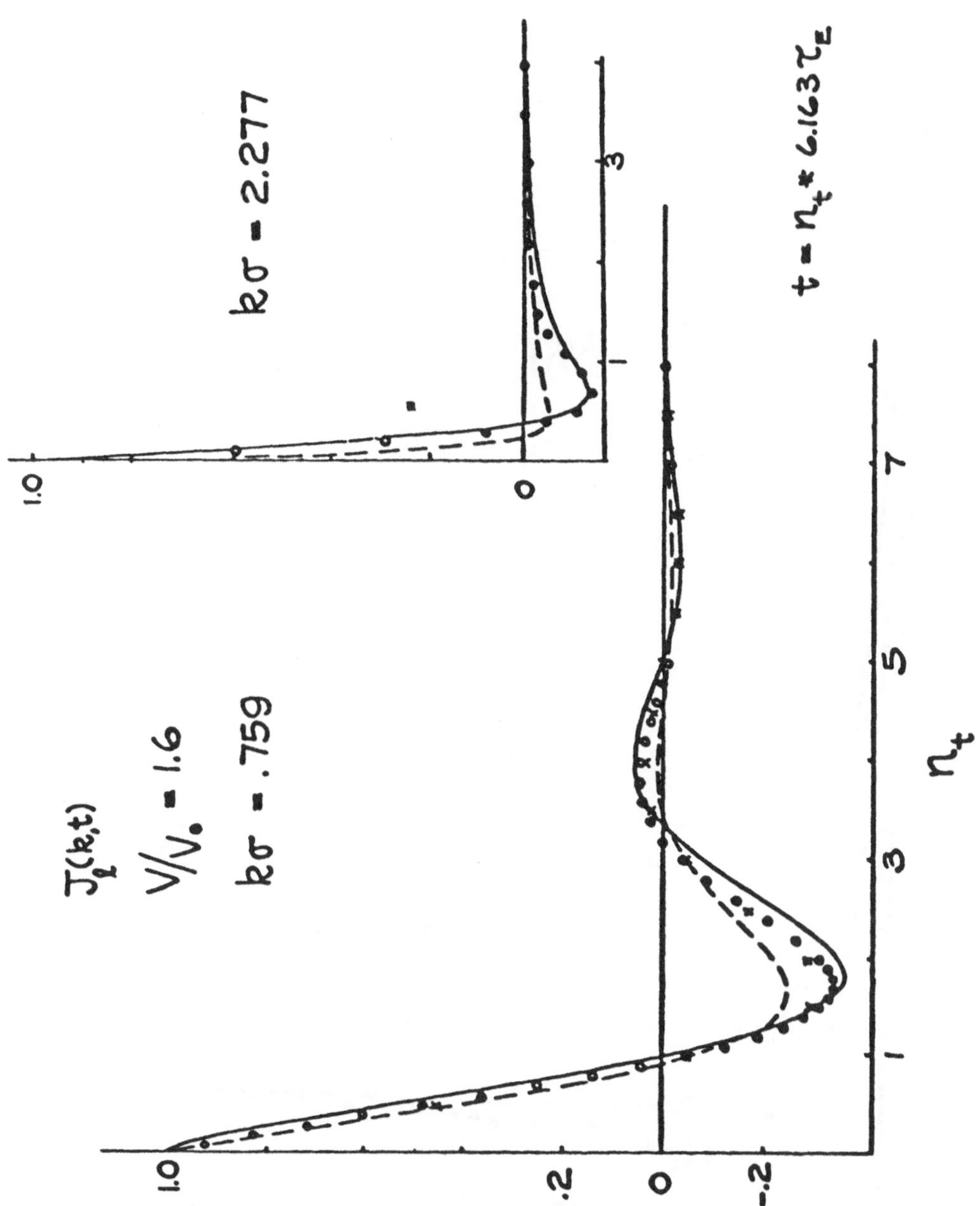


Fig. 6 : The longitudinal-current autocorrelation function (as in Fig. 2) . Time, $n_t$ , is given in terms of 6.163 Enskog mean collision times. The various symbols have the same meaning as in Fig. 4 . The insert gives the same results at a higher $k\sigma$ value.

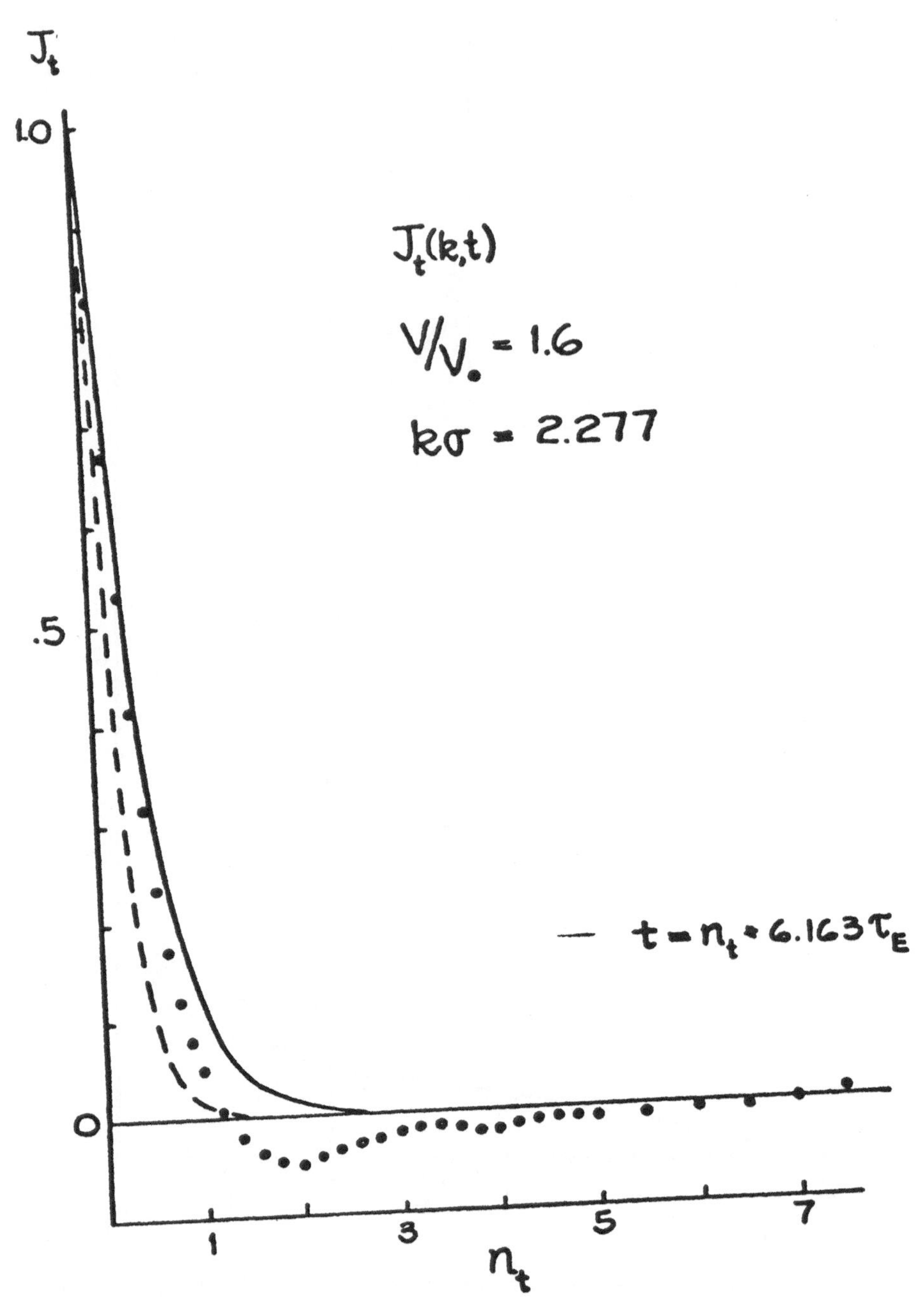


Fig. 7 : The transverse-current autocorrelation function as a function of time in the same notation as Fig. 6.

$J_x := m \sum_{\ell=1}^{N} \dot{x}_\ell \, e^{ikZ_\ell}$, which obeys the conservation law $\dot{J} + ikS = 0$. Thus,

$$\eta(k,\omega) = \int_0^\infty e^{-i\omega t} \langle S^k_{xy}(0) S^k_{xy}(t) \rangle \, dt ,$$

where now

$$S^k_{xy} = m \left[ \sum_\ell \dot{x}_\ell \dot{y}_\ell + \sum_{\ell,p} \ddot{x}_{\ell p} y_{\ell p} \left( \frac{e^{ikZ_{\ell p}} - 1}{ikZ_{\ell p}} \right) \right] e^{ikZ_\ell} .$$

The factor in the round brackets is such that at $k = 0$ its value is unity, so that the entire expression reduces to the earlier k-independent expression, while at $k = \infty$ the bracket term vanishes, so that only the first (the kinetic) term survives, as it should in this limit. This function is readily calculable by molecular dynamics, although the results are not as yet sufficiently digested to be presentable here.

Once the function $\eta(r,t)$ is known, it must be introduced in the Navier-Stokes equations and these must, in general, be solved numerically. One practical suggestion is to fit $\eta(r,t)$ by a sum of exponentials defined by a set of relaxation times $\lambda_\alpha^{-1}$: $\eta(r,t) = \sum_\alpha \eta_\alpha(r) \, e^{-\lambda_\alpha t}$. The advantage of this is that the stress now becomes a sum of elements $S(r,t) = \sum_\alpha S_\alpha(r,t)$ such that the time rate of change of $S_\alpha(r,t)$, required to predict the stress for the next time step in such numerical calculations, $S_\alpha(r,t+\Delta t) = S(r,t) + \Delta t(dS_\alpha(r,t)/dt)$ can be expressed by

$$\frac{dS_\alpha(r,t)}{dt} = \int_V \eta_\alpha(r-r') \, \varepsilon(r',t) dr' - \lambda_\alpha S_\alpha(r,t),$$

which is a differential equation for $S_\alpha(r,t)$ that requires only the value of $S_\alpha(r,t)$ at the present-time step. If such an exponential fit were not made, it would be necessary, in general, to remember $S(r,t)$ many time steps back, leading to much larger memory requirements for the computation. Such calculations are still in the future, but the principle has been successfully demonstrated[3)] with an $\eta(t)$ that was fitted numerically by a single exponential.

It is certainly true that the calculation of $\eta(r,t)$ is much more complex than that of the velocity autocorrelation that we are trying to explain by means of the generalized viscosity. Hence, one might wonder what has been gained. There are two things to be said.

One is that with the same function, $\eta(r,t)$, many other hydrodynamic problems can be solved. Secondly, theories and models can be worked out for this generalized viscosity, for example, by the generalized Enskog theory or more complex kinetic models. Furthermore, this generalized viscosity is subject to direct experimental study. Finally, it should be pointed out that so far we have considered only the linear generalized viscosity and it may very well be that nonlinear effects due to high-amplitude gradients may spoil the quantitative applicability of the linear theory to molecular processes, most probably at extremely short times. The nonlinear theory, both microscopically and macroscopically, is vastly more complex, and we shall not go into it here.

## 6. THE LORENTZ GAS

The Lorentz gas, in which a single particle moves through stationary objects, can clearly not have hydrodynamic modes. Since the single moving particle cannot generate either velocity or spatial correlations in the surrounding medium, the Lorentz gas is easier to analyse theoretically, and can be run by molecular dynamics about ten times faster than the equivalent fluid. Here the dynamical properties of the Lorentz gas have been studied by computer, in the hope that rigorous mathematical analysis will be possible. The Lorentz gas would then be the first nontrivial system for which such properties were rigorously established; however, even this task is difficult. So far only the central-limit theorem[8)] and the validity of the Fourier law of heat conduction in the Boltzmann-Grad limit[9)] appear to have been proven for the Lorentz gas, although a number of mathematicians are actively working in this field, so that there is hope of further exact results.

The nontriviality of the Lorentz gas is indicated by the graph-theoretical result[10)] that the velocity autocorrelation function in two dimensions and in the low-density limit is non-Markovian, namely, it decays as $\alpha_0/t^2$ , where $\alpha_0$ is a constant. These methods also suggest that the next-higher order velocity autocorrelation function, whose time integral leads to the Burnett coefficient, is of the form $\alpha_B/t$ , where $\alpha_B$ is another constant, and hence the Burnett coefficient diverges. If correct, this would imply that the Chapman-Enskog expansion of the Boltzmann equation[11)] is also not valid for this much simpler case. A version of the Lorentz model has already been studied on the computer, namely, the Ehrenfest tree model, and abnormal diffusion[12] was observed. In this case the moving point-particle is scattered among overlapping squares placed in such a manner that the particle can only travel in the x and y directions. The slow decay of the velocity autocorrelation function led to a predicted vanishing diffusion coefficient. Another computer study[13] of the two-dimensional Lorentz gas, where the moving point particle is scattered among overlapping discs, confirmed the theoretical prediction that the diffusion coefficient cannot be simply expanded as a power series in the density. A term logarithmic in the density with a predicted constant coefficient was found to be needed to fit the data.

The disc Lorentz gas is investigated here further, not only at low density but at all densities. The interest is in establishing the percolation density, the transport properties at that density, and the qualitatively different transport properties beyond that density. The percolation point is the lowest density for which the randomly placed discs trap the moving particle, and hence the diffusion coefficient vanishes at that density and higher densities. At the percolation density, the transport properties should show unique behaviour, which, unfortunately, is difficult to determine on the computer because of the large statistics required to sample the rare events that could lead the moving particle out of its trapped state. The percolation density has been estimated[14] to be at $n^* = 0.37 \pm 0.02$, where $n^* = nR^2$, the number density multiplied by the square of the radius of the discs.

The versatility of the computer studies allows investigations of the cause of the non-Markovian behaviour. For example, by studying the non-overlapping Lorentz gas, the role of the topology of the space in which the particle moves can be studied. The non-overlapping case is experimentally generated by freezing an instantaneous configuration of either a hard-disc fluid or solid and then letting a point particle move in the free space. In this way the percolation problem is avoided and a certain order is created in the scattering centres. Whether the moving particle is a point and the scatterers are discs or vice versa leads to completely equivalent results. Another variant of the Lorentz model investigates the effects of velocity persistence between successive collisions. That persistence can be removed by diffusive scattering at each collision rather than specular reflection. In diffusive scattering, the particle is scattered randomly in any of the possible directions rather than into the equal-angle direction called for by the dynamics. A further variant is to ignore the velocity and the collisions entirely and let the particle move, by the standard Monte Carlo sampling technique, through the fixed scatterers.

In this procedure a random move is accepted or rejected depending on whether it leads to an overlap with one of the fixed scatterers. The displacement of the particle is then recorded as a function of the number of moves and the asymptotic behaviour for large numbers of moves can be determined.

## 7. WAITING-TIME DISTRIBUTION

An attempt will be made to interpret the results on the Lorentz gas in terms of a more general than usual random walk in which, after each random step, the particle waits for a time as sampled from a waiting-time distribution before it makes the next random move[15]. In physical terms the waiting-time distribution can be thought of as corresponding to the different times it might take for the particle to escape from the various nearly-trapping regions it might find itself in. In practical terms the waiting-time distribution has to be introduced since the usual random walk cannot give a power-law decay of the velocity autocorrelation function. In mathematical terms the waiting distribution is, in fact, uniquely determined by the requirement that the experimentally determined velocity autocorrelation function be reproduced. This statement is equivalent to the statement that the waiting-time distribution is such that the second moment of the distribution of displacements is reproduced for all times. The question then is : can this more general random walk give the correct higher moments of the distributions, that is, the distribution itself, from a knowledge of the second moment only, at least at large times? The distribution of displacements for this more general but still random walk must still be Gaussian in the long-time limit. Thus, the limited but important objective is to find whether the random walk with a waiting-time distribution that gives the observed second moment at all times can reproduce at least the rate at which the higher moments approach their long-time Gaussian limit. On that question hinges the long-time behaviour of the Burnett coefficients. For example, the first linear super-Burnett coefficient measures the asymptotic rate of the second cumulant of the distribution.

Here, the random walk consists in sampling from a probability density in configuration space, $p(x)$, that a particle makes a jump and a probability density in time, $\psi(t)$, that the particle waits before making the next jump. Thus, the probability density for the walk to reach the position $x$ when the last transition occurred at time $t$ is simply

$$Q(x,t) = \sum_{n=0}^{\infty} p_n(x)\ \psi_n(t) \quad .$$

To bring the probability density up to the current time, account must be taken of the waiting time after the last jump. Thus the probability density of being at $x$ at time $t$ is

$$F(x,t) = \int_0^t \Psi(t-t')\ Q(x,t')dt',$$

where $\Psi(t) = 1-\int_0^t \psi(t)dt$ is the probability that the particle remains fixed, after the last jump, for a time $t$. The Fourier-Laplace transform of $F(x,t)$, indicated by $\tilde{F}(k,s)$, equals

$$\tilde{F}(k,s) = \frac{1-\tilde{\psi}(s)}{s}\ \tilde{Q}(k,s) \ ,$$

where

$$\tilde{Q}(k,s) = \sum_{n=0}^{\infty} \lambda^n(k)\tilde{\psi}^n(s) = \frac{1}{1-\lambda(k)\ \tilde{\psi}(s)}$$

and

$$\lambda(k) = \int_{-\infty}^{\infty} e^{-ikx}p(x)dx \ .$$

Finally, separating out the initial (n=0) term, which locates the particle at the origin, $p_0(x) = \delta(x)$, leads to

$$\tilde{F}(k,s) = \frac{1-\tilde{\psi}(s)}{s}\left[\delta(x) + \frac{1}{2\pi}\int_{-\infty}^{\infty}\frac{e^{ikx}\ \lambda(k)\ \tilde{\psi}(s)dk}{1-\lambda(k)\ \tilde{\psi}(s)}\right],$$

from which the moments of the distribution can be readily determined:

$$\langle \tilde{x}^{2n} \rangle = (-1)^n \frac{\partial^{2n}}{\partial k^{2n}}\ \tilde{F}(k,s) \ ,$$

where the derivatives are to be taken in the $k=0$ limit. Hence,

$$\langle \tilde{x}^2(s) \rangle = \langle \ell^2 \rangle \frac{1}{s}\frac{\tilde{\psi}(s)}{1-\tilde{\psi}(s)}$$

and

$$\langle \tilde{x}^4(s) \rangle = \langle \ell^4 \rangle \frac{1}{s}\ \frac{\tilde{\psi}(s)}{1-\tilde{\psi}(s)} + 6\langle \ell^2 \rangle^2\ \frac{\tilde{\psi}^2(s)}{(1-\tilde{\psi}(s))^2} \ ,$$

where

$$\langle \ell^n \rangle = \int_{-\infty}^{\infty} x^n \, p(x)dx$$

is the appropriate moment of the step-size distribution for which the odd moments vanish.

The relation between the second moment and the velocity auto-correlation function expressed in Laplace notation is

$$\tilde{\rho}(s) = s^2 \langle \tilde{x}^2(s) \rangle / 2 \langle \dot{x}^2 \rangle ,$$

so that the waiting-time distribution can be determined from a knowledge of the velocity autocorrelation function by

$$s\,\tilde{\psi}(s) = \frac{2\langle \dot{x}^2 \rangle}{\langle \ell^2 \rangle} \, \tilde{\rho}(s)(1-\tilde{\psi}(s)) \, .$$

Practically, the waiting-time distribution was obtained numerically by solving the above equation expressed in actual time, that is,

$$\psi(t) = \int_0^t K\,(t-t')\,[1-\psi(t')]\,dt',$$

where

$$K\,(t) = \frac{2\langle \dot{x}^2 \rangle}{\langle \ell^2 \rangle} \int_0^t \rho(t')dt'.$$

The analogous formula for the fourth moment

$$\langle x^4(t) \rangle = \langle \ell^4 \rangle \, \langle x^2(t) \rangle / \langle \ell^2 \rangle \; + 12 \int_0^t D(t-t') \langle x^2(t') \rangle \, dt',$$

where

$$D(t) = \int_0^t \rho(t')dt'$$

shows that this walk necessarily expresses the higher moments in terms of the second moment. Since the higher moments generally require higher correlations, this expression cannot be expected to be universally valid, except perhaps in the long-time limit, if the higher-order correlations die out sufficiently rapidly.

For the asymptotic analysis[16)] to make sense, the velocity auto-correlation function must first of all have a negative tail so that the waiting-time probability distribution is positive. Secondly, the diffusion coefficient must exist, so that if $\rho(t) \propto -t^{-\beta}$ , $\beta$ must be greater than one. If $\beta$ is less than 2 , then the dominant term in

$\langle x^4 \rangle$ is the second term in the above equation. Under these circumstances, the choice of the moments of the step-size distribution, $\langle \ell^n \rangle$, involved in the first term, becomes irrelevant. These moments could, however, be adjusted arbitrarily to fit the early behaviour of the fourth moment, or, equivalently, the Burnett coefficient. The asymptotic behaviour of the Burnett autocorrelation function[17], that is,

$$\dot{B}(t) = \frac{1}{24} \frac{d^2}{dt^2} \left[ \langle x^4(t) \rangle - 3 \langle x^2(t) \rangle^2 \right],$$

is then easily shown to be of the form $t^{-\beta+1}$ for $1 < \beta < 2$. Similarly,

$$K_2(t) = \frac{\langle x^4(t) \rangle - 3 \langle x^2(t) \rangle^2}{3 \langle x^2(t) \rangle^2}$$

behaves asymptotically as $t^{-\beta+1}$. This means that, if the low-density prediction of $\beta = 2$ is correct, the Burnett coefficient (the integral of the autocorrelation function) diverges, while the K function goes to zero asymptotically—a necessary condition for the distribution to be Gaussian.

If the diffusion coefficient vanishes, as it does above the percolation limit, the asymptotic analysis differs. The Burnett autocorrelation function then still diverges, as $t^{-2\beta+2}$ if $1 < \beta < 3/2$, and converges for $3/2 < \beta < 2$, as $-t^{-2\beta+2}$. Finally, for a vanishing diffusion coefficient and $\beta > 2$, the Burnett and velocity autocorrelation functions have identical behaviour, namely, as $-t^{-\beta}$. In the next section we shall compare these results with the ones found by computer.

## 8. RESULTS

The behaviour of the low-density velocity autocorrelation function is checked against the prediction[10] of $\rho(s) = -(n^*/\pi)s^{-2}$, as shown in fig. 8 and table I. At the lowest feasible densities, the $s^{-2}$ behaviour is indeed confirmed within the rather large experimental uncertainty, in spite of the very long runs, after about 10 collision times, s . The figure shows that the two different density runs lead to similar results, indicating that we are in the low-density regime. Furthermore, comparison with the Dutch result, using non-overlapping discs, is also very favourable—a further indication that we are at sufficiently low density. Emphasis is given to these facts,

TABLE 1. The long-time behaviour, represented by $\alpha t^{\beta}$, of the velocity autocorrelation functions that lead to the diffusion (D) and Burnett (B) coefficients of a two-dimensional Lorentz gas, and the diffusion coefficient itself.

| $n^{*}$ (a) | $-\alpha_D$ (b) | $-\beta_D$ | $D/D_E$ (e) | $\alpha_B$ (c) | $-\beta_B$ | $-\beta'_B$ |
|---|---|---|---|---|---|---|
| 0.736 | $0.88_8$ | $2.1_1$ | $0.013_2$ - $0.014_2$ | $-0.6_1$ | $1.7_3$ | $2.2_2$ |
| 0.654 | $0.42_5$ | $1.7_1$ | $0.04_1$ - $0.06_1$ | $-0.14_2$ | $1.2_1$ | $1.4_2$ |
| 0.654(S) | $0.40_4$ | $1.7_1$ | $0.08_1$ - $0.09_1$ | $-0.4_2$ | $1.2_1$ | |
| 0.477 | $0.19_2$ | $1.40_4$ | $0.14_1$ - $0.20_4$ | 0 | -- | |
| 0.370 | $0.18_1$ | $1.34_5$ | $0.21_1$ - $0.25_3$ | $(0.06_2)$ | $(0.75_4)$ | $0.68_9$ |
| 0.318 | $0.20_2$ | $1.40_5$ | $0.20_1$ - $0.16_2$ | $(0.11_2)$ | $(0.71_4)$ | $0.78_8$ |
| 0.260(N) | $0.27_2$ | $1.6_1$ | $0.55_1$ - $0.07_1$ | $0.15_5$ | $0.45_5$ | $0.6_1$ |
| 0.200 | $0.23_1$ | $1.59_5$ | $0.43_1$ - $0.09_1$ | $0.09_1$ | $0.60_1$ | $0.59_5$ |
| 0.200(S) | $0.17_4$ | $1.54_5$ | $0.36_5$ - $0.08_1$ | $0.08_2$ | $0.59_5$ | |
| 0.200(MC) | -- | -- | | -- | $0.65_7$ | |
| 0.050 | $0.07_1$ | $2.0_1$ | $0.81_1$ | $0.09_3$ | 1.0(d) | $1.0_1$ |
| 0.030 | $0.02_1$ | $2.0_1$ | $0.87_1$ | $0.05_2$ | 1.0(d) | $1.0_1$ |

(a) The first two entries use 90, the last two 1968, and the rest 504 particles. Runs were $5\times10^7$ collisions long, except at the lowest two densities where the velocity autocorrelation function results represent $4\times10^8$ collisions. Every $10^4$ collisions a new random scattering configuration was generated. The entry marked (S) stands for diffusive scattering, (N) for the no-overlap case, and (MC) for the Monte Carlo run.

(b) The magnitude is determined by normalizing the autocorrelation function initially to unity and by a fit of the data over a range of $t$ from 15 to 50 mean collision times, except at the lowest two densities, where the range is 10 to 20 collisions.

(c) The magnitude is determined by dividing $d^2(tB)/dt^2$ by $D^2$ ($D = 3v^2/8\Gamma$, where $\Gamma$ is the collision rate), and by a fit of the data over a range of times comparable to the diffusion data. At densities 0.370 and 0.318 the autocorrelation function changes at late times, of the order of several hundred mean collision times, and this is attributed to boundary-condition effects since the change depends on the number of particles used. Hence, the data that are given correspond to the early-time, trapped-particle region.

(d) The uncertainty in the last significant number is given in the small number following the entry, except in the case where the Burnett coefficient diverges logarithmically in the range of 10 to 50 mean collision times.

(e) The second number to be subtracted represents the tail correction.

---

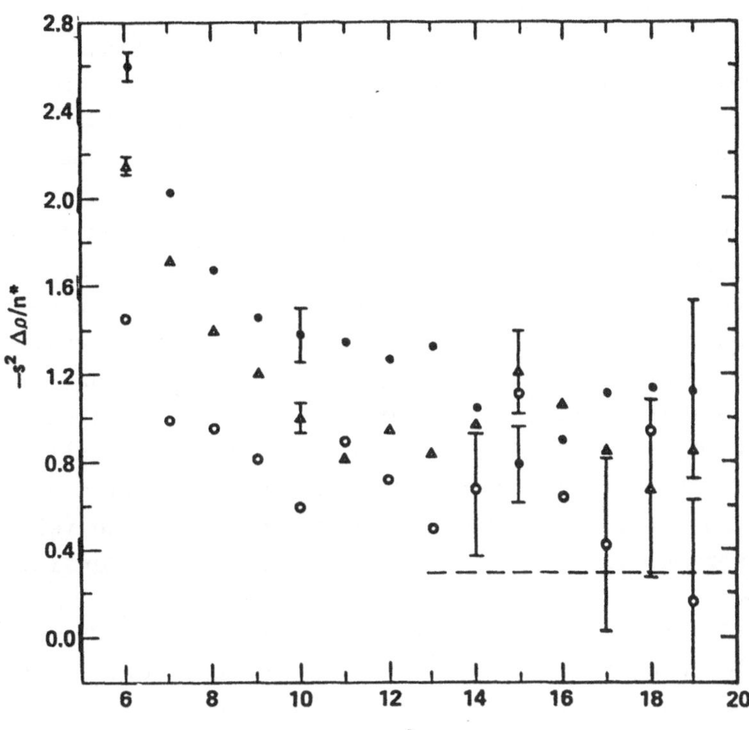


Fig. 8 : The low density velocity autocorrelation function for the Lorentz gas as a function of mean collision time, s . The deviation of the velocity autocorrelation function , $\rho(s)$, normalized so that $\rho(0) = 1$ , from an exponential $\rho_0(s) = \exp(-4s/3)$ , that is : $[\rho_0(s) - \rho(s)]s^2/n^*$ is plotted at $n^* = 0.03, 0.05$ . The solid circles represent a run of about $4 \times 10^8$ collisions at $n^* = 0.05$. The triangles represent a run of about $5 \times 10^8$ collisions at $n^* = 0.03$. Also plotted is a curve for the non-overlapping Lorentz gas at $n^* = 0.05$ which was taken from another paper. The vertical lines represent the estimated uncertainty in the results. The dashed line represents the low-density theoretical prediction.

because the experimental results lead to a coefficient of the $s^{-2}$ term about a factor of $\pi$ higher than the theoretical result, so that $\rho(s) \sim -n^* s^{-2}$. The discrepancy may still be due to the density not being low enough, or the asymptotic time regime not having been reached, but the theoretical result also needs to be checked to determine whether some graphs contributing in lowest order in the density have been omitted.

Fig. 9 shows the effect of density on the velocity autocorrelation function. As the density increases, back-scattering becomes more prominent, and the time at which back-scattering is most pronounced occurs later. This indicates that it takes longer for the Lorentz particle to escape its neighborhood the higher the density. Furthermore, as the density is increased, the figure indicates that the velocity autocorrelation function has a stronger negative-tail contribution, until the percolation density is reached. At this density the integral of the velocity autocorrelation becomes zero and, therefore, the stronger earlier back-scattering is compensated by a weaker tail to make the sum of the two regions a constant which cancels the contribution, independent of density, of the positive exponential part. As Table I shows, the tail of the velocity autocorrelation function could be represented over the entire density range by $\rho(s) = -\alpha_D t^{-\beta}$ and, as indicated above, the value of $\beta$ reaches an extremum of about 4/3 at the percolation density ($n^* \sim 0.37$) .

Figure 10 illustrates that the tail of the velocity autocorrelation function does not depend on whether the particle scatters diffusively or specularly at a given density. In the diffusive-scattering case, the back-scattering is initially longer and subsequently smaller, while the power-law decay as shown in Table I is, within experimental error, the same. Fig. 11 shows that similar slopes are also obtained for the tail of the Burnett coefficient at the same density for diffusive and specular scattering. In addition, fig. 11 shows that a Monte Carlo walk through the lattice also yields the same slope when the Burnett coefficient is plotted against the number of steps in the Monte Carlo walk. The appropriate numbers can also be found in Table I.

Figure 11 also illustrates that the Burnett coefficient never reaches a plateau value but appears to increase without bound as s increases; that is, it diverges. It is also clear from these results that this divergence, or asymptotic behaviour of the autocorrelation function, is a result of the topology of the scattering centres and that the kinematics only affects the early part of the correlation functions. There is also the question of whether the existence

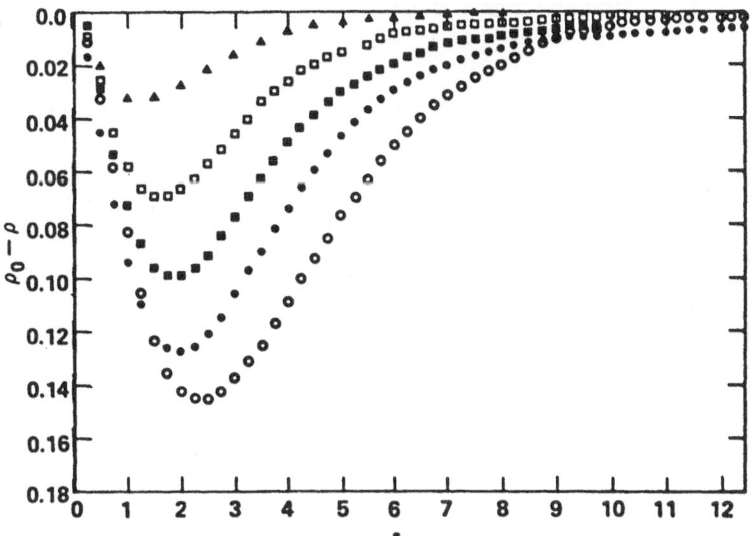


Fig. 9 : The deviation of the velocity autocorrelation function for the Lorentz gas from an exponential $\rho_0(s) = \exp(-4/3s)$, for a series of densities, as a function of mean collision time, s . The triangles Δ are for $n^* = 0.03$; the open squares □ are for $n^* = 0.10$ ; the closed circles ● are for $n^* = 0.37$ ; and the open circles O are for $n^* = 0.74$ .

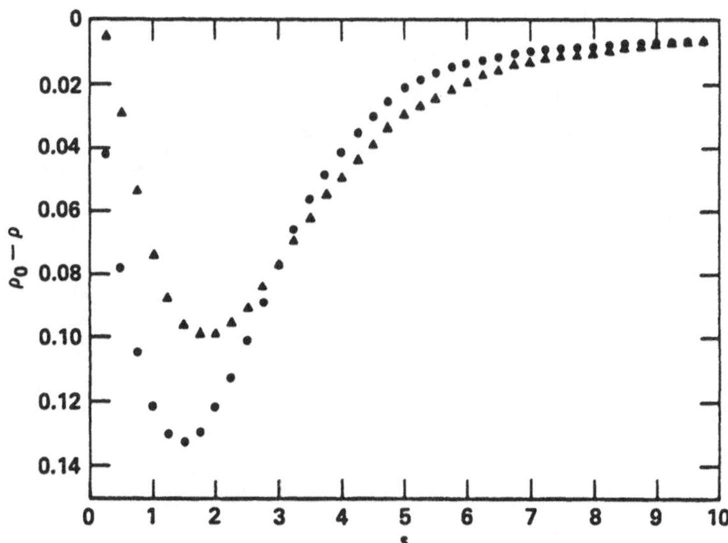


Fig. 10 : The deviation of the velocity autocorrelation function from an exponential $\rho_0(s) = \exp(-4s/3)$, for the Lorentz gas at $n^* = 0.20$ . The triangles are for a system in which the Lorentz particle undergoes specular reflection from the scatterers, and the circles are for the diffusive case.

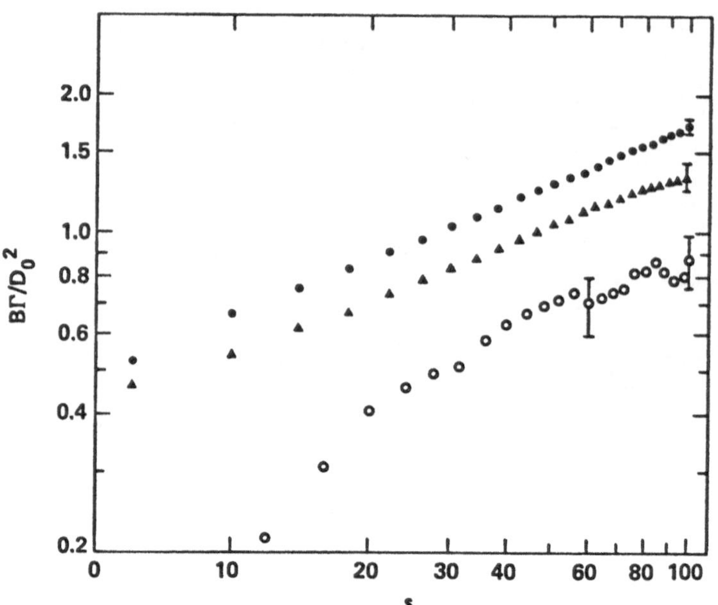


Fig. 11 : The reduced Burnett coefficient, $B\,\Gamma/D_0^2$ , as a function of mean collision time s for the Lorentz gas at a reduced density $n^* = 0.20$ . The solid circles are for the specular case, the triangles are for the diffusive case, and the open circles represent the results of a Monte Carlo walk through the scattering centres. In the latter case s represents the number of steps in the Monte Carlo walk.

of trapped regions is responsible for the long negative tail of the correlation functions of the Lorentz gas. Hence, the fact that the non-overlapping Lorentz gas also shows a tail (see Table I) in agreement with the overlapping case when the density is properly adjusted answers the previous question in the negative, since in the non-overlapping case no trapped regions exist and the percolation density is approached only in the close-packed limit. Hence, the long tails must be ascribed to a long sequence of back-scattering events that lead to a higher-than-random probability of return of a particle to its starting position. The fact that the velocity autocorrelation function has a negative tail while the Burnett one has a positive tail can be interpreted as being due to those particles that are not back-scattered diffusing for larger distances because of the presence of channels, that is, passages of less-than-average density. The particles which diffuse for larger distances are weighted more heavily in the fourth moment.

In fig. 12 the waiting-time distribution derived from the specular and diffusive velocity autocorrelation function is demonstrated. The fact that this distribution is negative at intermediate times indicates that it is physically meaningless except perhaps at long times. At long times, the K function, with a slightly different normalization than in the formula given earlier, and hence designated $K'$, extrapolates smoothly to zero, indicating that the distribution is consistent with a Gaussian distribution (see fig. 13). The two solid lines refer to different assumptions about the step-size distribution, which, as pointed out earlier, do not matter in the asymptotic limit. It is interesting to note that the earlier-time behaviour is fitted better by a Gaussian step-size distribution. In contrast , the $K_2$ function at a density beyond the percolation limit does not approach zero, as shown in fig. 14, and hence the distribution cannot be Gaussian. In fact, it can be shown that in the high-density limit of the overlapping Lorentz gas the distribution is exponential. Another non-Gaussian distribution at long times is indicated in fig. 15. This is for the two-dimensional hard-disk fluid. In this case the hydrodynamic modes cause the Burnett coefficients to be negative, indicating that the wings of the Gaussian distribution are reduced[18)]. This reduction in the number of particles that diffuse to large distances could be ascribed to the vortex mode which leads particles back to the centre.

The cumulant involving the sixth moment, for which the accuracy of the information is very limited, nevertheless also indicates consistency with a Gaussian limiting distribution and an early-time be-

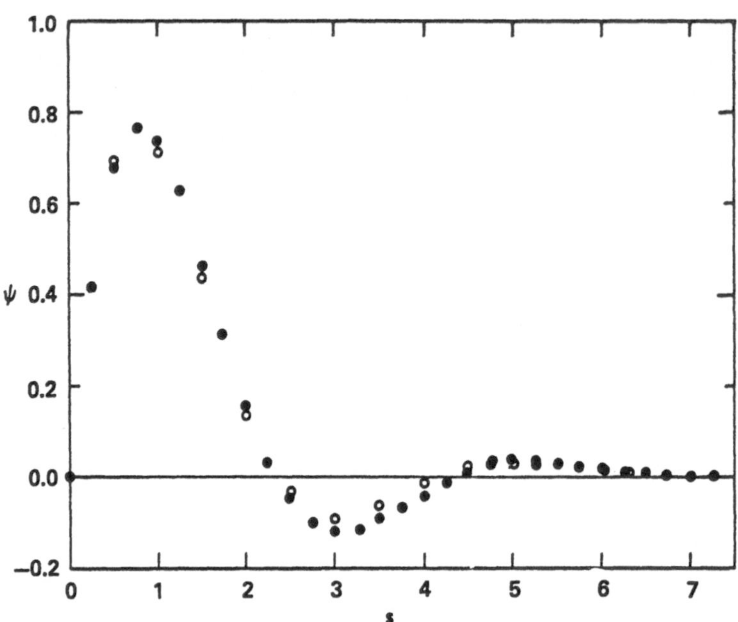


Fig. 12 : The waiting-time distribution as calculated from the velocity autocorrelation function of the overlapping Lorentz gas at $n^* = 0.20$ . The hollow circles represent a system in which the Lorentz particle scatters diffusively and the solid circles are for a system in which the Lorentz particle scatters specularly. s is in units of mean collisions.

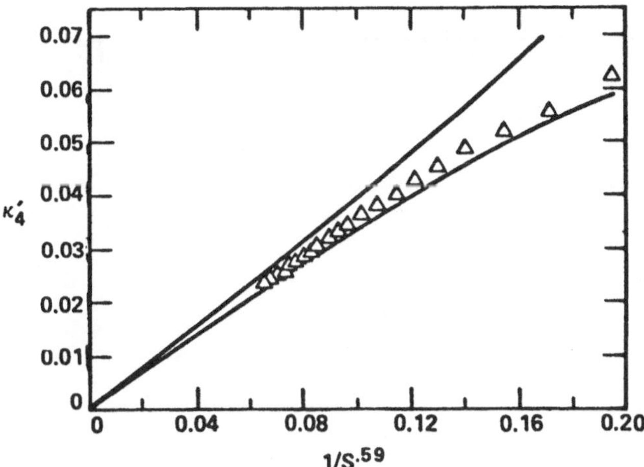


Fig. 13 : A plot of $K_4' = [\langle x^4(s)\rangle - 3\langle \Delta x^2(s)^2\rangle]/3(D_0 s)^2$ versus $1/s^{0.59}$, where s is time measured in collision times, for the overlapping Lorentz gas at a reduced density of $n^* = 0.20$ . The upper solid curve is the prediction of the continuous random-walk model if the distribution of step sizes is exponential. The lower solid line is the prediction if the distribution is a Gaussian. The statistical error is no larger than the plotting symbols (triangles) for the molecular-dynamics results.

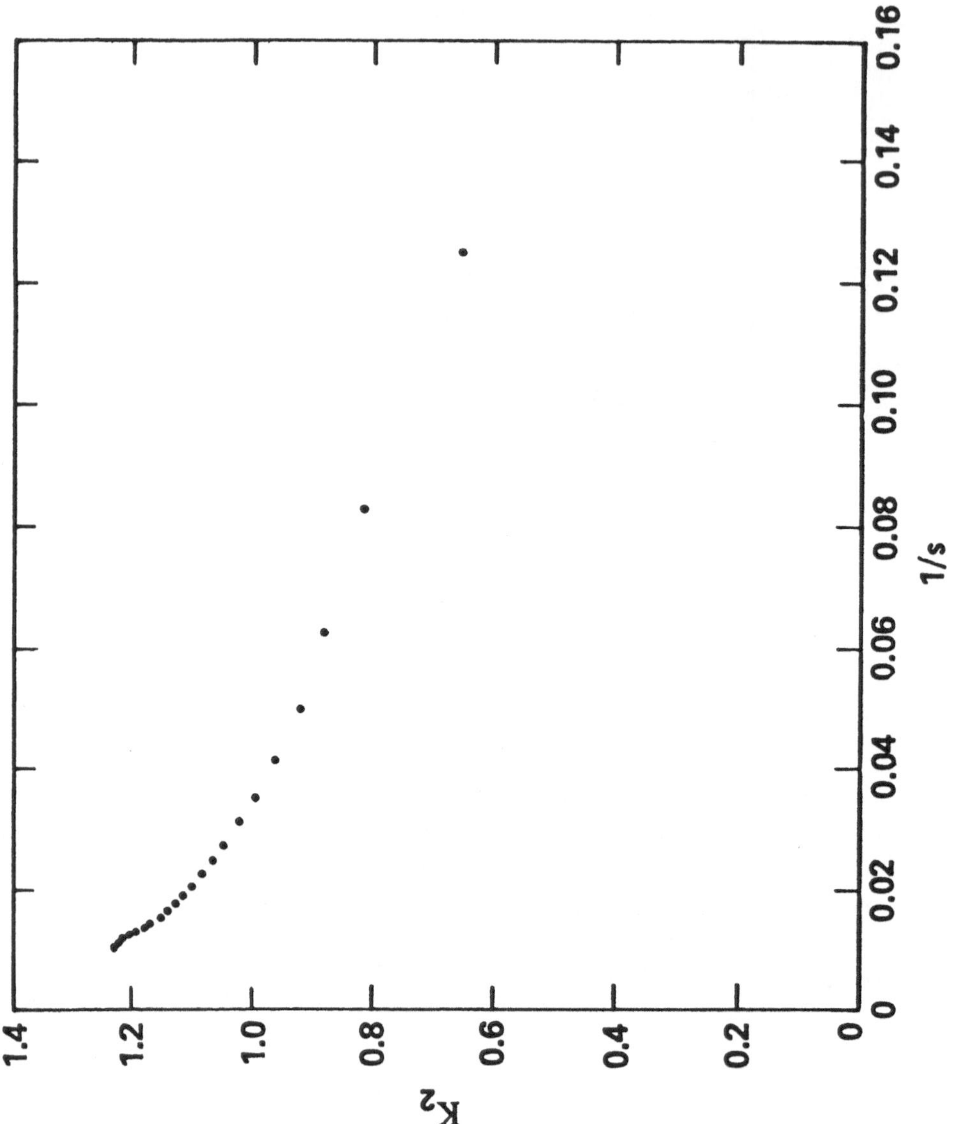


Fig. 14 : The fourth cumulant $K_2 = [\langle \Delta x^4(s)\rangle - 3\langle \Delta x^2(s)\rangle^2]/3\langle \Delta x^2(s)\rangle^2$ as a function of $1/s$ for a Lorentz gas at a density $n^* = 0.654$ .

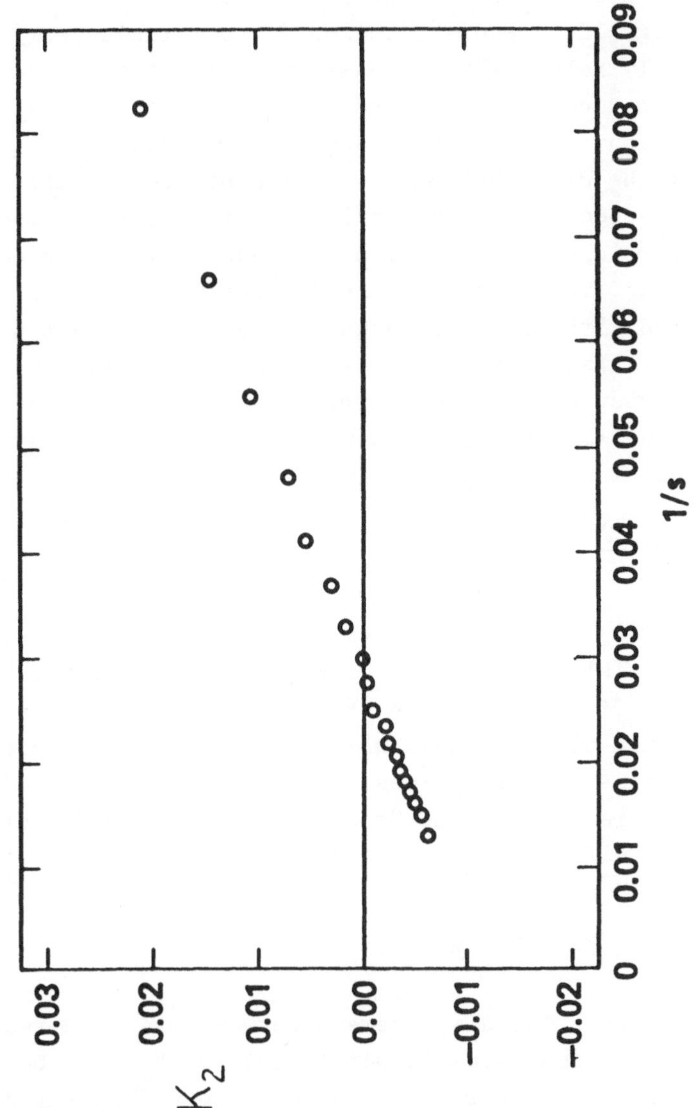


Fig. 15 : The fourth-order cumulant $K_2(s) = [\langle \Delta x^4(s)\rangle - 3\langle \Delta x^2(s)\rangle^2] / 3\langle \Delta x^2(s)\rangle^2$ for a hard-disc fluid at $A/A_0 = 3.0$ plotted against $1/s$ where $s$ is in mean particle collisions. The statistical error is smaller than the plotting characters.

haviour of the Gaussian step-size distribution for the Lorentz gas as shown in fig. 16. The distribution for a Lorentz gas itself is shown in fig. 17 relating to a Gaussian of the same half-width, as a function of reduced distance and at three times. It can be seen that the distribution does not change in character with time. The early peak in the distribution at small distances, which is indicative of the trapped particles, gradually disappears as the probability weight of particles in the traps is reduced by the spreading, diffusing particles. The large rise in the far wings of the distribution is also reduced with time, although even at the longest time the situation is still far from being Gaussian. The solid line merely indicates near-consistency of the distribution with that obtained from the Burnett equation where the known diffusion and Burnett coefficients are introduced. The Burnett equation

$$\frac{\partial F}{\partial t} = D \frac{\partial^2 F}{\partial x^2} + B \frac{\partial^4 F}{\partial x^4}$$

can be solved for $F$ by perturbation methods. To be sure, the expansion does not exist, because $B$ diverges. Nevertheless, a solution can be obtained if we substitute the value of $B$ appropriate to the time the solution of the equation is sought. If $B$ is small compared with $D$, as it is at these times, an expansion of $F=F_0+(B/D)F_1+\ldots$ can be made, where $F_0$ is the Gaussian distribution function. The solution of the differential equation then becomes, in a straightforward manner

$$F/F_0 = 1 + K_2(3/8 - 3\xi^2/4 + \xi^4/8) ,$$

where $\xi = x^2/2Dt$ . The disagreement between the observed and calculated distribution functions is partly due to $K_2$ not being quite small enough for the perturbation to converge, and partly due to higher moments that have been neglected. These higher moments are particularly needed to account for the cusp representing trapped particles, near the origin.

For the disc fluid a similar plot, shown in fig. 18, shows much better agreement with the perturbation solution. This is partly due to a smaller value of the Burnett coefficient and partly due to the absence of trapped particles (and, hence, a cusp) for a diffusing fluid. The change of sign in the Burnett coefficient with time, indicated in fig. 15, also leads to a change in the character of the distribution with time. The early-time behaviour is similar to that of the Lorentz

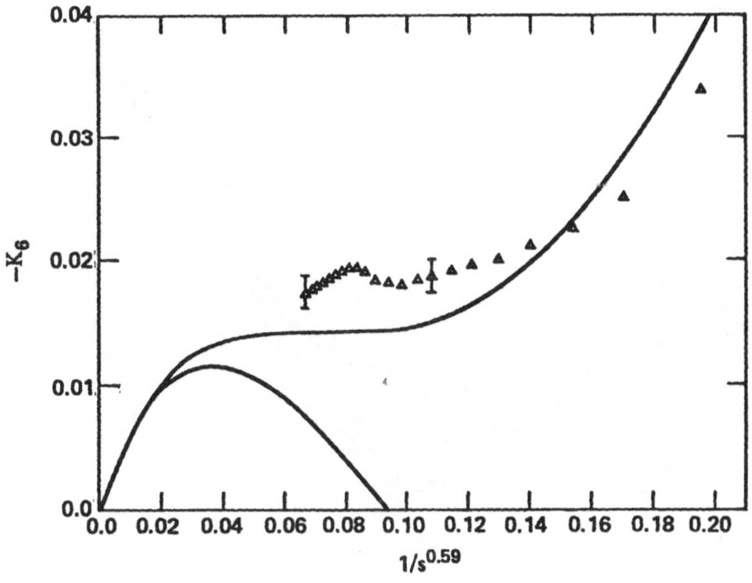


Fig. 16 : $K_6 = [\langle \Delta x^6(s)\rangle - 30\langle \Delta x^4(s)\rangle \langle \Delta x^2(s)\rangle + 15\langle \Delta x^2(s)\rangle^3 / 15\langle \Delta x^2(s)\rangle^3$ for the Lorentz gas at $n^* = 0.20$ . The upper solid curve gives the prediction from the random walk when the step-size distribution is a Gaussian; the lower solid line if the distribution is exponential. The molecular-dynamics data are given by the triangles.

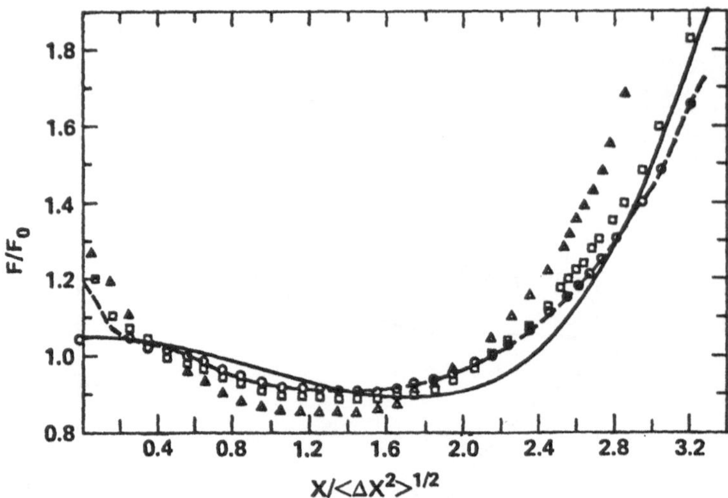


Fig. 17 : The distribution of displacements for a Lorentz gas at a reduced density $n^* = 0.20$ at three different times. The distribution, F, has been divided by a Gaussian, $F_0$, with the same second moment. The triangles are at 8 , the squares at 48 and the circles at 100 collisions, through which a dashed line has also been drawn. The solid line is the prediction from the solution of the Burnett equation at 100 collisions.

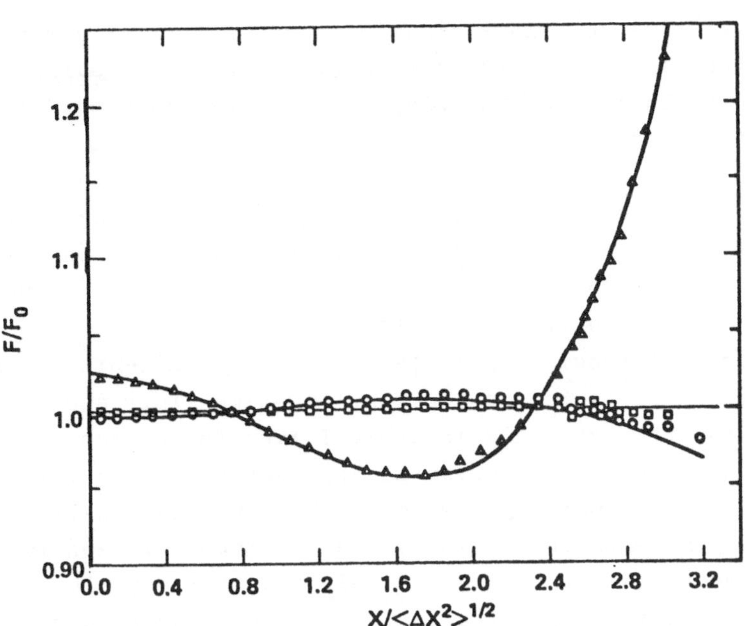


Fig. 18 : The distribution of displacements for a disc fluid at $A/A_0 = 3.0$ , at three different times. The distribution, F, has been divided by a Gaussian, $F_0$, with the same second moment. The triangles are at 3 , the squares are at 36 and the circles are at 75 collisions. The solid lines are the prediction for the solution of the Burnett equation at 3 and 75 collisions.

gas. The longest-time behaviour gives evidence for fewer particles in the wings than for a Gaussian distribution, as noted earlier. A plot of the distribution for the wings in fig. 19 contrasts clearly the difference between the Lorentz gas and the disc fluid. In the Lorentz case the distribution in the wings extrapolates to a Gaussian reasonably well, while in the fluid case the extrapolation is to less than the Gaussian value. The figure also gives evidence that the non-Gaussian distribution for the fluid is not due to the finite number of particles used in the computer calculation.

These graphical results indicate that for the Lorentz gas a Gaussian distribution is reached in the long-time limit and the numerical results for the exponents in Table I lead to the conclusion that the rate at which the Gaussian is approached at long times is correctly given by the waiting-time-distribution random walk. In the diffusing region the exponent of the Burnett autocorrelation function $\alpha_B t^{-\beta_B}$ is indeed larger by one than $\beta_D$, within experimental error. The exception is near the percolation density and can be ascribed to the limitations of the computer results. At $n^* = 0.318$ and 0.37 the diffusion coefficient, as shown in Table I, vanishes or nearly vanishes when the tail correction is taken into account. The long-time behaviour under these circumstances could not be reliably established because of the long time required for the particle to escape from its trapped state; this meant that during that time some particles had diffused to the boundaries of the periodic box used in the molecular-dynamics calculation. This was observed by a change of slope of the Burnett coefficient at long times that depended on the system size used. Hence, only the early-time results at these two densities are reliable and are indicative of the trapped state; that is, the Burnett coefficient is consistent with the $2\beta_D + 2$ prediction of the waiting-time distribution in the nondiffusing case. The waiting-time distribution thus correctly predicts the power of the diverging Burnett coefficient in the entire diffusing region.

In fact, the waiting-time distribution correctly predicts the divergence of the Burnett coefficient even in the nondiffusing region, that is, between the percolation density and an $n^*$ of about 0.477. Once $\beta_D$ exceeds 3/2 in the nondiffusing region, the waiting-time distribution correctly predicts that $\alpha_B$ changes sign and the Burnett coefficient converges. In that region, the Burnett coefficient can be obtained by numerical integration of the autocorrelation function and, once the tail is added, is found to vanish.

The major conclusion from this is that in the diffusing region

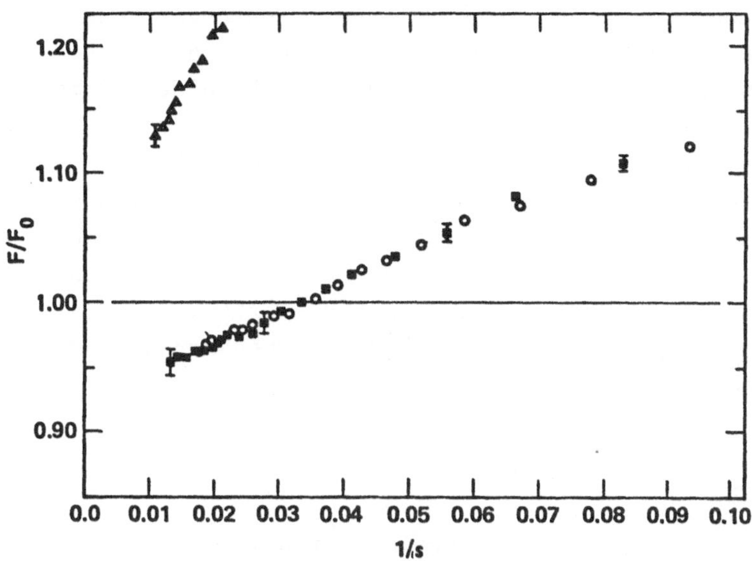


Fig. 19 : Comparison between a system of hard discs and a two-dimensional Lorentz gas. The distribution of displacements, normalized by a Gaussian of the same half width, $\sigma = \langle \Delta x^2(s) \rangle^{1/2}$ , in the interval between $3.1\sigma$ and $3.3\sigma$ . The triangles represent a Lorentz gas at $n^* = 0.1$ , while the circles correspond to a disc fluid of 1968 particles and squares to one of 4012 particles at an area relative to the close-packed area of $A/A_0 = 3$ . The error bars are indicated for the about $10^8$ collision runs.

the Burnett coefficient diverges for the Lorentz gas, and, therefore, the Burnett equation is not the way to expand to obtain k-dependent transport properties. This is also true for the disc and sphere fluids, and in generalized hydrodynamics we completely avoided expanding the transport properties in $k$. For the Lorentz gas it looks as if the waiting-time distribution leads to the correct long-time behaviour of the displacement distribution, so that we can learn how to expand this function so as to avoid the divergent coefficients in the Taylor expansion. This seems the simplest possible extension of a Markovian description and appears to work for one of the simplest nontrivial transport models.

ACKNOWLEDGEMENT

We wish to thank Mary Ann Mansigh for help with the computer programming. The work presented on the neutron scattering function is being done jointly with Professor S. Yip of M.I.T.

REFERENCES


1) B.J. ALDER and T.E. WAINWRIGHT, Phys. Rev. A1, 18 (1970).

2) B.J. ALDER, D.M. GASS and T.E. WAINWRIGHT, J. Chem.Phys. 53, 3813 (1970).

3) T.E. WAINWRIGHT, B.J. ALDER and D.M. GASS, Phys. Rev. A4, 233 (1971).

4) R. ZWANZIG and M. BIXON, Phys. Rev. A2, 2005 (1970).

5) W.E. ALLEY and B.J. ALDER, J. Chem. Phys. 66, 2631 (1977); R.D. MOUNTAIN, Rev. Mod. Phys. 38, 205 (1966).

6) J. SYKES, J. Stat. Phys. 8, 279 (1973).

7) E.P. GROSS and E.A. JACKSON, Phys. Fluids 2, 432 (1959); P.M. FURTADO, G.F. MAZENKO and S. YIP, Phys. Rev. A14, 869 (1976).

8) L.A. BUNIMOVICH, Th. Prob. and Appl. 19, 65 (1974).

9) H. SPOHN and J. LEBOWITZ, to be published.

10) M.H. ERNST and A. WEYLAND, Phys. Lett. 34A, 39 (1971).

11) S. CHAPMAN and T.G. COWLING, The Mathematical Theory of Non-Uniform Gases (Third Edition), Cambridge University Press (1970), Chapter 8.

12) W.W. WOOD and F. LADO, J. Comp. Phys. 7, 528 (1971).

13) C. BRUIN, Physica 72, 261 (1974).

14) S.W.HANN, Thesis, Univ. of Maryland (1977).

15) E.W. MONTROLL and G.H. WEISS, J. Math. Phys. 6, 167 (1965).

16) M.F. SHLESINGER, J. Stat. Phys. 10, 421 (1974).

17) J.A. McLENNAN, Phys. Rev. A8, 1479 (1973).

18) W.W. WOOD, Fundamental Problems in Statistical Mechanics, ed. E.G.D. COHEN, page 331, North Holland, Amsterdam (1975).

# KINETICS OF PHASE TRANSITIONS

O. Penrose
Faculty of Mathematics, Open University
Milton Keynes, MK7 6AA, England

## 1. INTRODUCTION

There are many different types of phase transition, and each one can happen under various different circumstances; so the subject of phase-transition kinetics appears at first sight frighteningly complex. However, with the help of a few basic concepts that apply to many of the different kinetic phenomena we can bring some of the complexity under control.

Let us follow the usual classification of phase transitions: first-order transitions, at which some or all of the properties of the substance change discontinuously, and second-order, at which they all change continuously. For a first-order transition, the kinetics of the transition are determined by two factors. These are, first of all, transport of matter and energy to or from the interface where the transition is taking place, and, secondly, the energy of the interface itself. For a second-order transition, no energy or matter need be supplied or taken away, so the effect of the interface energy is seen on its own.

In a first-order phase transition, the relative importance of the two factors, transport and surface energy, varies according to what stage the phase-transition process is at. It will be convenient to distinguish three régimes in the progress of a phase-transition process: <u>nucleation</u>, <u>growth</u>, and <u>coarsening</u>. Not all of them occur in every process, but if they do occur they occur in the order stated. By 'nucleation' I mean the initial stage of the phase transition, in which small nuclei of the new phase are formed. Sometimes nucleation is very rapid; sometimes it is exceedingly slow, in which case the old phase is said to be metastable; sometimes it does not exist at all, because a substantial amount of the new phase is present from the start. By 'growth' I mean the process by which the nuclei grow to macroscopic size; at the end of this stage, which again may be very rapid or very slow, the amount of new phase present is close to its final equilibrium value. Finally, by 'coarsening' I mean the (very slow) process by which the grown nuclei rearrange themselves and coagulate, gradually moving towards the equilibrium state predicted by thermodynamics, in which the interface between the two faces has as small an energy as possible.

The régime in which the surface energy predominates is the nucleation régime. When the nuclei of the new phase first form, they are very small; hence their ratio of surface area to volume is large and so their surface energy is a large fraction of their total energy. Since the surface energy is positive, a small nucleus of the low-temperature phase may have a higher (free) energy than no nucleus at all. There is an energy barrier to be overcome; nuclei smaller than a certain critical size will tend to contract rather than to grow, and the formation of the new phase cannot start in earnest until at least one nucleus of super-critical size is present. The super-critical nucleus may be formed at the irregular surface of some foreign body such as a speck of dust or the container holding the system - this is called heterogeneous or inhomogeneous nucleation - or it may be formed within the supercooled phase itself as the result of a fluctuation - this is called homogeneous nucleation. The rate at which these nuclei are formed will play a very important part in determining when and how the process of phase transformation begins, and in particular whether there can be a metastable state such as a supercooled vapour, a liquid under tension, or a persistent current in a superconductor.

Once the new phase is present in a significant amount, the phase transition has reached the growth stage. Surface energy is now of secondary importance; what is important is the rate at which energy and matter are supplied to or removed from the interface where the transformation is taking place. Transport of energy and matter usually requires gradients of energy and chemical potential, and so we must expect the progress of the transformation to be strongly influenced by these gradients.

As an example of the importance of these gradients, let us consider the freezing of a molten metal. Since the volume change in this transition is small, very little matter needs to be transported; the main transport process to consider is the one that removes the heat produced by the freezing. One way to remove this heat is to cool the solid, so that the liquid is warmer than the solid, and so the phase boundary is moving up the temperature gradient. The boundary advances in an orderly way into the liquid as the entire system cools. This is the method used in growing single crystals of metal.

But metals do not usually solidify as single crystals. In an iron foundry, say, the metal is solidified by cooling a mass of liquid with no solid present. In this case the crystal does not begin to grow until the liquid has cooled below the transition temperature (the reason for this delay is the delay in nucleating the solid phase). Once the crystal has formed, the heat of fusion warms it and the neighboring liquid, but the bulk of the liquid is still cooler than the crystal, in contrast to the case we considered previously.

The phase boundary is now moving down the temperature gradient instead of up it, and so there can be an instability similar to the instability of a cyclist riding downhill with no brakes. The part of the phase boundary that has advanced the furthest is in the best position to extract heat from the liquid (like a lightning conductor) and hence will advance fastest: the growth is unstable, and the crystal tends to grow as needles rather than as flat faces. The directions of these needles are determined by a surface-energy effect, which is that growth takes place much more easily on some crystal faces than others. The net result of these two effects in combination is that the crystal grows, to start with, as a complicated tree-like structure called a dendrite. The dendrite stops growing when the heat of fusion has warmed the remaining liquid back to the thermodynamic freezing temperature. After this temporary equilibrium has been reached, if further heat is now removed from the system the liquid between the dendrites will also freeze.

The formation of snowflakes appears to be due to a similar process, but in this case there are two transport effects to consider instead of one: in addition to heat transport there is now the diffusion of water through the air to the surface of the growing crystal. It is typical of phase transitions where one or both of the phases are mixtures that diffusion plays an important part in controlling the progress of the phase transition. In this case, however, diffusion and heat conduction together produce an effect similar to that of heat conduction on its own.

Once the system has reached a constant temperature and chemical potential this part of the process of growth is at an end. What happens now depends on what we do to the system. In practical examples, such as the cooling of a metal in an iron foundry, the temperature of the boundary of the system may be changed: this leads to further growth controlled by transport as before. But if the system is isolated, it will now enter the 'coarsening' phase, in which the interface energy once again becomes important. The shapes and positions of the regions occupied by the two phases will change so as to reduce the energy of the interface (which is proportional to its area). To do this, however, the system will have to shift matter from one place to another, and so transport processes will also be important. Thus in this third stage, coarsening, we have to consider both surface energy and transport processes.

In a second-order phase transition such as the ordering of an antiferromagnet or the formation of a super-lattice in an alloy, the sequence of events is simpler. No nucleation is necessary since the transition is continuous, but there are two kinds of ordered state and different parts of the system will find themselves ordered in these two different ways. The boundary between the two ordered regions will have a very complicated sponge-like structure, and the subsequent development of the system will be a coarsening process arising from the efforts of the system to move the boundary between the two ordered regions so as to reduce the area of the surface of this 'sponge', until eventually it disappears altogether and there is only one phase left. A similar sequence of events occurs in the so-called 'spinodal decomposition' of an alloy containing two constituents which mix at high temperatures. If such an alloy is cooled, the newly unmixed constituents separate out into two regions, not necessarily connected, and very complicated in shape. The later development of the alloy will be a coarsening process in which the regions change shape by interdiffusion of the two unmixed components, in such a way as to reduce the area of the interface.

In the rest of these lectures I shall be discussing theories that have been put forward for describing some of these phenomena, and how these theories can be tested against experiment and against computer simulations.

## 2. METASTABLE STATES IN MEAN-FIELD THEORY

Under some conditions (we shall see later what they are) it is extremely difficult for nuclei of the new phase to form. In such cases the old thermodynamic phase is said to be metastable and it behaves in most respects like a normal thermodynamic phase. For example, it can, in principle, be used as the working substance in a heat engine, and so we can apply normal thermodynamic reasoning to it.

The simplest theory of metastable states or phases is due to Maxwell (1875?) as a by-product of his description of the liquid-vapour transition. This description is based on the equation of state put forward by van der Waals (1873)

$$p = \frac{kT}{v-b} - \frac{a}{v^2}$$

where p is the pressure, k is Boltzmann's constant, T is the temperature, v is the volume per particle, a is a constant representing the strength of the attractive forces between molecules, and b is a constant representing the size of their repulsive cores. For temperatures less than the critical temperature given by $kT_c = 8a/27b$, the van der Waals isotherms look like the looped curves in the diagram.

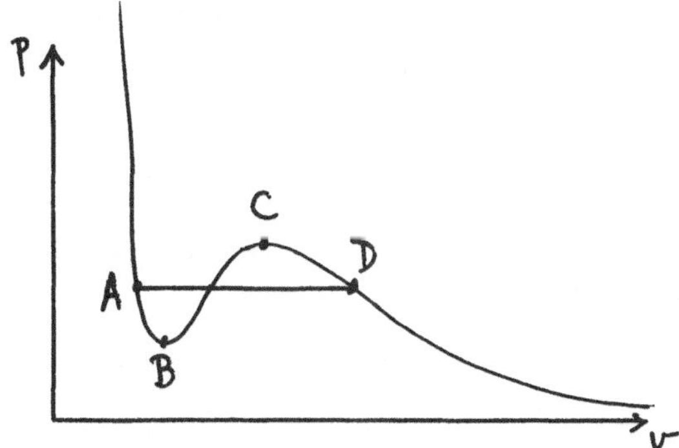


Maxwell proposed to replace the graph by a horizontal straight line, between the points A and D defined by the equal-area rule

$$p_A = p_D \; , \; \int_{v_A}^{v_D} (p-p_A)dv = 0 \, .$$

The points A and D represent the liquid and vapour phases which can be in equilibrium with each other at the temperature of this isotherm. Maxwell also proposed to interpret the arcs AB and CD as metastable states: AB describes superheated liquid; CD describes supercooled vapour. On these arcs the slope $dp/dv$ is negative, which corresponds to positive compressibility.

The arc BC corresponds to a state for which the compressibility is negative. The implications of this fact can be seen from the linearized transport equation for this system, which (neglecting viscosity) is the sound propagation equation

$$\frac{\partial^2 \rho}{\partial t^2} = \frac{dp}{d\rho} \nabla^2 \rho \, ,$$

where $\rho$ is the density and $dp/d\rho$ is calculated from the equation of state. With a negative compressibility, the solutions of this equation would grow exponentially, indicating that the spatially uniform unperturbed solution upon which it is based would be mechanically unstable against small perturbations of the density. On the other hand the metastable thermodynamic phases corresponding to the arcs AB and CD have positive $dp/d\rho$ and are therefore stable against small perturbations. It takes a large perturbation, namely, the nucleation of a new phase, to upset these states; that is why we call them 'metastable'.

A more flexible form of the van der Waals theory can be obtained by considering f, the free energy per particle, as a function of v. The van der Waals approximation gives

$$f = -kT \log(v-b) - \frac{a}{v} ,$$

from which the equation of state can be recovered using the thermodynamic relation $p = -df(v)/dv$.

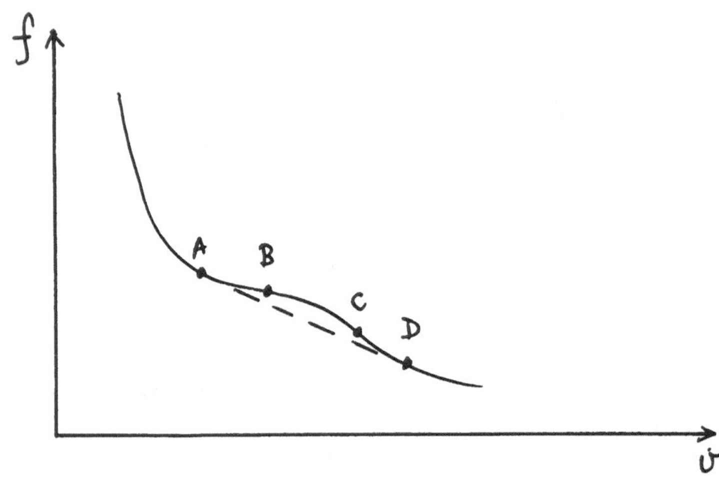


Maxwell's condition $p_A = p_D$ tells us that the slopes at A and D are equal and his equal-area condition $\int(p-p_A)dv = 0$ tells us that the tangents at A and D are the same line. The points B and C at which the metastable phases become unstable are points of inflection on the free-energy curve.

Various other phase transitions can be treated by approximate theories similar to that of van der Waals; they are known as mean-field theories. Examples are the Weiss theory of ferromagnetism and the Bragg-Williams theory of alloys. In each case the free energy can be plotted as a function of some order parameter analogous to v, the double-tangent construction gives the two phases that can coexist, and the metastable states are given by the upwardly curving arcs AB and CD. On the other hand the interpretation to be given to the downwardly curving arc BC varies from case to case, since the type of instability we get depends on the transport properties of the particular model and cannot be deduced from a diagram which contains only thermodynamic information.

As an example of the great qualitative success of van der Waals' ideas in describing metastable states, let us look at a typical isotherm for a temperature below $\frac{27}{32}T_c$ (about $0.84\ T_c$).

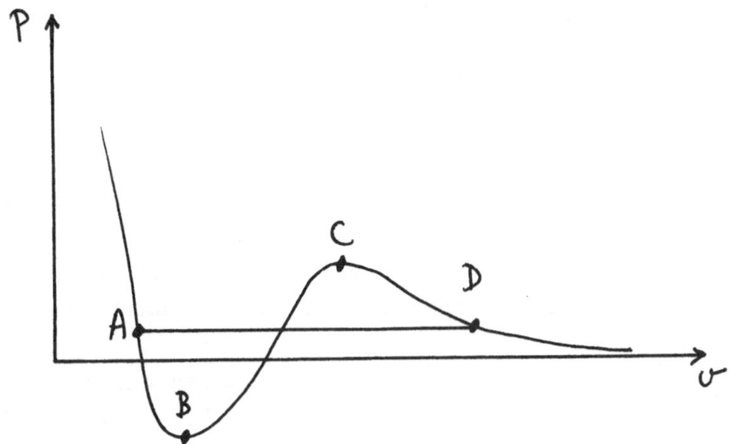


It contains a portion describing states with negative pressures — that is, a liquid under tension. According to Maxwell's construction the arc AB should describe metastable states, some with negative pressures, and the limit of metastability or 'ultimate tensile strength' should be given by the point B and should correspond to a negative pressure (i.e. a true tension) if $T < 27T_c/32$.

These theoretical predictions agree quite well with experimental findings on the tensile strength of liquids (Temperley, 1947; for later work see Temperley and Trevena, 1977).

The van der Waals method applies to transitions between phases that do not differ qualitatively (i.e., in their crystal symmetry) and therefore disappear above some critical temperature, but a similar method can be applied to two dissimilar phases, such as liquid and solid or two solids with different crystal symmetry. In that case the free-energy diagram consists of two distinct curves, one for each phase

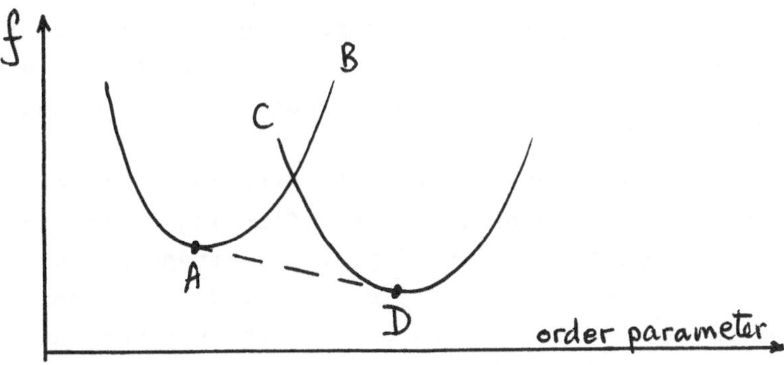

The double-tangent construction still gives the values of the order parameter (e.g., volume per particle, for the case of a liquid-solid transition) for two coexisting phases A and D. The arcs to the left of A and to the right of D still describe equilibrium states, and the arcs AB and CD still describe metastable states. The connecting arc BC is now absent, however, since one cannot continuously connect phases with different symmetries. Diagrams like the one shown, with empirically determined free-energy curves, are useful to metallurgists in studying the complicated families of phase transitions in the alloy systemsthey deal with.

Instead of obtaining the free energy curves empirically, or from approximate theories like that of van der Waals, we would like to know how to calculate them from partition functions according to the standard methods of statistical mechanics. It is a fundamental theorem, however, that the thermodynamic free energy per particle must be a convex function of the volume per particle. That is, the second derivative of the function f(v) must be non-negative, and the graph of this function must be either straight or curving upwards. It follows that a non-convex free-energy function, such as the one given by the van der Waals equation when $T < T_c$, cannot possibly be the 'true' thermodynamic free energy as given by the partition function. Indeed, in 1963, Kac, Uhlenbeck and Hemmer succeeded in finding a model system which exactly satisfies van der Waals' equation of state with Maxwell's rule incorporated in it. That is to say, the true free-energy curve for this system is the so-called <u>convex envelope</u> of the van der Waals free energy, obtained by replacing the part of the curve between A & D by the common tangent at these points. In 1964 van Kampen gave an argument showing that the partition function for more general liquid-vapour systems might also be expected to give a free energy obtainable by the double-tangent construction. Lebowitz and Penrose (1966) showed that van Kampen's argument could be made rigorous in a particular limiting case of systems with a very weak attractive force of very long range (of which the Kac-Uhlenbeck-Hemmer model, a one-dimensional hard-rod system with an attractive force of this kind, is a particular case).

To obtain any information about the metastable states from partition functions, therefore, some modification of the conventional methods is necessary. One possibility is to extrapolate the 'stable' part of the curve, e.g., by analytic continuation. This analytic continuation can easily be carried out for the van der Waals equation and the models with infinitely weak infinitely long-range forces, but for short-range interactions the free energy function is probably nonanalytic at the phase-transition point. This was suggested by Fisher (1967) and proved, for lattice systems with finite-range interactions, by Lanford and Ruelle (1969). It was suggested by Langer (1967) that despite this singularity an analytic continuation of the free energy is possible and that it might give a complex free-energy function whose real part can be interpreted as the free energy of the metastable state. A related idea has been explored by Newman and Shulman (1977). However, even if a good method of extrapolation is found, one still has to justify the assumption that the extrapolated thermodynamic functions do describe metastable states. That is to say, we should study things like the lifetimes of these states, and for this it is necessary to pay some attention to the kinetic or dynamic aspects of metastability.

## 3. ENSEMBLES FOR METASTABLE STATES

The problem considered in this section is how to calculate the free energy and other properties of metastable states from an ensemble formalism. The theorem that the free energy is convex rules out the direct use of the standard ensembles of statistical mechanics for this purpose. The reason is that the metastable state has a higher free energy per particle than the stable (two-phase equilibrium) state. Therefore in canonical averages taken over the whole ensemble the metastable states will in general be overwhelmed by the stable equilibrium states, whose contribution to the partition function is much larger. One way to avoid this is to replace the standard canonical ensemble by a restricted ensemble from which the unwanted states have been removed. The 'unwanted' states are those which contain an appreciable amount of the new phase - that is, the ones in which nucleation may be said to have taken place.

For the phase transition caused by a very weak, very long-range attractive force, a suitable restricted ensemble was found by Penrose and Lebowitz (1971). In this work the spatial region occupied by the system was partitioned into imaginary cells, whose size was chosen small compared with the range of the weak attractive force but large compared with the range of the other parts of the force; the restriction made was that the number of particles in each cell should lie between specified bounds $\omega/v_2$ and $\omega/v_1$ where $\omega$ is the volume of a cell and $v_1, v_2$ are two numbers, one just less than the actual specific volume and the other just greater. This ensured that the density distribution, when viewed on a scale much larger than the cell size, would be approximately uniform. Small-scale density fluctuations were still permitted, but such fluctuations should in any case be part of the statistical description of the metastable state, as indeed they are part of the statistical description of any equilibrium thermodynamic phase. The important thing is that the restriction rules out nuclei of the new phase by ruling out any fluctuation on a large enough scale to have a significant effect on the energy of the weak long-range attractive potential. It turned out that, in a limit where the size of these cells become very large, while the range of the attractive forces became even larger, this restricted ensemble could be proved to describe a state that had the following set of properties, which provide a convenient characterization of a metastable state:

1. only one thermodynamic phase is present
2. the lifetime is very long
3. the free energy per particle is greater than for the stable equilibrium state.

Moreover the free energy per particle in the metastable state has exactly the value given by the van der Waals-Maxwell prescription.

This particular way of specifying the restricted ensemble works only because the range of the weak attractive force responsible for the transition can be made arbitrarily large. For forces of finite but very long range the results would still hold in some approximate sense, but in practice we are usually concerned with short-range forces. A type of restricted ensemble suitable for some problems involving short-range forces was found by Capocaccia, Cassandro and Olivieri (1974) in a paper about the square Ising ferromagnet. Later on the method was extended to the two-dimensional Widom-Rowlinson model (Cassandro and Olivieri 1977), and there is no reason why it should not be extended to other models, not necessarily in two dimensions.

The square Ising model consists of a finite plane square lattice $\Lambda$ consisting of L sites, with each of which is associated a 'spin' variable capable of only two values, say -1 and 1. The energy of a given configuration of spins $\{s_1,\ldots,s_L\}$ in a magnetic field H, is taken to be

$$E = - mH \sum_{i=1}^{L} s_i - J \sum_{(ik)} s_i s_k \tag{3.1}$$

where $s_i$ is the spin variable at the $i^{th}$ site $(i = 1,2,\ldots,L)$, m is the magnetic moment of each spin, J is a positive constant, and the second sum goes over all nearest-neighbour pairs of sites on $\Lambda$.

Since metastability is a dynamical phenomenon, it is not enough to define the energy of each state; we must also specify how the system gets from one state to another. Capocaccia et al. used the dynamical model introduced by Glauber (1963), in which the system proceeds by a Markov chain. At each step, a site is chosen at random and then the spin at that site is reversed, with a probability which depends on the old and new configurations according to a rule chosen to satisfy detailed balancing.

To study metastable states we use the boundary condition that the spins on the boundary of $\Lambda$ are held fixed at the value + 1. If the temperature T is low and the magnetic field is zero (or even positive), this boundary condition ensures that the equilibrium state will be one of positive magnetization in which most of the spin variables are positive. Negative spins are somewhat exceptional and, for any given configuration, can be classified into <u>clusters</u>.

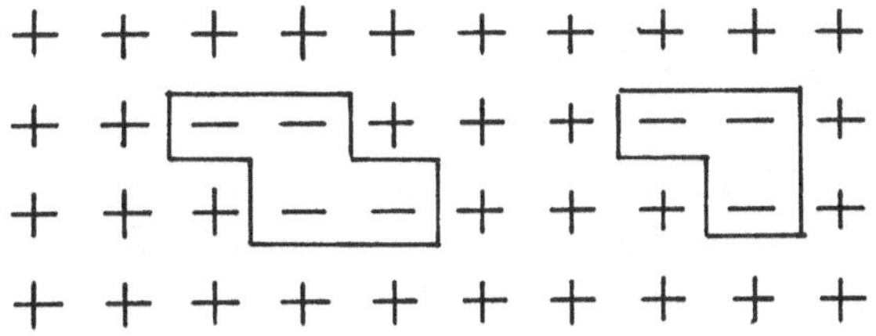

```
+ + + + + + + + + +
+ + - - + + + - - +
+ + + - - + + + - +
+ + + + + + + + + +
```

A cluster is a maximal set of negative spins connected by nearest-neighbour bonds. If we define the 'size' $\ell$ of a cluster as the number of negative spins in it, and its perimeter $p$ as the number of nearest-neighbour bonds joining these negative spins to positive spins, then the energy of the cluster is

$$-2m H\ell \quad - 2 J p$$

It was shown by Capocaccia et al that for $H \geqslant 0$ the canonical average of the number of clusters of a particular shape with size $\ell$ and perimeter p is at most

$$L\, e^{-(2mH\ell + 2Jp)/kT}$$

and, further, that the total number of clusters with a given size $a^2$ is at most

$$L 3^{4a} e^{-(2mHa^2 + 8Ja)/kT} \tag{3.2}$$

For $H \geqslant 0$ and $kT < 2J/\ell n 3$ this falls off rapidly with increasing p;

that is, there are very few large clusters.

The possibility of a metastable state arises if we make H negative, still keeping the spins on the boundary positive. Now the formula (3.2) no longer requires the number of large clusters to be few. These large clusters can be interpreted as regions of the new (negative-magnetization) phase. To describe the metastable state we can use a restricted ensemble from which such large clusters are excluded. Capocaccia et al did this by requiring all clusters to have sizes less than some specified value $a^2_{max}$. They showed that the formula (3.2) still applies to the restricted canonical ensemble (i.e., one from which all configurations containing clusters of sizes larger than $a^2_{max}$ have been removed, and the rest have the same statistical weight as before). If H is sufficiently small then it is possible to find a value for $a_{max}$ (the optimum value is $(2J - kT \ln 3)/ m|H|kT$) for which the restricted ensemble predicts only very few clusters with sizes comparable with $a^2_{max}$. Capocaccia et al were able to show that under these conditions (i.e., $kT < 2J/\ln 3$ and very small H) their restricted ensemble gave a state satisfying the three criteria of metastability we have mentioned earlier, viz. only one thermodynamic phase is present, the lifetime is very long and the free energy per site exceeds that for the stable equilibrium state.

In this work the free energy per particle in the metastable state is not uniquely defined - it depends on the choice of $a_{max}$. Such non-uniqueness seems to be inevitable in the theory of metastable states for short-range forces. However, under the conditions (small H and T) where the lifetime of the metastable state is long the dependence of the free energy on $a_{max}$ is very weak and so the non-uniqueness is not likely to lead to any consequences with practical importance.

A similar type of restricted ensemble can be used to describe metastable states for the lattice gas and for binary alloys, using the standard isomorphism between these systems and the Ising model.

## 4. THE THEORY OF NUCLEATION

Even if we can find a good restricted ensemble to tell us the equilibrium properties of the metastable state, it does not tell us the whole story, because metastability is a dynamic phenomenon. In particular, if we want to calculate the rate at which nuclei of the new phase will form we must look beyond the static description provided by the restricted ensemble. In this section we shall show how this nucleation rate may be estimated by considering the way clusters can change with time. The principles of the method are due to

Becker and Döring (1935) and it has been elaborated more recently by Binder and his collaborators (see Binder, 1977, for references).

For simplicity, and in order to make it as easy as possible to define the various quantities precisely, I shall describe the theory for a lattice model with nearest-neighbour interactions. This can be the Ising model of a ferromagnet, or an alloy model in which positive spins are treated as one kind of atom and negative spins as another. As a model of a real alloy this last has serious shortcomings, not least that the critical concentration has to be 50%; its usefulness is rather that it provides a mathematically convenient way of testing various theoretical predictions against computer simulations.

For the Ising model we use the Glauber dynamics already described in the previous section; for the alloy model we use the dynamics introduced by Kawasaki, in which a nearest neighbour pair of sites is selected at random and then the atoms on those sites may be interchanged, with a probability depending on the energies of the initial and final states in such a way that detailed balancing holds. This is a simplified model of interdiffusion in an alloy: real interdiffusion can also take place by other processes, such as the motion of vacancies and interstitial atoms.

In the alloy model, the energy of a configuration is taken to be

$$\sum_{(ij)} \left\{ \tfrac{1}{4}(1+s_i)\ (1+s_j)\ V_{AA} + \tfrac{1}{2}\ (1-s_i)\ (1-s_j)\ V_{AB} + \tfrac{1}{4}(1-s_i)(1-s_j)V_{BB} \right\}$$

$$= \text{const} + E$$

where

$$E \ = -\ V \sum_{(ij)} n_i n_j \tag{4.1}$$

with

$$V \ = V_{AA} - 2V_{AB} + V_{BB}$$

and

$$n_i = \tfrac{1}{2}\ (1-s_i) = \begin{cases} 1 \text{ if site i is occupied by a B atom} \\ 0 \text{ if not} \end{cases}$$

The 'constant' depends on $\sum n_i$, the total number of sites occupied by B atoms, but since this quantity is conserved by the Kawasaki dynamics we are justified in treating it as a constant. The formula (4.1) can be interpreted as the energy of a lattice gas in which the sites with $n_i = 1$ are occupied and those with $n_i = 0$ are empty, the interaction between neighbouring particles of the lattice gas being V.

The essence of the Becker-Döring method is to treat the system as a gas of clusters and to apply kinetic theory to them. The gas is a mixture of clusters of different sizes; we denote the concentration of clusters of size $\ell$ (that is the number of clusters per site summed over all possible shapes and perimeters for this size) by $c_\ell$.

It is not hard to calculate the equilibrium values of these concentrations for small $\ell$, to lowest order in the density of the lattice gas (that is, the number of B atoms per lattice site) or its fugacity z. The first few are given, for the simple cubic lattice, by

$$c_1 \simeq z$$

$$c_2 \simeq 3yz^2 \quad \text{where } y = e^{V/kT}$$

$$c_3 \simeq 15y^2z^3$$

$$c_4 \simeq (83y^3+3y^4)z^4$$

and, in general,

$$c_\ell \simeq Q_\ell \; z^\ell \tag{4.2}$$

where $Q_\ell$ is defined by

$$Q = \lim_{L\to\infty} \sum_K e^{-(\text{energy of cluster K})/kT}$$

where the sum goes over all $\ell$-particle clusters K on an L-site lattice. This 'cluster partition function' is a polynomial in y. Sykes (1976) has calculated $Q_\ell$ for the simple cubic lattice as far as $\ell = 10$ (and for the plane square lattice as far as $\ell = 20$). For example he finds

$$Q_5 = 486\, y^4 + 48y^5$$

$$Q_6 = 2967\, y^5 + 496\, y^6 + 18\, y^7$$

in the simple cubic lattice.

For larger values of $\ell$, nothing is known for certain, but we can interpret log $Q_\ell$ as - kT times a free energy. This free energy may be expected to include a 'bulk' term proportional to $\ell$ and a surface term proportional to $\ell^{2/3}$ (if a typical large cluster is approximately spherical). Frenkel (1946) suggested that the surface term might instead be proportional to $(\ell-2)^{2/3}$; in that case we should have approximately

$$\frac{Q_\ell}{Q_{\ell+1}} \simeq \exp\left(\frac{d}{d\ell}\log Q_\ell\right)$$

$$\simeq \text{const.} + \text{const.}\,(\ell-2)^{-1/3}$$

For $T = .59T_c$, the temperature at which the computer simulations I have been concerned with were done, it happens that

$$\frac{Q_\ell}{Q_{\ell+1}} \simeq .01053\left(1+\frac{2\cdot 415}{(\ell-2)^{1/3}}\right) \quad (3 \le \ell \le 9)$$

to 3-figure accuracy except when $\ell = 3$. This formula provides a method of extrapolating $Q_\ell$ to values of $\ell$ greater than 10.

The formula (4.2) has not been directly tested by computer simulations because at the densities where it holds the computations proceed too slowly. Higher-density corrections to (4.2) are considered by Lebowitz and Penrose (1977) and found to be in satisfactory agreement with the simulation results, at least for those values of $\ell$ where $Q_\ell$ is known.

The kinetic part of the Becker-Döring theory consists in the assumption that the only types of process altering the sizes of clusters are (a) collision between an $\ell$-cluster and a 1-cluster, and (b) evaporation of a 1-cluster from an $(\ell+1)$-cluster. It is further assumed that the rate of creation of $(\ell+1)$-clusters by process (a) is proportional to $c_1 c_\ell$, and that the rate of creation of $\ell$-clusters by process (b) is proportional to $c_\ell$. Therefore the net rate of conversion of $\ell$-clusters to $(\ell+1)$-clusters is (to lowest order in the density)

$$J_\ell = a_\ell c_\ell c_1 - b_{\ell+1} c_{\ell+1} \tag{4.3}$$

per site. For $\ell \geqslant 2$, the net rate of creation of clusters of size $\ell$ is therefore given by

$$\frac{dc_\ell}{dt} = J_{\ell-1} - J_\ell \tag{4.4}$$

This formula does not hold for $\ell = 1$ since clusters of size 1 have a special role; the simplest way to determine $c_1$ is from conservation of particle number:

$$c_1 = \rho - \sum_{2}^{\infty} \ell c_\ell \tag{4.5}$$

where $\rho$, a constant, is the total number of occupied sites (that is, the number of B-particles) per site.

Before we can use the Becker-Döring equations (4.3) and (4.4), together with the condition (4.5), to calculate such things as nucleation rates, we need to know the coefficients $a_\ell$ and $b_{\ell+1}$. Some information is given by the fact that the equilibrium distribution (4.2) should satisfy the equations. At equilibrium $J_\ell = 0$, by detailed balancing, and so (4.3) with (4.2) gives

$$a_\ell Q_\ell = b_{\ell+1} Q_{\ell+1} \, . \tag{4.6}$$

To the extent that the Q's are known, this tells us $b_{\ell+1}$ in terms of $a_\ell$.

The coefficients $a_\ell$ can be estimated by considering the kinetics of the interaction between an $\ell$-cluster and a monomer (a 1-cluster). For large values of $\ell$ (the ones most important in nucleation) the factor controlling the rate of interaction appears to be diffusion. This implies that the steady-state density of 1-body clusters near a given $\ell$-body cluster is not uniform but is given by the diffusion equation as

$$c_1 - \frac{A}{r}$$

where A is a constant. The average number of monomers reaching the $\ell$-body cluster per unit time is therefore

$$4\pi DA$$

where D is the diffusion constant for diffusion of monomers, and so we have

$$J_\ell = 4\pi DAc_\ell \tag{4.7}$$

To determine A we assume that $J_\ell$ satisfies a formula of type (4.3) with $c_1$ replaced by the density of monomers near the surface of the $\ell$-body cluster, which is $c_1 - A/R_\ell$ where $R_\ell$ is the radius of that cluster. It follows, by combining (4.3)with (4.7), that

$$J_\ell = 4\pi DAc_\ell = \bar{a}_\ell (c_1 - A/R_\ell) c_\ell - \bar{b}_{\ell+1} c_{\ell+1} \tag{4.8}$$

where $\bar{a}_\ell$ and $\bar{b}_\ell$ are the coefficients in (4.3) when the value of $c_1$ at the surface of the $\ell$-body cluster is used in place of its value averaged over the whole system. Eliminating A from (4.8) we obtain

$$J_\ell = \frac{\bar{a}_\ell c_\ell c_1 - \bar{b}_{\ell+1} c_{\ell+1}}{1 + \bar{a}_\ell / 4\pi R_\ell D}$$

For large $\ell$, we expect $\bar{a}_\ell$ to increase as the surface area of a cluster, that is as $\ell^{2/3}$, whereas $R_\ell$ will only increase as $\ell^{1/3}$; consequently, we expect to find $\bar{a}_\ell / 4\pi R_\ell D \gg 1$ and the formula to reduce to

$$J_\ell = 4\pi R_\ell D(c_\ell c_1 - (\bar{b}_{\ell+1}/\bar{a}_\ell) c_{\ell+1})$$

That is, the effective value of $a_\ell$ in equation (4.3) is proportional to $R_\ell$ and therefore to $\ell^{1/3}$:

$$a_\ell = \text{const.}\ \ell^{1/3} \tag{4.9}$$

It remains to work out an expression for the rate of nucleation. Becker and Döring do this by considering the solution of eqns. (4.3) and (4.4) in which $J_\ell$ is independent of $\ell$. This is (Becker and Döring 1935)

$$J_\ell = J = 1/\sum_1^\infty R_n \tag{4.10}$$

$$c_\ell = \frac{\sum_\ell^\infty R_n}{\sum_1^\infty R_n} Q_\ell c_1^\ell \tag{4.11}$$

where

$$R_n = \frac{1}{a_n Q_n c_1^{n+1}} = \frac{1}{b_{n+1} Q_{n+1} c_1^{n+1}}$$

There are some mathematical objections to this procedure, for example the series $\sum \ell c_\ell$ for the total number of particle diverges if (4.11) is substituted in it (because, by (4.3), $c_\ell > J/a_\ell c_1 =$ $=$ const. $\ell^{-1/3}$ if all the $J_\ell$ are equal). However, the result can probably be justified, with some suitable interpretation, under the conditions where the metastable state has a long lifetime.

There has been no very careful check of the Becker-Döring formula (4.10) for the nucleation rate against computer simulations but Kalos et al (1978) find order-of-magnitude agreement for a related estimate of the lifetime of a metastable state.

For a full discussion of the Becker-Döring equations as applied to the liquid-gas system, including comparison with experiment, see Abraham (1974).

## 5. GROWTH IN THE LATTICE MODEL

The Becker-Döring model provides a method of estimating the rate at which nuclei are produced. In order to relate it to experiment we would like to know what happens to these nuclei afterwards: how long does it take them to grow large enough to be observed, like the droplets you see when the steam issuing from a kettle cools? The growth of these nuclei will be controlled by transport processes, in accordance with the general principles outlined in the Introduction. In the lattice model we are considering, the temperature is held constant throughout, which is equivalent to making the heat conductivity infinite, and so we need not consider energy transport. Only the transport of matter need be considered, and the relevant transport mechanism in this model is diffusion.

In the preceding section we obtained equations which can be written

$$\frac{dc_\ell}{dt} = J_{\ell-1} - J_\ell \tag{5.1}$$

$$J_\ell = 4\pi D R_\ell (c_1 c_\ell - z_\ell c_{\ell+1}) \tag{5.2}$$

where $z_\ell = b_{\ell+1}/a_\ell$

$$= Q_\ell / Q_{\ell+1} \tag{5.3}$$

Eq. (5.2) can be rewritten as

$$J_\ell = v_\ell c_\ell + 4\pi D R_\ell z_\ell (c_\ell - c_{\ell+1}) \tag{5.4}$$

where

$$v_\ell = 4\pi DR_\ell (c_1 - z_\ell) \tag{5.5}$$

The first term in (5.4) represents the avcrage rate at which the clusters of size $\ell$ are growing; the second term describes the effect of fluctuations about this average which produce an additional contribution to $J_\ell$ (and therefore to $dc_\ell/dt$) which can be thought of as a diffusion along the 'gradient' of c in $\ell$-space. For small $\ell$ this 'diffusion' is very important, indeed it is the mechanism that makes nucleation possible; but for large $\ell$ the relative change in $\ell$ when we go from $\ell$ to $\ell+1$, and therefore in c when we go from $c_\ell$ to $c_{\ell+1}$, is small. As a first approximation we may neglect it and approximate (5.4) by

$$J_\ell = v_\ell \; c_\ell \tag{5.6}$$

This equation says that, in the approximation we are now using, the rate at which a cluster of size $\ell$ grows is $v_\ell$ where $v_\ell$ is given by (5.5). Eq. (5.5) says that the rate at which monomers are arriving at the surface of a large cluster of size $\ell$ is proportional to the difference between $c_1$, which is the concentration of monomers a long way from the large cluster, and $z_\ell$, which is the concentration of monomers at the surface of this cluster. This latter is the concentration of monomers that can be in equilibrium with a cluster of size $\ell$. According to the formula for the vapour pressure over a curved surface this will, at the very low monomer densities we are considering here, be simply related to $z_s$, the concentration of monomers in the saturated vapour. The relationship is

$$z_\ell = z_s + \frac{2\sigma}{R_\ell} \tag{5.7}$$

where $\sigma$ is the surface tension; according to a formula in section 4 it has the form

$$z_\ell = z_s(1 + \gamma/(\ell - 2)^{1/3})$$

where $\gamma$ is a constant. For large $\ell$, this last formula is approximately

$$z_\ell = z_s(1+ \gamma/\ell^{1/3})$$

Since we take $R_\ell$ proportional to $\ell^{1/3}$, the two ways of estimating $z_\ell$ are consistent.

Equation (5.5) can be brought to a more useful form if we define $\ell^*$, the size of a cluster that is in equilibrium with monomers of concentration $c_1$, by

$$c_1 = z_s \; (1+ \gamma/\ell^{*\,1/3}) \tag{5.8}$$

Then, if we also assume that $R_\ell$ is proportional to $\ell^{1/3}$, eq. (5.5) reduces to

$$v_\ell = A\left[(\ell/\ell^*)^{1/3} - 1\right] \qquad (5.9)$$

where A is a constant. This equation indicates, as we might expect, that clusters larger than the critical size $\ell^*$ will grow, and clusters smaller than $\ell^*$ will shrink.

According to (5.9) the rate of growth of a cluster of size $\ell$ is given, on average, by

$$\frac{d\ell}{dt} = A\left[\left(\frac{\ell}{\ell^*}\right)^{1/3} - 1\right] \qquad (5.10)$$

Just after nucleation, $\ell$ is greater than $\ell^*$ and so the cluster grows. Later on, as the growing large clusters take monomers out of the 'vapour' phase, $\ell^*$ will also increase (see eq. (5.8)) and the right-hand side of (5.10) will fall again to zero or below. (It must do this, since the cluster cannot grow indefinitely).

For example, if all the growing clusters were the same size, and if there were $\nu$ of them per unit volume, and if the total density were $\rho$, we should have approximately (neglecting all clusters of sizes other than 1 and $\ell$)

$$\rho = c_1 + \nu\ell \qquad (5.11)$$

which combined with (5.8) and (5.10) gives

$$\frac{d\ell}{dt} = \frac{A}{\gamma}\left\{\left(\frac{\rho}{z_s} - 1\right)\ell^{1/3} - \frac{\nu\ell^{4/3}}{z_s} - \gamma\right\} \qquad (5.12)$$

The right-hand side of this equation is plotted schematically in the diagram below.

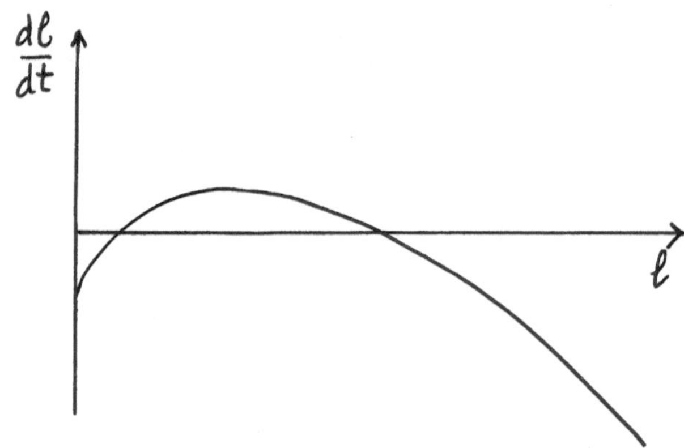

The first intercept on the right gives the size of the droplets at the moment when nucleation starts the growth process; the second gives the ultimate size of the droplets at the end of the growth process. The duration of the growth process depends in a rather complicated way on the parameters in the equation (5.12).

## 6. COARSENING

Our simple model of droplet growth given by eq. (5.12) is unrealistic because it assumes that all the droplets are the same size. In practice they are not. The ones that are larger than average will grow faster than average; those that are smaller than average will grow more slowly, or may even shrink. This process of coarsening follows, or even accompanies, the growth process outlined in the preceding section. To describe it properly we now go back to the more detailed description using clusters of many different sizes, which we first used in section 4.

Since we are now considering very large values of $\ell$, we can simplify the equations we used in section 4 in two ways, first by using the approximation $J_\ell = v_\ell c_\ell$ (eq. (5.6)) and secondly by treating $\ell$ as a continuous variable and replacing $J_{\ell-1} - J_\ell$ by a partial derivative. The resulting equations are

$$\frac{\partial c}{\partial t} \approx \frac{\partial}{\partial \ell}(-vc) \qquad \text{(for large } \ell\text{)} \tag{6.1}$$

$$v = A\left[\left(\frac{\ell}{\ell^*}\right)^{1/3} - 1\right] \tag{6.2}$$

with $\ell^*$ determined from eq. (5.8):

$$c_1 = z_s\,(1 + \gamma/\ell^{*\,1/3}) \tag{6.3}$$

and $c_1$ (on the simplifying assumption, good at very low densities, that all the small clusters are monomers)from the condition that the total number of particles is conserved:

$$c_1 = \rho - \int_0^\infty \ell c \, d\ell \tag{6.4}$$

The equations can be simplified further by making use of the suggestion made by Lifshitz and Slyozov (1961, see also Wagner, 1961) that in the later stages of the coarsening process

$$\frac{d\ell^*}{dt} = \text{constant, say } K \tag{6.5}$$

This assumption is confirmed as a good approximation both by computer simulation at density $\rho = 0.075$ (Penrose et al, to appear) and by a numerical integration of the Becker-Döring equations (Buhagiar and Penrose, unpublished). Making this assumption, it is possible to treat $\ell^*$ instead of t as the time variable and solve the partial differential equation (6.1) by the method of characteristics. The solution has the form

$$c = \partial g/\partial \ell \tag{6.6}$$

where

$$g(\ell, \ell^*) = \psi\left[\ln \ell^* + \phi\left(\frac{\ell}{\ell^*}\right)\right] \tag{6.7}$$

with $\psi$ arbitrary and $\phi$ defined by

$$\phi(x) = \int_0^x \frac{dx}{x - C\, x^{1/3} + C} \tag{6.8}$$

where $C = A/K$.

Lifshitz and Slyozov showed how the function $\psi$ can be determined, to lowest order in $c_1$, from the requirement that $\int_0^\infty \ell c \, d\ell$ should be (to this order) equal to the constant $\rho$.
This requirement gives

$$\psi(x) = \text{const. } e^{-x} \tag{6.9}$$

so that

$$g(\ell, \ell^*) = \text{const. } \frac{e^{-\phi(\ell/\ell^*)}}{\ell^*}$$

They then argued, from the fact that the integral in (6.4) must converge, that C should have the value which gives the integrand in (6.8) a double pole. This value is $C = 27/4 = 6.75$, and the double pole is then at $x = 27/8 = 3.375$ and it has the consequence that

$$g(\ell, \ell^*) = 0 \text{ for } \ell > 3.375\ell^* . \tag{6.10}$$

Our computer simulations were done at a higher density than the one for which the formula (6.9) should hold, but they confirmed an analogous formula, obtained by generalizing the Lifshitz-Slyozov theory to somewhat higher densities. On the other hand they did not confirm the prediction $C = 6.75$; our value was, to a good approximation,

C=4. Apart from this discrepancy, which remains to be explained, the ideas of Lifshitz and Slyozov give a remarkably good picture of the coarsening process in these simulations.

Experimentally, the Lifshitz-Slyozov predictions can be tested by measuring the average particle size in a suitable alloy that has been quenched from a high temperature to a low one. The mean particle diameter $\lambda$ can be estimated using x-ray diffraction or by looking at the metal with an electron microscope. According to the Lifshitz-Slyozov rule (6.5), $\lambda^3$ should vary linearly with time, and this has been observed, for example in a Cu-Ti alloy (Datta & Soffa, 1976) and in a Ni-Al alloy (Ardell et al.,1969). Ardell et al also checked that the excess concentration of minority atoms (Al, in their case) in the 'vapour' phase varied as $t^{1/3}$, as predicted by the theory (see eq. (5.8), with $\ell^*$ assumed proportional to t), and they gave histograms of the particle sizes. These histograms differ somewhat from the Lifshitz-Slyozov prediction. It may be that the agreement could be improved by taking C = 4 as given by the simulations, in place of C = 6.75 as predicted by Lifshitz and Slyozov.

## REFERENCES

ABRAHAM, F.F.,(1974) Homogeneous Nucleation Theory (Academic Press, NY)

ARDELL, A.J., NUTALL and NICHOLSON, R.B.,(1969) The mechanics of phase transitions in solids (institute of metals monograph series, London) 33, 111

BECKER, R., and DÖRING (1935) Ann. der Phys. 24, 719

BINDER, K. (1977) Phys. Rev. B15, 4425.

CAPOCACCIA, D., CASSANDRO, M., and OLIVIERI, E. (1974) Commun. Math. Phys., 39, 185

CASSANDRO, M., and OLIVIERI, E. (1977) J. Stat. Phys. 17, 229.

DATTA, A., and SOFFA, W.A., (1976) Acta Metall. 24, 987.

FISHER, M.E.,(1967) Physics 3, 255.

FRENKEL, J., (1946) Kinetic Theory of Liquids (Dover, NY), p. 381

GLAUBER, R., (1963) J. Math. Phys. 4, 294.

KAC, M., UHLENBECK, G.E., and HEMMER, P., J. Math. Phys. 4, 216

KALOS, M., LEBOWITZ, J.L., PENROSE, O., and SUR, A., (1978) J. Stat. Phys. 18, 39.

van KAMPEN, N.G. (1964) Phys. Rev. 135A, 362.

KAWASAKI, K., (1966) Phys. Rev. 145, 224

LANFORD, O. and RUELLE, D., (1969) Comm. Math. Phys., 13, 194.

LANGER, J.S. (1967) Ann. Phys. (N.Y.) 41, 108.

LEBOWITZ, J.,and PENROSE, O., (1966) J. Stat. Phys. 7, 98.

LEBOWITZ, J.,and PENROSE, O., (1977) J. Stat. Phys. 16, 321.

LIFSHITZ, I.M.,and SLYOZOV, V.V., (1961) J. Phys. Chem. Solids, 19, 35.

MAXWELL, J.C., (1874 or 5) Scientific papers, p. 425,(Dover reprint).

NEWMAN, C., and SHULMAN (1976) preprint.

PENROSE, O., and LEBOWITZ, J.L., (1971) J. Stat. Phys. 3, 211.

SYKES, M., (1976) private communication.

TEMPERLEY, H.N.V., and TREVENA, D.H., (1977) Proc. Roy. Soc. A357, 395.

TEMPERLEY, H.N.V., (1947) Proc. Phys. Soc., 59, 199.

van der WAALS, J.D., (1873) Ph.D. Thesis, University of Leiden.

WAGNER (1961), Z. Electrochem. 65, 581.

# STOCHASTIC BEHAVIOR OF SIMPLE DYNAMICAL SYSTEMS

Y. Pomeau
Commissariat a L Energie Atomique
Centre d Etudes Nucléaires de Saclay
91190 Gif-sur-Yvette, France

# STOCHASTIC BEHAVIOR OF SIMPLE DYNAMICAL SYSTEMS

Y. Pomeau

Commissariat à L'Energie Atomique
Centre d'Etudes Nucléaires de Saclay
91190 Gif-sur-Yvette, France

## 1. *INTRODUCTION*

Since the early days of statistical mechanics, there has been a lot of work to understand the connection between the apparent irreversibility of the macroscopic laws of motion (such as, say, the Navier-Stokes equations) and the deterministic and reversible character of the classical motion of systems of particles According to many physicists, including Landau, the irreversibility is basically due to the presence of infinitely many degrees of freedom (or of infinitely many particles) : the initial data involve the choice of countably many real numbers, and this should explain the practical unpredictability of the motion. A good example of such a situation is given by the infinite perfect gas : if one concentrates one's attention on what happens in a finite region of space, say $\Omega$ , during the history of the system, one sees particles which started at $t=0$ farther and farther from $\Omega$ as time goes on. And the fluctuations in $\Omega$ arise from the randomness in the initial data of the infinite system.

The question of the irreversibility of infinitely many *interacting* particles is not so clear. Of course molecular dynamics (that is integration on computers of the equations of the motion) is able to recover the thermodynamic properties of systems of a few thousand hard spheres, hard discs, Lennard-Jones particles, and so on.... This proves, at least implicitly, the validity of the ergodic hypothesis[(1)]. However, numerical simulations[(2)] of anharmonic lattices seem to show the existence of a critical energy per particle : at lower energies the system is not ergodic; on the contrary, it becomes ergodic at energies larger than this critical energy. However, numerical simulations of a 1d - plasma (i.e., particles interacting through a potential law $e_i e_j |x_i - x_j|$ , $x_i$, $x_j \in \mathbb{R}$) have shown[(3)] that for other systems the ergodicity sets in when the number of particles increases beyond, say, 5 or 6. At the present time there is no clear understanding of the way in which the ergodicity (and the stronger mixing and Bernouilli properties) is connected with the existence of "many" or "infinitely many" degrees of freedom. There does not seem to be any well-defined way of attacking this problem, except by computer simulations.

So far I have only mentioned the problem of the irreversibility of classical interacting particles. Another important domain wherein irreversibility comes from reversible equations is the hydrodynamic turbulence. Ruelle and Takens[(4)] have explained the transition from regular to stochastic or turbulent flows as the Reynolds number increases, without requiring the notion of an infinite number of degrees of freedom.

Giving up the idea of an infinite number of degrees of freedom, it is possible to construct deterministic dynamical systems that have, in a certain sense, a "random behavior". This is not too surprising, since a real number (which may be regarded as one of the initial data of a dynamical system with a few degrees of freedom) has an infinite number of "degrees of freedom" : its binary expansion (for instance) involves the free choice of countably many digits. If, as time goes on, the dynamics depends on digits ever further from the first few, it is natural to say that the evolution is unpredictible, as it depends for large times on an uncontrollable parts of the initial data.

The mathematical techniques that have been invented for studying this sort of problem are often rather subtle, using for instance nontrivial results in algebraic number theory. It must be also recognized that one is far from understanding the dynamics of many simple systems (the Hénon mapping [5] is probably the best example of such a simple system which is very difficult to understand). From that point of view of classical statistical mechanics, the major recent achievement in this field is Sinai's proof of the ergodicity of the hard-sphere gas in a finite box[6]. For instance, no such proof exists for smooth potentials, even purely repulsive. On the other hand, despite convincing experimental evidence[7], it is not yet proved that beyond a finite Reynolds number a hydrodynamic flow becomes "turbulent".

Before going into more details, I shall explain first what I mean by "dynamical system". One wants to study the behavior "at large times" of solutions of ordinary differential equations. It is convenient to consider the "Poincaré map" generated by the differential equation; the differential equation defines trajectories in a d-dimensional space and we may assume that these trajectories are not too far from a closed trajectory. Hence, at each "turn" each trajectory cuts a given (d-1)-dimensional plane (or surface) orthogonal to the closed trajectory (see Fig. 1).

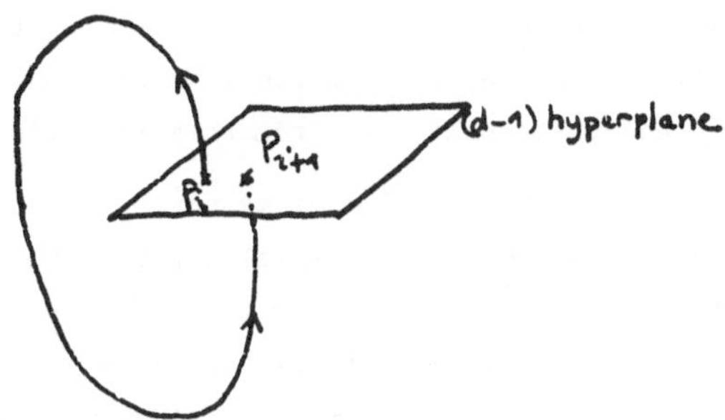


This defines a mapping (the Poincaré map) of this (d-1) hyperplane into itself : starting from point $P_i$ in this plane, one gets after one turn to the point $P_{i+1}$ in this plane, and so on.... The mapping $P_i \rightarrow P_{i+1}$ is one-to-one; it is possible to trace back trajectories, as the velocity is uniquely defined at each point. It is obvious, however, that this Poincaré map might not exist for a given differential system : for instance $P_i$ has no image in the (d-1) hyperplane if the trajectory running through $P_i$ reaches a fixed point lying outside of the cut plane.

We are thus led to consider the iteration of maps of spaces into themselves. We shall take first a rather general point of view to define what is called an "abstract dynamical system" and then come to more realistic systems.

## 2. MEASURE SPACES - DYNAMICAL SYSTEMS

### A. *Measure spaces.*

In this subsection I give definitions only and consider some examples. More details on measure theory can be found in many books; the classical one in this field is Halmos' "Measure Theory"[(8)].

To define a measure space one has to introduce first a so called σ-ring (loosely speaking, the prefix σ refers to the possibility of defining a property by countably many operations). A σ-ring $\underset{\sim}{S}$ is a collection of parts of a nonvoid set X such that

(i) $E \in \underset{\sim}{S}$, $F \in \underset{\sim}{S} \Rightarrow E - F \in \underset{\sim}{S}$ $(E - F = E \cap F'$, $F'$ = complement of F). The void subset $\emptyset = E - E'$ belongs to $\underset{\sim}{S}$,

(ii) Let $\{E_i, i \in \mathbb{N}\}$ be a denumerable collection of subsets of X; all of them belong to S; then $\bigcup_{i=1}^{\infty} E_i \in \underset{\sim}{S}$ (in other terms, $\underset{\sim}{S}$ is closed w.r. to countable unions). As $\bigcap_{i=1}^{\infty} E_i = E - \bigcup_{i=1}^{\infty} (E - E_i)$, S is closed under countable intersections too.

A measure μ is a positive function defined on the σ-ring S which is countably additive : Let $\{E_i \,;\, i \in \mathbb{N}, E_i \in \underset{\sim}{S}, E_i \cap E_j = E_i \delta_{ij}\}$ be a set of disjoint subsets in S; thus,

$$\mu\left(\bigcup_{i=1}^{\infty} E_i\right) = \sum_{i=1}^{\infty} \mu(E_i)$$

This implies, of course, $\mu(\emptyset) = 0$.

This idea of measure extends the usual notion of length, surface, volume,... In what follows we shall consider normalized finite measures only, such that the entire space X belongs to $\underset{\sim}{S}$ and that $\mu(X) = 1$. From this it follows that $\mu(E) \leq 1$ $\forall E \in \underset{\sim}{S}$.

Let us give two well-known examples of measure spaces :

1. On X = [0,1], one defines as the Borel algebra the smallest σ-ring including all the intervals [a,b] $0 \leq a \leq b \leq 1$; the measure μ of this interval is ([a,b]) = (b-a) and for any set in $\underset{\sim}{S}$ it is defined by means of the countable additivity. This is the well-known Lebesgue measure. (However, as explained by Halmos[(8)], there is a difference between the Borel and the Lebesgue measure. I shall not consider here this rather difficult point). Any countable union of points in [0,1] (such as the rationals) has zero measure, being a countable sum of zero-measure sets. In this sense, the well-known triadic Cantor set is of zero measure too, although it is not countable. To find non measurable sets on [0,1] , one has to use the axiom of

choice[9], in order to construct sets by non-countably-many operations, otherwise the set would belong to $\underset{\sim}{S}$, that is, would be closed under countably many operations.

2. Consider now the set of all doubly infinite sequences of finitely many symbols, say p symbols $(a_1,\ldots,a_p)$ :

$$X = \{x = \{\alpha_i \ ; \ i \in \mathbb{Z}, \ \alpha_i \in \{a_1,\ldots,a_p\}\}\} \quad ;$$

the $\sigma$-ring $\underset{\sim}{S}$ is the smallest $\sigma$-ring including all the cylindric subsets : an elementary cylinder is made of all the elements of X such that a given $\alpha_i$ has a well defined value. This sort of system is important for dynamics: if one plays dice (or roulette...) an infinite number of times, one gets such a doubly infinite sequence of outcomes ; in the case of usual dice the measure of an elementary cylinder is just the probability of the corresponding outcome (1/6 in principle !) at the $i^{th}$ game. More generally, if one specifies (q) outcomes, the corresponding subset has the measure $(1/6)^q$.

B. *Dynamical systems.*

To define an "abstract" dynamical system I consider a measure space, say X, a $\sigma$-ring $\underset{\sim}{E}$ of subsets of X, and a measure $\mu$ on $\underset{\sim}{E}$ s.t. $X \in \underset{\sim}{E}$ and $\mu(X) = 1$ (this assumption can, of course, be deleted but we keep it for the moment). Furthermore we consider on X what is called a measure-preserving transform $\phi$ ; by definition,

(i) if $A \in \underset{\sim}{E}$; then $\phi^{-1}(A) \in \underset{\sim}{E}$

(ii) $\mu(A) = \mu(\phi^{-1}(A))$.

We shall say that, $\phi$ given, $\mu$ is an "invariant measure".

At this step, we have modified (and generalized) the original problem of studying ordinary differential equations (or flows), since we do not require any continuity property for $\phi$ , although one expects that the images of two close starting points are close to each other, if the flow is sufficiently smooth. However, this is *not* always the case : the Poincaré map may have discontinuities, for instance, whenever some trajectories run to a fixed point, outside the cut-plane.

The measure-theoretic properties of dynamical systems are the ones which exhibit best their stochastic properties; moreover they allow one (in some simple cases at least) to predict in some sense the behavior of "almost all" the trajectories. When I will study the endomorphism of the real line, I hope to illustrate what we can learn from the topological approach to dynamical problems[10]. An important review on topological dynamics has been written by Smale[11].

Let me give a few dynamical systems, which are the classical examples studied in many text-books.

1. The rotation on a circle (with a unit circumference, to avoid awkward $\pi$ factors!).

The measure space X is the unit interval [0,1] with the Lebesgue measure; this measure is invariant under the transform

$$T_\alpha \ : \ x \to (x+\alpha)$$

where, by definition $(x) = x - [x]$ , $[x]$ being the integer part of x, and $\alpha$ a given real number.

If $\alpha$ is irrational, the only invariant measure s.t. $\mu([0,1]) = 1$ is the Lebesgue measure. If $\alpha$ is rational, say $\alpha = p/q$, then $T_\alpha^q(x) = (x + q\,\frac{p}{q}) = (x+p) = (x)$, so that each point is periodic of period q (that is, the $q^{th}$ transform of x is x, and no $r^{th}$ transform $1 < r < q$, is equal to x). Thus any sum like

$$\frac{1}{q} \sum_{i=1}^{q} \delta(x - \frac{ip}{q}) \quad (\delta = \text{Dirac distribution})$$

is an invariant measure; so is any linear combination with positive coefficients of total mass 1 of such functions with various x. This elementary example shows that for a given transformation many invariant measures may exist.

2. The $\beta$-transform[12] is defined on [0,1] by $x \to (\beta x)$, $\beta > 1$. It is not invertible (contrary to what one would expect from the "philosophy" of the Poincaré map); x and $x + \frac{1}{\beta}$ have the same transform if $0 < x < 1 - \frac{1}{\beta}$ . However, there is a "natural" extension of this $\beta$-transform to make it invertible.

An invariant measure for this $\beta$ transform has been found by Parry[13]; it is absolutely continuous w.r. to the Lebesgue measure : that is, any set of nonzero (positive) measure (Lebesgue measure) is of nonzero measure w.r. to this invariant measure, and conversely.

A very simple case of a $\beta$-transform is the dyadic transform : it corresponds to $\beta = 2$, and is written at once in binary notation : let $x = .i_1 i_2 \ldots . i_{k-1} i_k i_{k+1} \ldots$ be the binary expansion of a number ($i_k \in \{0,1\}$); then $(2x) = .i_2 i_3 \ldots . i_{k-2} i_{k-1} i_k \ldots$ is obtained by shifting the row "x" one step on the left and erasing the first digit $i_1$. The natural extension of the dyadic transform to an invertible transform is the *shift*. It acts on the doubly infinite sequences of two symbols :

$$\phi(\ldots . i_{-1}, i_0, i_1, i_2, \ldots) = (\ldots . j_{-1}, j_0, j_1, j_2; \ldots)$$

where $j_0 = i_1$, $j_1 = i_2, \ldots$

They are many invariant measures for the dyadic transform : the most obvious one is the Lebesgue measure, for which any elementary cylinder has the mass 1/2.

Other invariant measures (which are *not* absolutely continuous w.r. to the Lebesgue measure) exist e.g. those having as support all numbers whose binary expansion does not involve a given sequence – say, 00. This is allowed : the sequence is absent in any iterate if it does not appear in the starting sequence. This support is of zero Lebesgue measure. It is generated in a way very similar to the triadic

Cantor set

(i) First step : exclude any number in [.00,.01] (in binary notation),

(ii) second step : exclude [.0,.0001] (already excluded), [.100,.101] ,

(iii) third step : exclude $I_1$ = [.0,.00001], $I_2$ = [.0100,.0101] , $I_3$ = [.1000,.1001] and $I_4$ = [.1100,.1101] , $I_1$ and $I_3$ being already excluded.

Let us assume that before the $n^{th}$ step any number in $I_n$ is excluded; then at the $n^{th}$ step any number $x/2$ and $(1+x)/2$, $x \in I_n$ is excluded, except, of course, if it is already in $I_n$.

Proceeding further and further, one excludes in this way almost all the numbers in [0,1]; this is a simple consequence of the fact that the successive digits of almost all numbers are at random, so that in any very long sequence of digits one may find almost surely a given finite sequence of digits.

Again there are many invariant measures on this peculiar invariant subset. One of these measures has an interesting connection with the golden number, and it is another way of looking at a particular two-state Markov process : let A and B be these two states. If at the $n^{th}$ step the system is in state A (corresponding to the digit 0), then at the $(n+1)^{th}$ step it must be in state B, although if it starts from B, it jumps equiprobably to A or B.

To show the connection between measure and probability in this simple example, let us compute the probability of being in A or B at any given time. Let us consider all possible paths of the system of a given length, the allowed paths of length 4 are

ABAB, ABBA, ABBB, BABA, BABB, BBAB, BBBA and BBBB

(or, in binary notation .0101, .0110, ...., these numbers being written in increasing order). The measure is constructed by assuming that these paths have the same probability.

Let $\mathcal{A}_n$ and $\mathcal{B}_n$ be the number of sequences ending (on the right) with A and B, respectively. These quantities satisfy the recursion relations :

$$\mathcal{A}_{n+1} = \mathcal{B}_n$$

$$\mathcal{B}_{n+1} = \mathcal{A}_n + \mathcal{B}_n$$

with $\mathcal{A}_1 = \mathcal{B}_1 = 1$.

This has the solution, where the golden number $\frac{1+\sqrt{5}}{2}$ appears :

$$\mathcal{A}_n = \mathcal{B}_{n-1}$$

and

$$\mathcal{B}_n = \left(\frac{\sqrt{5}+5}{10}\right)\left(\frac{1+\sqrt{5}}{2}\right)^n + \left(\frac{5-\sqrt{5}}{10}\right)\left(\frac{1-\sqrt{5}}{2}\right)^n$$

The probability of being in the state A at the $n^{th}$ step is $\frac{\mathcal{A}_n}{\mathcal{A}_n+\mathcal{B}_n}$. When one considers path of arbitrary length, the probability of a given state is dominated by the very long paths, and the probability of being in A is

$$\lim_{n\to\infty} \frac{\mathcal{A}_n}{\mathcal{A}_n+\mathcal{B}_n} = \frac{3-\sqrt{5}}{2}$$

and in B, $\frac{\sqrt{5}-1}{2}$ .

This gives an invariant measure of the elementary cylinder in the initial problem : if one specifies the $k^{th}$ digit, and does not specify anything else (except, of course, that the sequence .00 nowhere appears), the measures of the two cylinders are $\frac{3-\sqrt{5}}{2}$ and $\frac{\sqrt{5}-1}{2}$, depending if one specifies the digit 0 or 1. But, contrary to the case of the usual Lebesgue measure, for which the successive digits are independently chosen, in the present case the successive digits are correlated. The measure of more-complicated cylindric sets, where one specifies more than one digit in the infinite sequence, is not completely obvious. If two digits only are specified, this measure can be computed rather simply from $\mathcal{A}_n$ and $\mathcal{B}_n$.

3. I shall end this series of examples with the so-called Gauss measure. It is connected with the continued-fraction expansion, which plays a basic role in many problems of dynamics (for instance in the Kolmogoroff - Arnol'd-Moser theorem (14)).

To explain in short what is the continued-fraction expansion (c.f.e.), let consider a number $0 < x < 1$ and the set of integers $X_n$ defined by the recursive operations

$$x_n = T_n(x) = \left(\frac{1}{x_{n-1}}\right)$$

$$\left(\frac{1}{x_{n-1}}\right) = \frac{1}{x_{n-1}} - \left[\frac{1}{x_{n-1}}\right]$$

$$x_1 = x \quad \text{and} \quad X_n = \left[\frac{1}{x_n}\right]$$

(we do not consider the case where $\frac{1}{x_{n-1}}$ is an integer, so that the continued fraction stops).

When no $x_n$ is zero, which is always the case when $x$ is irrational, $x$ can be written as

$$x = \cfrac{1}{X_1 + \cfrac{1}{X_2 + \cfrac{1}{X_3 + \dots}}}$$

Any rational has a finite c.f.e. (contrary to the case of binary or n-ary expansions which are only periodic for rationals) and the fractional part of any real solution of a second-degree equation with rational coefficients has a *periodic* c.f.e. The Euler constant has a remarkable c.f.e.(15).

The transform $x \to T(x)$ has infinitely many discontinuities, at the points $1/2, 1/3, 1/4, \ldots, 1/k$, $k \in N_+$. $T(x)$ is defined analytically as follows :

$$\begin{array}{lll} \frac{1}{2} < x < 1 & T(x) & = \frac{1}{x} - 1 \\ \frac{1}{3} < x < \frac{1}{2} & T(x) & = \frac{1}{x} - 2 \\ \vdots & \vdots & \\ (\frac{1}{k+1}) < x < \frac{1}{k} & T(x) & = \frac{1}{x} - k \end{array}$$

Let us look for the invariant measure of this transform. As, on $]0,1[$, the derivative of $T(x)$ is either indefinite or larger than 1, and as $x = 1$ is not an attracting point, a general theorem by Lasota and Yorke(16) tells us that an invariant measure exists for T which is absolutely continuous w.r. to the Lebesgue measure.

This measure has been found long ago by Gauss. From the general definition of an invariant measure,

$$\mu(A) = \mu(T^{-1}(A))$$

for any measurable set A. Restricting oneself to sets A made of a small interval dy around y $(0 < y < 1)$, one gets :

$$\mu(y)\, dy = \sum_{i=1}^{\infty} \mu(x_i)\, dx_i$$

where $x_i$ is the set of all points s.t. $y = (\frac{1}{x_i})$ and $0 < x_i < 1$ :

$$y = \frac{1}{x_i} - i \quad \text{or} \quad x_i = \frac{1}{y+i} .$$

Thus $dx_i = -\frac{1}{(y+i)^2}\, dy$ and $\mu$ must satisfy the functional relation

$$\mu(y) = \sum_{i=0}^{\infty} \frac{1}{(y+i)^2}\, \mu(\frac{1}{y+i}) .$$

Gauss has found its solution :

$$\mu(y) = \frac{C}{y+1} .$$

C is found from the normalization :

$$\int_0^1 dy\, \mu(y) = 1 \quad \text{or} \quad C = \frac{1}{\ln 2}$$

It is striking that $\mu$ is perfectly smooth, although the initial transform is full of discontinuities. The mean value of a digit in the c.f.e. of a number should be given by the integral $\frac{1}{\ln 2} \int_0^1 \frac{dy}{1+y} [\frac{1}{y}]$. This diverges logarithmically at $y = 0$ and makes rather subtle the statistical properties of the digits of the c.f.e.(17).

## 3. ERGODIC SYSTEMS. THE BIRKHOFF AND OSELEDEC THEOREMS

As is well known, and as I shall try to explain , the stochasticity of a dynamical system can be defined in a number of ways. There is a "one-dimensional" hierarchy of properties : each of them (except of course the "end" properties) is just weaker than one property and just stronger than another one. The weakest property of stochasticity is the so-called "minimal property". It is of a topological nature (contrary to all the others) ; a transform $\phi$ is minimal on X if the set $\{\phi^n(x), n \in \mathbb{N}\}$ is dense on X, or, equivalently, given an open set in X there is at least an $n \geq 0$ s.t. $\phi^n(x)$ is in this open set. There are transforms which are only minimal, and do not possess the stronger property of ergodicity(18). (This statement is rather strange, of course, since one might imagine dynamical systems where the metric and topological properties are disconnected; we refer the interested reader to ref. (18) for more details).

The ergodic systems are very important for physics, since the ergodic property allows one to replace the *time average* by the equilibrium average over the Gibbs ensemble. The basic property of ergodic systems that we shall explain in some detail is given by the Birkhoff theorem.

Before explaining the main ideas of its proof, I define first an ergodic system. A dynamical system [remember that this is a "quadruple " $(X,\underset{\sim}{S},\mu,\phi)$ : X = basic set, $\underset{\sim}{S}$ = $\sigma$-ring of subsets of X, $\mu$ = measure on $\underset{\sim}{S}$, $\phi$ = measure-preserving map of X into itself] is ergodic iff $A \in \underset{\sim}{S}$ and $\phi(A) \subseteq A \Rightarrow \mu(A) = 0$ or 1. That is, in the sense of the measure $\mu$ , the only invariant sets are $\emptyset$ and X. The Birkhoff theorem states among other things ) that in an ergodic system a measurable subset B in S is visited by a trajectory "in proportion" to its measure, that is, for almost all x (in the sense of the invariant measure $\mu$ — in short, $\mu$-almost) :

$$\lim_{n \to \infty} \frac{1}{n} \sum_{i=1}^{n} \chi_B(\phi^i(x)) = \mu(B) ,$$

where $\chi_B(.)$ is the characteristic function of B.

The Birkhoff theorem is proved in two steps.

### A.1. *The Hopf maximal ergodic theorem.*

Let us state first the theorem :

Let $f \in L^1(X,S,\mu)$ be an integrable function on X (the construction of such functions is explained in ref. (7)). Let $\phi$ be a map of X into itself which preserves the measurability. Consider the sum

$$\sigma_n(\omega) = \sum_{i=0}^{n-1} f(\phi^i(\omega))$$

(by convention $\phi^o(\omega) = \omega$)

and the set $N = \{\omega : \limsup_n \sigma_n(\omega) > 0\}$.

Thus the maximal ergodic theorem states $\int_N f\, d\mu \geq 0$. (For simplicity we shall write the integral of any function f with the measure $\mu$ as $\int f$ .)

*Proof*

Consider the measurable functions of X

$$M_n(\omega) = \max \{0, \sigma_1(\omega), \ldots, \sigma_n(\omega)\}$$

$$M_n^*(\omega) = \max \{\sigma_1(\omega), \ldots\ldots, \sigma_n(\omega)\}$$

From the definition of $\sigma$,

$$\sigma_k(\phi(\omega)) = \sigma_{k+1}(\omega) - f(\omega) \ ,$$

and one gets

$$M_{n+1}^*(\omega) = \max \Big\{0 + f(\omega), \sigma_1(\phi(\omega)) + f(\omega), \ \ldots, \ \sigma_n(\phi(\omega)) + f(\omega)\Big\}$$

$$= f(\omega) + \max \Big\{0, \sigma_1(\phi(\omega)), \ldots, \sigma_n(\phi(\omega))\Big\} \ .$$

Also

$$f(\omega) = M_{n+1}^*(\omega) - M_n(\omega) \geq M_n^*(\omega) - M_n(\omega) \qquad (*) \ ,$$

since

$$M_{n+1}^*(\omega) = \max \Big\{M_n^*(\omega) \ , \ \sigma_{n+1}(\omega)\Big\} .$$

Consider now the (measurable) set $A_n = \{\omega : M_n(\omega) > 0\}$ .

From the above inequality (*) :

$$\int_{A_n} f \geq \int_{A_n} M_n^*(\omega) - \int_{A_n} M_n(\phi(\omega)) \qquad (**)$$

Consider now separately the two terms on the r.h.s. of (**). For any $\omega \in A_n$, $M_n^* = M_n$ [as $M_n^* = \max \{0, M_n\}$ and $M_n(\omega) > 0$ if $\omega \in A_n$] and

$$\int_{A_n} M_n^*(\omega) = \int_{A_n} M_n(\omega) \geq \int M_n(\omega) \quad ,$$

as $M_n$ is negative or zero outside of $A_n$.

Furthermore $\int_{A_n} M_n(\phi(\omega)) = \int_{\phi^{-1}(A_n)} M_n(\omega)$ is less than $\int_{A_n} M_n(\omega)$ , since $\int_A M_n(\omega)$ is maximal w.r. to A when $A = A_n$.

Recapitulating,

$$\int_{A_n} f \geq \int_{A_n} M_n^*(\omega) - \int_{A_n} M_n(\phi(\omega)) \geq \int_{A_n} M_n(\omega) - \int_{A_n} M_n(\omega) \ .$$

Thus $\int_{A_n} f \geq 0$ and, as $\lim_{n \to \infty} A_n = N$, $\int_N f \geq 0$, which is the final result.

## A.2. *Birkhoff individual ergodic theorem*

Let again f be in $L^1(X,S,\mu)$. The Birkhoff theorem states that a function $f^* \in L^1$ exists s.t.

$$\lim_{n \to \infty} \frac{1}{n} \sum_{i=0}^{n-1} f(\phi^i(\omega)) = f^*(\omega) \qquad \mu\text{-almost everywhere}$$

and $\int f^* = \int f$.

*Proof*

Let $a < b$ be two different numbers and

$$A(a,b) = \left\{\omega : \liminf_n \frac{1}{n} \sigma_n(\omega) < a < b < \limsup_n \frac{1}{n} \sigma_n(\omega)\right\}$$

First we shall prove, by using the previous Hopf theorem, that A(a,b) is of zero measure.

A(a,b) is invariant under $\phi$, since $\sigma_n(\omega)$ and $\sigma_n(\phi(\omega))$ differ by a finite quantity. Applying now the maximal ergodic theorem to $g(\omega) = f(\omega) - b$, one gets

$$\int_N g = \int_N f - b\mu(N) \geq 0,$$

where, by definition,

$$N = \left\{\omega : \sup_n \frac{1}{n} \sum_{k=0}^{n-1} \mu(f(\phi^k(\omega)) - b) > 0\right\}$$

$$= \left\{\omega : \sup_n \frac{1}{n} \sigma_n(\omega) > b\right\} .$$

From the definition of A(a,b), any point in A(a,b) is in N too and $b\mu(N) \geq b\mu(A(a,b))$. Thus

$$\int_{A(a,b)} f \geq b\mu(A(a,b))$$

Considering (-f) instead of f and replacing a by b, one proves similarly

$$\int_{A(a,b)} f \leq a\mu(A(a,b))$$

Comparing these two inequalities, one gets

$$\mu(A(a,b)) = 0.$$

Given a pair of real numbers a,b $(a < b)$ it is possible to find a pair of rationals a',b' s.t. $a < a',b' < b$ and $A(a,b) \subseteq A(a',b')$. Thus, if a point $\omega$ belongs to a set $A(a,b)$ ,a,b real, it belongs too to the countable union of the sets $A(a',b')$, a',b' rational. This union is of zero measure, as any set $A(a',b')$ is of zero measure.

Thus $\lim_{n\to\infty} \frac{1}{n}\sigma_n(\omega)$ exists $\mu$-almost everywhere. [ Summarizing the proof:any measurable set of $\omega$ s.t. $\lim_{n\to\infty} \sup \frac{1}{n}\sigma_n(\omega) \neq \lim_{n\to\infty} \inf \frac{1}{n}\sigma_n(\omega)$ is of zero measure.] Furthermore,

$$\left| \frac{1}{n} \int \sigma_n(\omega) \right| < \frac{1}{n} \sum_{k=0}^{n-1} \int \left| f(\phi^k(\omega)) \right| < \int |f(\omega)| \, .$$

Thus $\frac{1}{n}\sigma_n(\omega)$, which has a bounded integral in absolute value, tends $\mu$-almost everywhere to an integrable function $f^*$ by Fatou's lemma.

It remains to prove that $\int f^* = \int f$ . This is straightforward, since

$$\int \frac{1}{n} \sigma_n(\omega) = \frac{1}{n} \sum_{k=0}^{n-1} \int f(\phi^k(\omega)) = \int f = \int f^* \, .$$

A.3. *Corollary*

1. An important corollary is that the Birkhoff sums are $\mu$ -almost everywhere constant if the underlying dynamical system is ergodic. This comes from the remark that $f^*$ is $\phi$-invariant :

$$f^*(\phi(\omega)) = f^*(\omega) \quad \mu\text{-almost everywhere } (\mu\text{-a.e.})$$

More generally, $f^*$ is a.e. constant in each ergodic component of X [an ergodic component of X is a measurable subset Y s.t. $\phi(Y) \subseteq Y$ , $\mu(Y) > 0$, and if $Y_1 \subset Y$, $\phi(Y_1) \subseteq Y_1 \Rightarrow \mu(Y_1) = \mu(Y)$ or $\mu(Y_1) = 0$] . If $f^*$ were not a.e. constant in an ergodic component, one may split this ergodic component in two parts, say, the part where it is strictly larger than its mean value and the part where it is smaller or equal. Both parts are invariants, which contradicts the assumption, unless one of them is of zero measure.

An immediate consequence of this corollary is

$$\lim_{n\to\infty} \frac{1}{n} \sum_{k=0}^{n-1} \mu(A \cap \phi^{-k}(B)) = \mu(A)\,\mu(B) \quad ,$$

A and B being any two measurable subsets. To prove this formula, one takes $\chi_B$ as the function f in the Birkhoff theorem and writes

$$\mu(A \cap \phi^{-k}(B)) = \int \chi_A \chi_{\phi^{-k}(B)}$$

$\frac{1}{n} \sum_{k=0}^{n-1} \chi_{\phi^{-k}(B)}$ tends $\mu$-a.e. to a constant, and this constant is $\mu(B)$. Thus

$$\lim_{n\to\infty} \frac{1}{n} \sum_{k=0}^{n-1} \mu(A \cap \phi^{-k}(B)) = \int \chi_A \mu(B) = \mu(A)\,\mu(B).$$

2. Remarks

(i) If $\lim_{n \to \infty} \frac{1}{n} \sum_{k=0}^{n-1} \mu(A \cap \phi^{-k}(B)) = \mu(A)\mu(B) \; \forall \; A$ and $B \in S$, the system is ergodic : let us assume $\phi^{-1}(B) = B$ and $A = B'$ (complementary of B); the left hand side of the above equality is zero, and one must have either $\mu(B) = 0$ or $\mu(B') = 0$, which proves the ergodicity.

(ii) The Birkhoff theorem is not constructive and it may be extremely difficult to show for a given function and a given element $\omega$ that the Birkhoff sum tends to the limit value given by the Birkhoff theorem. As an example, consider the dyadic transform $x \to (2x)$ on $[0,1]$ . It is ergodic with aspect to the Lebesgue measure. This is shown (for instance) by proving $\frac{1}{n} \sum_{k=0}^{n-1} f(\phi^k(\omega)) \xrightarrow[(n \to \infty)]{}$ constant for any function $\exp(\frac{2i\pi p}{q} \omega)$ (p,q integers) and for $\mu$- almost any $\omega$. But, except for very special cases, it is impossible to prove that the Birkhoff limit is reached for a given number. Choosing the function f to be, e.g. the characteristic function $\chi_{[0,1/2[}$ of the half interval, one knows that the corresponding Birkhoff sum for a.e. number x has the limit 1/2 (which is the probability of the digit 0 in the binary expansions), but one does not know [20] if $\sqrt{2} - 1$, $e - 2$, $\pi - 3$, ln2, ... are "normal" numbers, that is, if the probability of occurrence of a given digit in their expansion is the same as for almost any number in $[0,1]$ .

(iii) Halmos[21] has defined a property that is, in principle, stronger than ergodicity. It may be called "double" ergodicity and states that

$$\frac{1}{m} \sum_{k'=0}^{m-1} \left[ \frac{1}{n} \sum_{k=0}^{n-1} \mu(A \cap \phi^{-k}(B) \cap \phi^{-k'}(C)) \right] \underset{\substack{n \to \infty \\ m \to \infty \\ m \gg n}}{\to} \mu(A)\, \mu(B)\, \mu(C) \quad .$$

It is generalizable to "n-tuple" ergodicity and is stronger than ergodicity. It is not yet known whether ergodic systems exist that are not n-ply ergodic.

(iv) An interesting question is the one of the *discrepancy*[22]. It refers to the manner in which a Birkhoff sum reaches its limit. The quantity

$$F_n(\omega) = \frac{1}{n} \sum_{k=0}^{n-1} (f(\phi^k(\omega)) - f^*)$$

is a "fluctuating" quantity : its average (from the Birkhoff theorem) is zero : $\int F_n(\omega) = 0$. For physical systems, the fluctuations of $F_n$ are connected with the transport properties[23]. For the dyadic transform, and for any $\beta$-transform actually, $F_n(\omega)$ tends to a normal distribution of zero average and with a width of order $1/\sqrt{n}$ (if $f \in L^2$). One might imagine this is "generally" the case, as $F_n(\omega)$ is $\frac{1}{n} \times$ a sum of n terms, each of them being of zero average and of finite variance.
But in general there are correlations among the various terms of the Birkhoff sum, so that the central-limit theorem may not be true.

For the continued fraction expansions, Gauss made the conjecture, which was proved much later on by Levy and Kuzmin[(24)], that $\int nF_n^2(\omega) \underset{n\to\infty}{\to} 0$. The discrepancy of the irrational rotation involves rather subtle questions of algebraic number theory. Let us only say that, in this case, the correlations among the various terms are very important, and the distribution of $F_n(\omega)$ depends in a rather complicated way on n, depending on the continued-fraction expansion of the irrational angle[(25)].

B. *Rotations of an irrational angle are ergodic*

This is the map of $[0,1]$ into itself defined by $x \to (x+\alpha)$, where $\alpha$ is the angle of rotation. This rotation is ergodic w.r. to the Lebesgue measure iff $\alpha$ is irrational.

To prove this, we need the

*Lemma*:

$\forall\varepsilon > 0$ and $\forall 0 < r < 1$, $\exists m$ (integer) s.t. $(m\alpha) < \varepsilon$ and, more generally, $\exists n$ s.t. $|(n\alpha - r)| < \varepsilon$.

*Proof*.

Consider the numbers $(\alpha),(2\alpha),\dots,(k\alpha)$, which are all different and included between 0 and 1. Two numbers $k_1$ and $k_2$ exist s.t. $1 \le k_1 < k_2 \le k$ and $((k_2-k_1)\alpha) = (m\alpha)$ is less than $1/k = \varepsilon$.

The second part of the lemma is proved by considering the lattice of points $(m\alpha)$, $(2m\alpha)$, ... $([\frac{1}{m\alpha}]\, m\alpha)$ which are all within $\varepsilon$ of each other and disposed on $[0,1]$ so that no point of $[0,1]$ is more than $\varepsilon$ from one of these points.

This proves that the irrational rotations are minimal, which is known as the Kronecker theorem.

It remains to prove that the irrational rotations are ergodic. Let us first consider a rotation by an angle $1/p$ ($p$ integer $> 1$) and two sets A and B, each being made of a finite union of segments ("subunits" in short, hereafter) $[\frac{i}{p}, \frac{i+p}{p}]$ $0 \le i\,(\text{integer}) \le p-1$, A being made of $j_A$ segments, B of $j_B$ segments. After p rotations of angle $1/p$ each subunit has covered the entire circle; thus,

$$\sum_{i=0}^{p-1} \mu(R^i_{1/p} A \cap B) = \frac{j_A j_B}{p}$$

(By definition $R^i_\beta C = \{(x+i\beta)\ ;\ x \in C\}$.)

As the left-hand side of this equality is a sum of p positive terms, at least one of them is larger than or equal to $(j_A j_B)/p^2 = \mu(A).\mu(B)$. Let $i_o$ be the index of one of these terms:

$$\mu(R^{i_o}_{1/p} A \cap B) \ge \mu(A).\mu(B)$$

But $R^{i_o}_{1/p} = R_{i_o/p}$, and from the second part of the above lemma this rotation can be approximated arbitrarily closely by a rotation of angle $(n\alpha)$, $\alpha$ being irrational. The non-zero part of $R_{i_o/p}A \cap B$ is made of subunits in $R_{i_o/p}A$ that cover exactly a subunit of B. When the rotation of angle $i_o/p$ is approximated by $(n\alpha)$ at a distance $\varepsilon$, the common part of two subunits in $R_{(n\alpha)}A$ and B is $(\frac{1}{p} - \varepsilon)$, instead of being $\frac{1}{p}$ in $R_{i_o/p}$ A and B. Thus,

$$\mu(R_{n\alpha}A \cap B) \geq (1 - p\varepsilon)\ \mu(A).\mu(B) \quad .$$

It remains to extend this inequality to any pair of measurable subsets.

If C and D are finite unions of intervals, one may repeat the previous reasoning by considering a partition of $[0,1]$ in p subunits, p being large enough to make negligible the "end effects" in C and D. This proves the existence of n s.t.

$$\mu(R_{n\alpha}C \cap D) > K\cdot\mu(C).\mu(D) \qquad 0 > K > 1.$$

To extend this reasoning to any pair of measurable subsets, one considers again a partition in subunits, say $s_1, s_2 \ldots s_p$, and the restriction $\tilde{C}$ of a given measurable set C to those subunits "bearing" the measure of C :

$$\tilde{C} = \bigcup_{j=1}^{q} (C \cap s_j) \qquad (q \geq 1)$$

$$\text{s.t.} \quad \mu(S_j \cap C) \geq \frac{k\mu(C)}{p}$$

$$( \geq k\ \mu(C)\mu(S_j)) \quad ,$$

$0 < k < 1$ given.

Consider now D as a finite union of $S_i$ : since $\tilde{C} \subseteq C \quad \forall n$ ,

$$\mu(R_{n\alpha}C \cap D) \geq (R_{n\alpha}\tilde{C} \cap D).$$

Applying now the same reasoning as before, one gets : $\exists n$ s.t.

$$\mu(R_{n\alpha}\tilde{C} \cap D) \geq k\ \mu(C)\mu(D) \left[\sum_{j=1}^{q} \mu(S_j)\right]$$

which implies

$$\mu(R_{n\alpha}C \cap D) \geq K\ \mu(C)\mu(D) \quad , \quad 0 < K < 1$$

This sort of reasoning can be extended to any measurable D. The ergodicity is proved as follows : $D = C'$ (complement of C). If $R_\alpha(C) \subseteq C$, $R_{n\alpha}(C) \subseteq C$ and $R_{(n\alpha)}(C) \cap C' = \emptyset$ , which is compatible with the above inequality if $\mu(C) = 0$ or $1$.

This proves the ergodicity of the irrational rotation, and is known as the Weyl theorem.

*Remark*

The more general case of diffeomorphism of the circle (= continuous invertible mapping of [0,1] into itself, 1 being identified with zero) has been studied for a long time, especially in connection with the properties of flows on the torus $T^2$. Let us give a few important results in this field.

Let f(x) be one such mapping of [0,1] into itself; it must satisfy f(0) = f(1) and $f^{-1}(0) = f^{-1}(1)$.

Poincaré and Denjoy [(26)] have defined what is called the rotation number of this diffeomorphism. It is usually noted $\rho(f)$, and, roughly speaking, it describes the average speed of rotation of a point around the circle, under successive applications of f. In our notation, this rotation number is connected with the number of times a point jumps over an arbitrary point, say $x_o$. This is the number of times it falls in $I_o = [f^{-1}(x_o), x_o]$ . This set could be made "apparently" of a few pieces, owing to the "apparent" discontinuity of f at 1 and $f^{-1}(0)$. The number of rotations is

$$\rho(f) = \frac{1}{n} \lim_{n \to \infty} \sum_{i=1}^{n-1} \chi_{I_o}(f^i(x)),$$

where $\chi_{I_o}$ is the characteristic function of $I_o$. This rotation number is independent of $x_o$ and of x, and depends continously on f. When this rotation number is irrational and sufficiently far from any rational [which is expressed by a condition on the c.f.e. of $\rho(f)$] , and when f is smooth enough, f is conjugat with a rotation of angle $\rho(f)$, that is a one to one mapping h of the circle into itself exists s.t.

$$R_{\rho(f)} = h^{-1} \circ f \circ h$$

where $R_{\rho(f)}$ is the rotation of the angle $\rho(f)$ and o denotes the usual functional composition law. It can be shown that h is sufficiently differentiable to map the Lebesgue measure, invariant under $R_{\rho(f)}$, into a measure invariant under f, which is absolutely continous w.r. to the Lebesgue measure. The differentiability of h is crucial for this property, as under the conjugacy h (when it is differentiable) the measure element dy is mapped into $|h'(x)|dx$ (this is nothing else but the usual formula of change of variable in an integral).Many important results on the diffeomorphisms of the circle have been proved by Hermann in his thesis[(26)].

## C.1. *An extension of the Birkhoff theorem : the Oseledec theorem.*

An important extension of the Birkhoff theorem, especially from the point of view of physics, is the non commutative ergodic theorems[(27)].

In these theorems, one considers, instead of functions f(x),matrices M(x) with real coefficients which depend on x, which is itself an element of the space of an ergodic system $(X,\underset{\sim}{S}, \mu,\phi)$ . We consider furthermore the product of matrices

$$M_n(x) = M^T(\phi^n(x))\, M^T(\phi^{n-1}(x)) \ldots M^T(x) M(x) \ldots\ldots M(\phi^n(x)) ,$$

where $M^T$ is the transpose of M (if M has complex elements, one must consider the hermitian conjugate instead). This matrix $M_n(x)$ has positive (or zero) eigenvalues, say $\lambda_1(x|n)$, $\lambda_2(x|n)$ ... $\lambda_k(x|n)$, which may be ordered as follows :

$$\lambda_1(x|n) \geq \lambda_2(x|n) \geq \lambda_3(x|n) \ldots \geq \lambda_k(x|n).$$

Let us consider the quantity $\frac{1}{2n} \ln_+ \lambda_1(x|n)$ ($\ln_+ x = \sup(0, \ln x)$).
If $M(x)$ is such that $\lambda_1(x|n)$ is in $L^1(X,S,\mu)$, the Oseledec theorem states that $\frac{1}{2n} \ln_+ \lambda_1(x|n)$ converges $\mu$-a.e. to a number that is independent of x.

The Birkhoff theorem, when restricted to positive functions, is a consequence of the Oseledec theorem, when applied to $1 \times 1$ matrices.

There is an important "practical" difference between the Oseledec theorem and the Birkhoff theorem : if one knows explicitly the matrix M(x) and the invariant ergodic measure, one has no explicit formula for computing $\frac{1}{2n} \ln_+ \lambda_1(x|n)$, unless the matrices M(x) commute. In order to illustrate this last point, I shall detail an application of the Oseledec theorem to a problem of many-body physics.

## C.2. *An application of the Oseledec theorem*[28].

We want to calculate the free energy of a random bond Ising chain in an uniform magnetic field H. The interaction energy between a pair of spins is +J with the probability (1 - x) and -J with the probability x.

The partition function of the chain is

$$Z_N = \mathrm{Tr}\; O_N$$

where $O_N$ is a $2 \times 2$ matrix

$$O_N = \prod_{i=1}^{N} M_i ,$$

the transfer matrices $M_i$ being randomly chosen : the matrix $\begin{pmatrix} z^{1+\alpha} & z^{-1+\alpha} \\ z^{-1-\alpha} & z^{1-\alpha} \end{pmatrix}$ is chosen with the probability $(1-x)$, and the matrix $\begin{pmatrix} z^{-1+\alpha} & z^{1+\alpha} \\ z^{1-\alpha} & z^{-1-\alpha} \end{pmatrix}$ is chosen with the probability x,

$$z = \exp \frac{J}{k_B T} \quad \text{and} \quad \alpha = H/J .$$

To find the free energy per spin, in the thermodynamic limit, one has to calculate

$$F(T,H) = \left\langle \lim_{N\to\infty} \frac{1}{N} \ln Z_N \right\rangle ,$$

the average being taken over all possible choices of matrices $M_i$. As the elements of these matrices are all positive, one can verify that

$$\lim_{N \to \infty} < \frac{1}{N} \ln \operatorname{Tr} O_N> \;=\; \lim_{N \to \infty} < \frac{1}{N} \ln \text{(any element of } O_N)> ,$$

and that $F(T,N)$ is $\frac{1}{2N} \ln_+$ (largest eigenvalue of $O_N$). The Oseledec theorem states that $F(T,H)$ exists for almost any choice of the + or − bond. Here the underlying dynamical system is the one built on all the doubly infinite sequences of two symbols (the + or − in front of the bond), $\phi$ being the shift mapping and the invariant measure gives to the elementary cylinders ...(+J)...and....(−J)... the weight $(1-x)$ and $x$ respectively.

As stated before, it is not possible to get a closed formula (with, say, a few quadratures) for $F(T,H)$. It is only possible to find the ground state energy ($F(T,H)$ at $T = 0$) and entropy $\left(-\frac{1}{k_B} \frac{\partial F(H,T)}{\partial T}\Big|_{T=0}\right)$.

For that purpose let us write $O_N = \begin{pmatrix} z^a e^u & z^b e^v \\ \cdots & \cdots \end{pmatrix}$ as $z \to \infty$ (or $T \to 0$). As $O_{N+1} = O_N M_N \underset{z \to \infty}{\simeq} \begin{pmatrix} z^A e^U & z^B e^V \\ \cdots & \cdots \end{pmatrix}$, one can find A, B and U,V knowing $M_N$ and a,b and u,v.

Let us call $C = U-V$, $c = a-b$, $W = U-V$, $w = u-v$. One has to consider the difference between the powers of z in the matrix elements of $O_N$ and $O_{N+1}$, since one keeps at each step (in N)the dominant term at $z \to \infty$ only in this matrix element. This dominant term in $O_{N+1}$ may come from any one of the elements of $O_N$ of the same row , depending on their relative order of magnitude.

With probability $1-x$, one has :

If $-2 \leq c < 2-2\alpha$ : $A = a+1+\alpha$ $\quad U = u$
$B = b+1-\alpha$ $\quad V = v$
$C = c+2\alpha$ $\quad W = w$

If $c = 2-2\alpha$ : $A = a+1+\alpha$ $\quad U = u$
$B = b+1-\alpha$ $\quad V = \operatorname{Log}(e^u + e^v)$
$C = c+2\alpha$ $\quad W = -\operatorname{Log}(1+e^{-w})$

If $2-2\alpha < c \leq 2$ : $A = a+1+\alpha$ $\quad U = u$
$B = a-1+\alpha$ $\quad V = u$
$C = 2$ $\quad W = 0$

With probability x, one has :

If $-2 \leq c < 2 - 2\alpha$ : $A = b + 1 - \alpha$ $U = v$

$B = a + 1 + \alpha$ $V = u$

$C = -c - 2\alpha$ $W = -w$

If $c = 2 - 2\alpha$ : $A = b + 1 - \alpha$ $U = \mathrm{Log}\,(e^u + e^v)$

$B = a + 1 + \alpha$ $V = u$

$C = -2$ $W = \mathrm{Log}\,(1 + e^{-w})$

If $2 - 2\alpha < c \leq 2$ : $A = a - 1 + \alpha$ $U = u$

$B = a + 1 + \alpha$ $V = u$

$C = -2$ $W = 0$

Let us assume $\frac{2}{r+1} < \alpha < \frac{2}{r}$, r being an integer.

From the above (random) recursion relations for c, it is easy to see that c can have values of the form $\pm(2 - 2i\alpha)$ $0 \leq i$ integer $\leq r$, and one gets a random walk "with boundaries" : if $1 \leq i \leq r-1$ and $c = -2 + 2i\alpha$ , one has a probability (1-x) to decrease i by one unit and a probability x to increase i by one unit. When the boundary is reached, at i = 0 for instance, at the next step either i stays at zero or increases.

Let $p_i$ and $q_i$ be the probabilities that $c = -2 + 2i\alpha$ and $c = 2 - 2i\alpha$ respectively. They become stationary in the large N limit and satisfy linear relations, which are deduced at once from the above recursion relations :

For $1 \leq i \leq r - 1$

$$p_i = (1-x)p_{i-1} + x\, q_{i+1}$$
$$q_i = (1-x)q_{i+1} + x\, p_{i-1} ,$$

and for the boundary probabilities :

$$p_r = (1-x)p_{r-1}$$
$$q_r = x\, p_{r-1}$$
$$p_o = x\,(p_r + q_o + q_1)$$
$$q_o = (1-x)\,(p_r + q_o + q_1)$$

One can solve this system :

$$p_i = \beta(i + 1 - r - 1/x) \qquad 0 \leq i \leq r$$
$$q_i = \beta(i - r - 1) \qquad 1 \leq i \leq r$$
$$q_o = \beta(1 - r - 1/x)(1-x)/x$$

with $\beta = x^2/[(1 + rx)(1 + rx - x)]$ to ensure the normalization : $\sum_i (p_i + q_i) = 1$.

Using these probabilities and the recursion relations, we can now obtain the ground state energy :

$$\frac{E}{J} = \lim_{N \to \infty} (\langle a\rangle_{N+1} - \langle a\rangle_N) =$$

$$(1-x)(1+\alpha) + x \sum_{i=0}^{r-1} p_i(2-2i\alpha+1-\alpha) + x\, p_r(\alpha-1)$$

$$+ x \sum_{i=1}^{r} q_i(-2+2i\alpha+1-\alpha) + x\, q_o(\alpha-1)$$

$$= \frac{r^2x^2 + rx(2-x) + (2x-1)(x-1) + \alpha(2rx(1-x)+1-x)}{(1+rx)(1+rx-x)}$$

One can verify that $\langle b\rangle_{N+1} - \langle b\rangle_N$ gives the same result.

The ground state entropy is connected with the growth of the factor of largest power of z in any matrix element of $O_N$ as $N \to \infty$ . If one writes this dominant term (in z) as $e^u z^a$ ,

$$\frac{S}{k_B} = \lim_{N \to \infty} (\langle u\rangle_{N+1} - \langle u\rangle_N )$$

is the entropy per spin in the ground state. This entropy has a different form, depending whether $\frac{2}{\alpha}$ is an integer or not. I shall explain the calculation in this last case only.

One notes first that the only possible values for (c,w) are :

$$\begin{cases} c = -2 + 2i\alpha \\ w = \text{Log } n \end{cases} \quad \text{or} \quad \begin{cases} c = 2 - 2i\alpha \\ w = -\text{Log } n \end{cases}$$

with $0 \le i \le r$ and n integer $\ge 1$.

Let $R_i(n)$ be the probability that $c = -2+2i\alpha$ , $w = \text{Log } n$,

and $S_i(n)$ be the probability that $c = 2-2i\alpha$ , $w = -\text{Log } n$.

One has the relations :

$$1 \le i \le r-1 \quad \begin{cases} R_i(n) = (1-x)\ R_{i-1}(n) + x\ S_{i+1}(n) \\ S_i(n) = (1-x)\ S_{i+1}(n) + x\ R_{i-1}(n) \end{cases}$$

$$R_r(n) = (1-x)\ R_{r-1}(n)$$

$$S_r(n) = x\ R_{r-1}(n)$$

$$S_o(1) = (1-x)\ (p_r + q_o)$$

$$R_o(1) = x\ (p_r + q_o)$$

$$n \geq 2 \quad \begin{cases} S_o(n) = (1-x)\ S_1(n-1) \\ R_o(n) = x\ S_1(n-1) \ . \end{cases}$$

The solution of this system is :

$$\left.\begin{aligned} R_i(n) &= p_i(1-\nu)\ \nu^{n-1} \\ S_i(n) &= q_i(1-\nu)\ \nu^{n-1} \end{aligned}\right\} \quad \text{for} \quad 0 \leq i \leq r$$

with $\nu = r\,x^2/(1 + rx - x)$ .

It is now possible to obtain the entropy :

$$\begin{aligned} \frac{S}{k_B} &= \lim(\langle u\rangle_{N+1} - \langle u\rangle_N) \\ &= x \sum_{n=1}^{\infty} \Big[ \sum_{i=0}^{r-1} R_i(n)\ (-\mathrm{Log}\ n) + \sum_{i=2}^{r} S_i(n)\ \mathrm{Log}(n) \\ &\qquad + S_1(n)\ \mathrm{Log}(n+1) \Big] \\ &= \frac{x(1-x)^2}{(1+(r-1)x)^2} \sum_{n=1}^{\infty} \nu^{n-1}\ \mathrm{Log}\ n \end{aligned}$$

Another important field of applications of the non commutative ergodic theorems is the question of stability of non periodic attractors. I refer to the recent work of Ruelle on this question(29).

## 4. WEAK AND STRONG MIXING : *The Kakutani couter example.*

In the introduction, I discussed in rather vague terms the connection between the apparent irreversibility of macroscopic phenomena and the reversibility of classical dynamics. In some sense, an ergodic system displays such an irreversible behavior : starting almost anywhere one reaches any measurable part of the system with a frequency proportional to the measure of this part.

As is well known, this property is enough for the foundation of equilibrium thermodynamics : an isolated ergodic system of particles in a closed box "reaches" the thermodynamic equilibrium, as any time average yields the same results as averages over the microcanonical ensemble.

It is well known too that this does not fully account for the irreversible macroscopic effects : in particular a system may be ergodic, although there is no growth of the errors in its evolution. This makes an important difference between the irrational rotations on the circle and the dyadic transform. In the case of irrational rotations, starting from two points close to each other, the $n^{th}$ iterates of these points will remain close to each other ; for the dyadic transform, on the contrary, the $n^{th}$ iterate depends on the $n^{th}$, $(n+1)^{th}$,... digits so that two numbers with the same binary expansion up to the $n^{th}$ digit remain close to each other after the first, second,... $(n-1)^{th}$ iteration, although their $n^{th}$, $(n+1)^{th}$,... iterates may be completely different. In this last case a small fluctuation in the initial data (that is a change in the large order digits) is "amplified" by the dynamics. In this sense, the dyadic transform is more stochastic than the irrational rotation.

In order to account for this stronger "stochasticity" the property of mixing has been invented. It refers to the following observation : at some time, say t, one does not know exactly the position of a point in phase space. This point belongs to some measurable set, say A, which may be seen as a small cube defined by the inaccuracies (or "error bars") of our knowledge of the initial conditions. One tries then to understand the evolution of A. It may happens that this small set A becomes diluted in the whole system as, say, a drop of wine is diluted uniformly in stirred water.

The mathematical definition corresponding to this observation is the (weak or strong) mixing property.

A dynamical system is weakly mixing iff

$$\mu(A \cap \phi^{-n}(B)) \underset{n \to \infty}{\to} \mu(A)\mu(B) \qquad \forall\, A,B \in \underset{\sim}{S} \tag{1}$$

$$n \notin \mathcal{J} \text{ of zero density in } \mathbb{N}$$

$\mathcal{J}$ is of zero density in $\mathbb{N}$ if *cardinality of* $\dfrac{\mathcal{J} \cap \{0,1,\ldots,N\}}{N} \underset{N \to \infty}{\to} 0$

A dynamical system is strongly mixing iff

$$\mu(A \cap \phi^{-n}(B)) \underset{n \to \infty}{\to} \mu(A)\mu(B) \tag{2}$$

Weak mixing is an obvious consequence of strong mixing and ergodicity is a consequence of weak mixing. That ergodicity does not imply weak mixing is obvious : consider an irrational rotation on the circle. It is ergodic w.r. to the Lebesgue measure, but not weak mixing, as $\mu(A \cap \phi^{-n}(B))$ is an oscillating function of n.

Many properties may be shown to be equivalent to weak or strong mixing. For instance

$$(*) \quad \lim_{n \to \infty} \frac{1}{n} \sum_{k=0}^{n-1} |\mu(A \cap \phi^{-k}(B)) - \mu(A)\ \mu(B)| = 0 \ \forall\, A \text{ and } B \in S$$

is equivalent to weak mixing. This property strongly reminds us of a similar one valid for ergodic systems, but without the absolute value.

To prove that (*) is equivalent to (1), one has to show the general statement : let $\{u_n\}$ be a sequence of uniformly bounded positive numbers s.t. $u_n \to u$ $(n \to \infty)$ $(n \notin \mathcal{J}$ of zero density).

This is equivalent to $\frac{1}{n} \sum_{j=0}^{n-1} |u_j - u| \underset{n \to \infty}{\to} 0.$

*Proof.*

Let $\mathcal{J}_N$ be the number of elements of $\mathcal{J}$ in $\{0,1,\ldots,N\}$ , thus

$$\frac{1}{N} \sum_{j=0}^{N-1} |u_j - u| \le \frac{1}{N} \sum_{\substack{j=0 \\ j \notin \mathcal{J}}}^{N-1} |u_j - u| + \frac{\mathcal{J}_N}{N} (b + u),$$

where b is the uniform upper bound : $|u_n| < b \quad \forall\, n.$

As $\frac{\mathcal{J}_N}{N} \underset{N \to \infty}{\to} 0$, this proves $\frac{1}{N} \sum_{j=0}^{N-1} |u_j - u| \to 0$ if $u_j \to u,\ j \to \infty$ and $j \notin \mathcal{J}$ .

To prove the converse, assume $\frac{1}{n} \sum_j |u_j - u| \to 0$ and $u_j \underset{j \to \infty}{\to} u$ for a subset of $\mathbb{N}$ of finite density, that is, $\exists \varepsilon > 0$ s.t. $\forall\, N$, $\exists \alpha(\varepsilon) > 0$ and a subset $\mathcal{J}(\varepsilon,N)$ of $\{0,1,\ldots,N\}$ with $\frac{\mathcal{J}(\varepsilon,N)}{N} \ge \alpha$ and $|u_j - u| \ge \varepsilon \ \forall \ j \in \mathcal{J}(\varepsilon,N)$. This implies at once $\frac{1}{N} \sum_{j=0}^{N-1} |u_j - u| \ge \varepsilon\alpha$, which contradicts the hypothesis.

Now it remains to prove a non trivial thing : weak mixing does *not* imply strong mixing. This is done by a counterexample : one constructs a system that is weak mixing, but not strong mixing. I shall give this counterexample in some detail since it is historically the first (and remain the simplest one to explain) of a set of counterexamples proving the non equivalence of various properties connected with stochasticity. Furthermore its study leads to the introduction of some important concepts.

## *The Kakutani Counterexample*

One defines first the transform $\phi$ of $[0,1] = X$ into itself as follows : Let $x = .x_1 x_2 \cdots\cdots x_k \cdots\cdots$ be the binary expansion of any number in X ($x_k$ = 0 or 1), (another possible notation is $x \in \{0,1\}^{\mathbb{N}}$).

If $x_1 = 0$ , $\phi(x) = .(x_1 + 1)x_2 \cdots\cdots x_k \cdots\cdots$

If $x_1, x_2 \cdots\cdots x_p = 1 (p \ge 1)$ and $x_{p+1} = 0$,

$$\phi(x) = .\underbrace{00\cdots\cdots 0}_{p \text{ zeros}}1\, x_{p+2} \cdots\cdots x_k \cdots\cdots$$

This transform has discontinuities at $x = 1 - 2^{-i}$ , $i = 1,2,\ldots$, leaves invariant the Lebesgue measure as $\frac{d\phi}{dx} = 1$ a.e. and is invertible.

This system [that is (X, $\underset{\sim}{S}$, Lebesgue measure, $\phi$)] is ergodic but not weak mixing. To prove this consider the action of $\phi$ on the first p digits of x. Let p = 3 for instance, and start from $x = .1010\, x_5 x_6 x_7 \cdots\cdots$ The beginning of the binary expansion of x , $\phi(x)$, $\phi^2(x)$ ...... $\phi^8(x)$ is given in the following table :

$$\begin{array}{ll}
x & .1010\ x_5\ldots\ldots \\
\phi(x) & .\underline{0110}\ x_5\ldots\ldots \\
\phi^2(x) & .1110\ x_5\ldots\ldots \\
\phi^3(x) & .\underline{0001}\ x_5\ldots\ldots \\
\phi^4(x) & .1001\ x_5 \\
\phi^5(x) & .\underline{0101}\ x_5\ldots\ldots \\
\phi^6(x) & .\underline{1101}\ x_5\ldots\ldots \\
\phi^7(x) & .\underline{0011}\ x_5\ldots\ldots \\
\phi^8(x) & .1011\ x_5,\ldots\ldots
\end{array}$$

The first three digits on the left of $\phi^8(x)$ are those of x. A slight extension of this construction (which may be called an adding machine) shows that from the point of view of the first p digits $\phi$ has period $2^p$ : after $2^p$ actionsof $\phi$ one recovers the starting pattern, each of the $2^p$ different patterns (or numbers between 0 and $2^p - 1$ in binary notation) being obtained once and only once in the period. If one considers all possible sequences of p digits, each of them with measure $2^{-p}$, the transform $\phi$ , acting on the p first digits, is ergodic. This implies by a straightforward extension that $\phi$ is ergodic on $X = [0,1]$ .

This transform is not weak mixing. Consider the two sets $A = \{.000 \text{ (anything)}\}$ and $B = \{ .100 \text{ (anything)}\}$ ; one has $\phi^{-1}(B) = A$ (see the above table) and more generally

$$\phi^{-1-8k}(B) = A\ , \quad k \in \mathbb{N}_+ \quad \text{and} \quad \phi^{-j}(B) \cap A = \phi \quad \text{iff} \quad j \neq 1 \pmod 8 .$$

This shows

$$\mu(A \cap \phi^{-1-8k}(B)) = \mu(A) = 1/8$$

and

$$\mu(A \cap \phi^{-j}(B)) = 0 \quad \text{iff } j \neq 1 \pmod 8)$$

which excludes that $\phi$ is weak mixing.

To generate from $\phi$ a weak but not strong mixing transform, one uses the idea of induced transform.

Let A be a measurable set in $\underset{\sim}{S}$ s.t. $\mu(A) > 0$; the induced transform is a mapping $\phi_A$ of A into itself defined as follows :

$$\phi_A | x \to \phi_A(x) = \phi^{n_A(x)}(x) ,$$

$n_A(x)$ being the smallest integer s.t. $\phi(x),\ldots, \phi^{n_A(x)-1}(x) \notin A$ and $\phi^{n_A(x)}(x) \in A$. Such an integer exists for almost any x, due to the well known Poincaré theorem.

*Proof of the Poincaré recurrence theorem.*

Let $\bar{A} = \left\{x\,;\, x \in A\ ,\ \text{no } n \geq 1 \text{ exists s.t. } \phi^n(x) \in A\right\}$ ; thus $\phi^{-1}(\bar{A})\ldots.\phi^{-k}(\bar{A})$

are pairwise disjoint, otherwise take $y \in \phi^{-k}(\bar{A}) \cap \phi^{-k'}(\bar{A})$ $k \neq k'$ and $\phi^k(z) = \phi^{k'}(z)$ , $z \in \bar{A}$ and if $k > k'$, $z = \phi^{k-k'}(z)$ which is impossible, by definition of A. But $\mu(x) < \infty$ and to avoid the indefinite growth of $\mu(\bar{A} \cup \phi^{-1}(\bar{A}) \ldots \cup \phi^{-k}(\bar{A})) = k\,\mu(\bar{A})$ as $k \to \infty$, one must have $\mu(\bar{A}) = 0$.

Let us now take for A the set of numbers $x = .x_1 x_2 \cdots$ s.t. the smallest n with $x_n = 0$ is odd. A number is in A if its binary expansion starts as $.0\cdots$, or as $.110\cdots$, or as $.11110\cdots$ and so on. Thus the measure of A is $\frac{1}{2} + \frac{1}{8} + \ldots + \frac{1}{2.4^n} + \ldots = \frac{1}{2}\,\frac{1}{1-\frac{1}{4}} = \frac{2}{3}$. In the above table $\phi(x)$, $\phi^3(x)$, $\phi^5(x)$, $\phi^6(x)$, and $\phi^7(x)$ belong to A. But if one restricts oneself to the first three digits, without knowing the fourth one, one does not know a priori if $\phi^3(x) = .111\cdots$ belongs to A or not. This depends on the digits on the right of first three ones. This dependence on the next order digits will be basically the reason why $\phi_A$ is weak mixing although $\phi$ is not.

To prove that $\phi_A$ is weak but not strong mixing, we shall need the :

*Lemma.*

Let $\psi_n(x)$ be the number of visits to A in the orbit $x, \phi(x), \ldots \phi^{n-1}(x)$. Then, for $n = 2^{2p}$ , p integer, $\psi_n(x)$ takes only two values, namely $b_p$ and $b_p + 1$ where $b_p = 4^p(\frac{1}{2} + \frac{1}{16} + \ldots + \frac{1}{4.2^{2p-1}}) = \frac{1}{3}(2.4^p - 1)$ and

$$\mu(x : \psi_{4^p}(x) = b_p) = 1/3$$

$$\mu(x : \psi_{4^p}(x) = b_p + 1) = 2/3.$$

*Proof of the Lemma*

As already seen, if one restricts oneself to the first 2p digits, all the different combinations of 2p digits 0 and 1 are found in $x, \phi(x) \ldots \phi^{4^p}(x)$.

The binary expansion of half of these $4^p$ numbers starts as $.0\cdots$ , of $1/8^{th}$ of these numbers as $.110\cdots$ ; thus the proportion of the $4^p$ sequences of length 2p belonging to A is

$$(\frac{1}{2} + \frac{1}{8} + \ldots + \frac{1}{4.2^{2p-1}}) = b_p$$

But we have not yet considered the iterate $\underbrace{.1111\cdots}_{2p \text{ digits } 1} x_{2p+1}\, x_{2p+3} \cdots$ . We only know that the sequence after the 2p first digits (that is $x_{2p+1}\, x_{2p+2} \cdots$) belongs to A with a probability 2/3 [from the definition of A, if $x_{2p+1}\, x_{2p+2} \cdots \in A$, $\underbrace{.1111}_{2p \times 1} x_{2p+1} \cdots$ is in A too ]. Thus $\psi_{4p}(x)$ takes the value $b_p$ $(\underbrace{.111\cdots 1}_{2p \times 1} x_{2p+1} \cdots \notin A)$ with the probability 1/3 and the value $b_p + 1$ $(\underbrace{.11\cdots 1}_{2p \times 1} x_{2p+1} \in A)$ with the probability 2/3.

$\phi_A$ *is not strong mixing ; Proof.*

If $\mu(\phi^{-j} A \cap B) \underset{j \to \infty}{\to} \mu(A)\mu(B)$, thus

$$q\, \mu(\phi^{-j} A \cap B) + (1-q)\mu(\phi^{-(j+1)} A \cap B) \underset{j \to \infty}{\to} \mu(A)\mu(B) \qquad \forall\, 0 \leq q \leq 1.$$

From the above lemma, we know that there is a sort of period for $\phi_A$ : after $b_p$ or $b_{p+1}$ applications of $\phi_A$ one recovers the same (2p) first digits. Let us consider the quantity

$$v = q\, \mu(\phi_A^{-b_p}(C) \cap C') + (1-q)\, \mu(\phi_A^{-(b_p+1)}(C) \cap C')$$

where the measurable subset C is in A and is defined by means of the pattern of the few first digits : for instance

$$C = \{x : x = .110\cdots\}$$

and C' is the complement of C in A.

Thus we know that with a probability 1/3 $\phi_A^{-b_p}(C) = C$ and that with a probability 2/3, $\phi_A^{-b_p}(C)$ = something else, say $C_1$. Similarly, with a probability 2/3 $\phi_A^{-(b_p+1)}(C) = C$ and a probability 1/3 $\phi_A^{-b_p}(C) = C_2$.

As $\phi_A$ is measure preserving : $\mu_A(C_1) \leq \frac{2}{3}\mu_A(C)$

$$\mu_A(C_2) \leq \frac{2}{3}\mu_A(C)$$

$$\mu_A(C_1 \cap C') \leq \frac{2}{3}\mu_A(C)$$

and $\mu_A(C_2 \cap C') \leq \frac{1}{3}\mu_A(C)$

We have introduced the induced measure $\mu_A$ : it is defined on the subset A as $\mu_A(C) = \frac{\mu(C \cap A)}{\mu(A)}$ , and is invariant under $\phi_A$.

The above inequalities show

$$v \leq \left(\frac{2q}{3} + \frac{1-q}{3}\right)\mu_A(C) \qquad \text{if} \quad 0 < q < 1$$

or

$$(*) \quad v \leq \frac{1+q}{3}\mu_A(C)$$

But strong mixing implies $v \underset{p \to \infty}{\to} \mu_A(C)\,(1 - \mu_A(C))$.

If one chooses $0 < q < 2 - 3\mu_A(C)$, which is possible if $\mu_A(C) < 2/3$, (*) is incompatible with strong mixing.

If one takes $C = \{.1100\cdots\}$ , then $\mu(C) = 1/8$, and

$$\mu_A(C) = \frac{\mu(C \cap A)}{\mu(A)} = \frac{1}{8} \cdot \frac{3}{2} = \frac{3}{16} < \frac{2}{3}$$

To prove that $\phi_A$ is weak mixing we need

*A brief account of spectral theory.*

Let us consider the space of functions $L^2(X,\underset{\sim}{S},\mu)$ and the operator $U_\phi$ defined by

$$U_\phi f(.) = f(\phi(.)) \quad , \qquad f \in L^2$$

$U_\phi$ is unitary, if the measure is $\phi$-invariant :

$$\int |(U_\phi f(.)|^2 = \int |f(\phi(.))|^2$$
$$= \int |f(.)|^2$$

Thus the spectrum of f is on the unit circle :

$$U_\phi f = \lambda f \Rightarrow \lambda\lambda^* = 1$$

A number of theorems[30] relate the structure of the spectrum of $U_\phi$ [that is the topology of the singularities of $(z - U_\phi)^{-1}$] to the statistical properties of the underlying dynamical system :

(i) 1 is an eigenvalue, corresponding to the a.e. constant eigenfunction.

(ii) if $\phi$ is ergodic, all the eigenvalues are simple and they form a subgroup of the multiplicative group of the complex number of unit modulus.

(iii) if $\phi$ is weak mixing, $U_\phi$ has a continuous spectrum on the complement of the space of constant eigenfunctions. This is a necessary *and* sufficient condition.

For the proof of the last statement which is rather lengthy and difficult, I refer again to Halmos[30].

$\phi_A$ *is weak mixing : Proof.*

The idea of the proof is to show that, if f is an eigenfunction of $U_{\phi_A}$ in $L^2(A,S_A,\mu_A)$, then the corresponding eigenvalue is necessarily 1. This is done by calculating in two different ways the quantity

$$\int_{[0,1]} |f(\phi^{4p}(.)) - f(.)|^2$$

when f is an eigenfunction of eigenvalue $e^{2i\pi\lambda}$ and $p \to \infty$.

For a given x $f(\phi^{4p}(x))$ is equal to $f(x)\, e^{2i\pi\lambda n(x)}$ where n(x) is the number of $\phi^j(x)$ , $1 \leq j \leq 4p$ in A. One knows that for $1/3^d$ of the x's (in measure) $n(x) = b_p$ and for $2/3^d$ $n(x) = b_p + 1$. Thus

$$\int_{[0,1]} |f(\phi^{4p}(.)) - f(.)|^2 = (\frac{1}{3} |e^{2i\pi\lambda b_p} - 1|^2 + \frac{2}{3} |e^{2i\pi\lambda(b_p+1)} - 1|^2) \int_{[0,1]} |f(.)|^2$$

Let us calculate now the same quantity by considering the spectrum of $U_\phi$ (take care that $U_\phi$ is *not* the operator as restricted to A, this is the original operator).

We have already seen that $\phi$ , as seen from the point of view of the first 2p digits is a permutation operator among the $4^p$ possible pattern of 2p digits 0 or 1. Thus its spectrum has the eigenvalues $\exp \frac{2i\pi}{4^p} k$, with $k = 1,\ldots, 2^{p-1}$.

This means, in particular that $(U_\phi)^{4^p}$ acting on any one of these eigenfunctions reduces to the identity operator.

Expanding f on the eigenfunctions of $U_\phi$ , one shows that one may neglect the eigenfunction associated with a large p and

$$(U_\phi)^{2p} f \underset{p \to \infty}{\to} f \text{ in the } L^2 \text{ topology.}$$

If one compares with the above relation, one must have :

$$e^{2i\pi\lambda b_p} \underset{p \to \infty}{\to} 1 \quad \text{and} \quad e^{2i\pi\lambda(b_p + 1)} \underset{p \to \infty}{\to} 1 ,$$

which implies $e^{2i\pi\lambda} = 1$, and proves that $\lambda = 0$ is the only eigenvalue of $U_{\phi_A}$ acting in $L^2$, and thus, from the spectral theorem (iii), $\phi_A$ is weak mixing.

*Final remarks on the mixing systems.*

The property of strong mixing is very commonly used by physicists. In statistical physics in particular, it is often assumed, after Landau, that in the ordered phase of a many body system an "order parameter" takes a finite value. From the point of view of dynamical systems[31], this may be easily understood by reference to the mixing property : when the "order parameter" is zero, the system has the mixing property . The choice of the magnetization for instance, at some site of a ferromagnet above the Curie temperature, does not determine the magnetization very far away from this point : correlations "decay" at large distances. Here the dynamical system is obtained by considering the translation of the underlying lattice, the invariant measure being the Gibbs measure.

It is striking to notice that mathematics tell us that statistical properties exist which are *stronger* than mixing. One may wonder, if phase transitions exist which could be characterized by the breaking of one of their properties, without breaking the mixing itself.

Perhaps this is a manner of explaining what happens in the controversial "spin glass" state, for which it is rather difficult to characterize the transition from the paramagnetic state by means of an "order parameter".

## 5. BERNOUILLI SHIFTS : *The Meshalkin isomorphism.*

In this chapter I consider Bernouilli shifts (B. Shift in short), which are the most "stochastic" dynamical systems. In many books on dynamics, an intermediate step appears between strong mixing and Bernouilli system, that is the K-systems (which are also called undeterministic sometimes). And it is known that K systems exist which are not Bernouilli and strong mixing systems which are not K. For the theory of these K systems I refer the interested reader to (32) and (37).

Roughly speaking, a Bernouilli shift is a dynamical system which is completely unpredictible from a "certain point of view ". This is the case for the roulette game : if the roulette is not biased, the successive outputs are at random* . To formulate this more precisely, consider a dynamical system $(X,\underset{\sim}{S},\mu,\phi)$. It is Bernouilli iff a countable (possibly finite, as for roulette, head and tail,...), partition $P = (P_1,P_2,\ldots)$ of X exists s.t.

(i) $P_j \in \underset{\sim}{S} \quad \forall j \in \mathbb{N}_+$

(ii) $\mu(P_j) = p_j \qquad 0 \leq p_j \leq 1$

(iii) $\bigcup_{n=-\infty}^{n=+\infty} \phi^n(P_j)$ generates $\underset{\sim}{S}$ **

(iv) $\mu\left(\bigcap_{j=1}^{\ell} \phi^{n_j}(P_{k_j})\right) = \prod_{j=1}^{\ell} p_{k_j}$ for each choice of $\ell$, $k_1,\ldots,k_\ell$ and

$n_1 < n_2 < \ldots < n_\ell$ .

To make these statements clearer, let us say that (iii) defines the Bernouilli shift as a symbolic system : any measurable set in $\underset{\sim}{S}$ is a countable union or intersection of elementary cylinders, each of these cylinders being specified by the outcome (w.r. to the partition P ) at a given step. In other terms the B. shift is the shift on the doubly infinite sequences of symbols $(\ldots, x_{-1},x_0,x_1,\ldots)$ where the $x_i$ are at random and can take a denumerable set of values with the probabilities $p_1,p_2\ldots p_k \ldots$ . This is what is stated in (iv).

---

*As is well known, this sentence, or similar ones, is at the basis of epistemological difficulties in the foundations of probability theory : it is tautological as "roulette is not biased" precisely means that the output is at random. Actually "roulette is not biased" could mean that this roulette is built to avoid a priori any breaking of its apparent rotational symmetry, and that the croupier is honest. Then there is no redundancy in the statement.

** $\underset{\sim}{S}$ is the smallest σ-ring including all the $\phi^n(P_j)$, $n \in Z$ , $j \in \mathbb{N}$ .

It is important to realize that the property of being a B. shift implies the choice of a particular partition. Consider, for instance, the B. shift (1/2, 1/2) [that is the elementary cylinders are $(P_1,P_2)$ and $\mu(P_1) = \mu(P_2) = 1/2$ ]. This is the dynamical system acting on the doubly infinite sequences of two symbols as, say, 0 and 1 (or $X = \{0,1\}^Z$) . The occurrence of these two symbols is at random, each of them with probability 1/2. If, instead of considering the partition generated by $(P_1,P_2)$ one considers the partition generated by the choice of consecutive pairs of symbols, one gets four elementary cylinders $Q_1,Q_2,Q_3$ and $Q_4$ (in order) corresponding to the four possible choices for the pair $(x_o,x_1) = (0,0)$ or $(0,1)$ or $(1,0)$ or $(1,1)$. Each of these cylinders has the measure 1/4, but the set $\phi(Q_1) \cap Q_4$ is empty [ shifting a sequence $(\ldots,x_o=0,\ x_1=0,\ldots)$ on the left, one never gets $(\ldots,x_o=1,x_1=1,\ldots)$] Thus the measure defined on the partition generated by the $Q_i$ $i = 1\text{-}4$, does not satisfy the above property (iv). In this example it is rather easy to build an isomorphism between the Markov shift defined from the Q partition and the original B.shift. This general question of building B. shifts from Markov shifts has been already considered[33].

*The Ornstein theorem : entropy of B. shift*

Given two B.shifts is it possible to build an isomorphism between these two systems? In other terms, does a $1 \leftrightarrow 1$ mapping exist between the two dynamical systems s.t.

(i) the measure is preserved by this mapping

(ii) the two shifts commute with the mapping

(iii) the points with more than one image or without image are of zero measure.

This problem has been solved in two steps. First Kolmogoroff[34] introduced what is called the K-entropy. In the case of B. shifts this K-entropy is

$$h(\phi) = -\sum_i p_i \ln p_i$$

As this entropy is invariant under an isomorphism mapping between two B. shifts, two B shifts may be isomorphic only if they have the same entropy. In 1970, Ornstein[35] proved by an explicit calculation that two B. shifts with the same entropy are isomorphic.

Before Ornstein's proof, Meshalkin[36] and Blum and Hanson[37] proved a restricted version of the Ornstein's theorem. I explain hereafter this form of the isomorphism theorem. It is interesting in itself and gives the idea of what is meant by isomorphims of two B. shifts.

*The Meshalkin isomorphism theorem.*

The two B. shifts under consideration are $(\frac{1}{4}, \frac{1}{4}, \frac{1}{4}, \frac{1}{4})$ and $(\frac{1}{2}, \frac{1}{8}, \frac{1}{8}, \frac{1}{8}, \frac{1}{8})$. Their common entropy is

$$h = -\sum_{(4)} \frac{1}{4} \ln \frac{1}{4} = -\frac{1}{2} \ln \frac{1}{2} - \sum_{(4)} \frac{1}{8} \ln \frac{1}{8} = \ln 4.$$

The first (second) B. shift is the shift transform acting on doubly infinite sequences of 4(5) symbols. The isomorphism theorem is proved by constructing a "coding device", that is a rule associating to any element of $\{0,1,2,3\}^{\mathbb{Z}}$ an element of $\{0,1,2,3,4\}^{\mathbb{Z}}$ ( a simple example of such a coding device is the expansions of a given number in $[0,1]$ on different bases. In this case the coding does not commute with the shift, this is rather natural, as the entropy of the shift is $\ln n$, n being the base ).

Let $x = (\ldots x_{-1}, x_0, x_1, \ldots)$ be a doubly infinite sequence of one of the four symbols 0,1,2,3 ($x \in \{0,1,2,3\}^{\mathbb{Z}}$). Replacing each number by its binary expansion, $0 \to 00$, $1 \to 01$, $2 \to 10$, $3 \to 11$ , one gets a doubly infinite sequence of ordered pairs (or "2-stacks") of symbols 0 or 1. For instance

$$x = (\ldots.1,0,3,2\ldots..) \text{ becomes}$$

$$(*) \qquad \begin{matrix} & & 0,0,1,1 \\ x & = & (\ldots.1,0,1,0\ldots..). \end{matrix}$$

More generally an element of $\{0,1,2,3\}^{\mathbb{Z}}$ is written as a doubly infinite sequence of 2-stacks :

$$(*') \qquad x = \left(\ldots \begin{matrix} x^1_{-1} \\ x^0_{-1} \end{matrix}, \begin{matrix} x^1_0 \\ x^0_0 \end{matrix}, \begin{matrix} x^1_1 \\ x^0_1 \end{matrix} \ldots\right)$$

where $x^0_i$ and $x^1_i \underset{i \in \mathbb{Z}}{\in} \{0,1\}$ .

Now let us consider the following binary coding of the five symbols of the B. shift $(\frac{1}{2}, \frac{1}{8}, \frac{1}{8}, \frac{1}{8}, \frac{1}{8})$ :

$$(**) \qquad 0 \to 0\ , \quad 1 \to \begin{matrix}0\\0\\1\end{matrix}\ , \quad 2 \to \begin{matrix}1\\0\\1\end{matrix}\ , \quad 3 \to \begin{matrix}0\\1\\1\end{matrix}\ , \quad 4 \to \begin{matrix}1\\1\\1\end{matrix}$$

We shall construct the isomorphism $\{0,1,2,3\}^{\mathbb{Z}} \to \{0,1,2,3,4\}^{\mathbb{Z}}$. For that purpose, one has to transform a sequence of 2-stacks such as in (*) into a sequence of three- or one-stacks, such as appear in (**).

The idea is that, in the 2-stack "expansion" of any element in $\{0,1,2,3\}^{\mathbb{Z}}$ some elements have a zero at the bottom, and some others have 1, thus one shifts the digit on the top of a zero to put it on a 2-stack with a 1 at the bottom, and the the result is something like

$$\begin{matrix} \sigma_2 & & \sigma_1 \\ 0 & \ldots. & 1\ldots. \end{matrix} \quad \to \quad \begin{matrix} & & \sigma_2 \\ & & \sigma_1 \\ \ldots.0\ldots. & & 1\ldots. \end{matrix} \quad , \text{ with } \sigma_1, \sigma_2 \in \{0,1\} \ .$$

Now the symbols $\overset{\sigma_2}{0}$ and $\overset{\sigma_1}{1}$ are decodable by means of the table (**).

We do not expect a unique rule for this coding. Let us explain one possible set of rules. One starts at a doubly infinite sequence, as written in (*').

(i) *First step*

Consider all the pairs $\ldots \begin{matrix} x_i^1 & x_{i+1}^1 \\ x_i^o & x_{i+1}^o \end{matrix} \ldots\ldots$ s.t.

$x_i^o = 1$ and $x_{i+1}^o = 0$ , then replace this pair by

$$\begin{matrix} x_{i+1}^1 & \\ x_i^1 & \\ \ldots\ldots\ldots\ldots x_i^o & x_{i+1}^o \ldots\ldots \end{matrix}$$

Both 1-stack $x_{i+1}^o$ and 3 stack $\begin{matrix} x_{i+1}^1 \\ x_i^1 \\ x_i^o \end{matrix}$ are decodable by (**), since $x_{i+1}^o = 0$ and $x_i^o = 1$.

(ii) *Second step.*

Once the above substitution has been done everywhere , one has a sequence like

$$\ldots\ldots\ldots\ldots \begin{matrix} x_j^1 \\ x_j^o \end{matrix} \underset{\text{coded 2-stacks}}{\underline{\qquad\qquad\qquad}} \begin{matrix} x_{j+\ell}^o \\ x_{j+\ell}^o \end{matrix} \ldots\ldots\ldots\ldots$$

One may now continue the process defined above, by forgetting the already coded 2-stacks : if, for instance, $x_j^o = 1$ and $x_{j+\ell}^o = 1$, then one gets from the above sequence :

$$\ldots\ldots\ldots\ldots \begin{matrix} x_{j+\ell}^1 \\ x_j^1 \\ x_j^o \end{matrix} \underset{\text{coded 2-stacks}}{\underline{\qquad\qquad\qquad}} x_{j+\ell}^o \ldots\ldots\ldots\ldots$$

And one repeats the process as manytimes as required. This process converges a.e.

For we have to prove that almost surely any 2-stack $\begin{matrix} x_j^1 \\ 1 \end{matrix}$ or $\begin{matrix} x_j^1 \\ 0 \end{matrix}$, is "mated" at the end of the process. Consider the first case (the second one is very similar).

The element $\begin{matrix} x_j^1 \\ 1 \end{matrix}$ will not be coded if, on its right, all the elements $\begin{matrix} x_{j+k}^1 \\ 0 \end{matrix}$ $k \geq 1$ are mated with $\begin{matrix} x_{j+k'}^1 \\ 1 \end{matrix}$ $1 < k' < k$. This happens if, among the symbols

$x^o_{j+1}, \dots, x^o_{j+k'}$, more zeroes than 1 appear, whatever k' is $1 \leq k' \leq k$. This is very unlike as $k \to \infty$ , as a consequence of the "drunk man" theorem : a drunk man doing equiprobably steps +1 or -1 on a line, and starting anywhere, goes almost surely anywhere else,at home in particular ! (It is well known too that this theorem is not true for any regular and fully connected lattice in any dimension larger than 2). This is proved by noticing that, if the drunk man walks left, than one chooses $x^o_{j+k'} = 0$ and $x^o_{j+k'} = 1$ if he walks right. As he goes almost surely to any point on the right of the starting point, $\exists$ k' almost surely s.t. more ones than zeroes appear in $x^o_{j+1} \dots x^o_{j+k'}$.

Now it remains to prove that the measure is preserved by the coding in $1 \leftrightarrow 1$ (the coding obviously commutes with the shift, as it is translationnaly invariant).

The measure of the elementary cylinder $\{\dots,0,\dots\}$ in $(\frac{1}{2}, \frac{1}{8}, \frac{1}{8}, \frac{1}{8}, \frac{1}{8})$ is, by definition $\frac{1}{2}$. This is precisely the sum of the measure of the two cylinders $\{\dots \overset{1}{0} \dots\}$ and $\{\dots \overset{0}{0} \dots\}$ in $\{0,1,2,3\}^{\mathbb{Z}}$ and, by the coding, these last two cylinders are mapped into the above cylinder.

An element $(\dots \overset{x^1_i}{1} \dots)$ of $\{0,1,2,3\}^{\mathbb{Z}}$ will be mapped equiprobably into $\left(\dots \begin{smallmatrix} 1 \\ x^1_i \\ 1 \end{smallmatrix} \dots\right)$ or $\left(\dots \begin{smallmatrix} 0 \\ x^1_i \\ 1 \end{smallmatrix} \dots\right)$, since the coding process is independent of the value of the digit added on the 2-stack to make a 3-stack, and the digits chosen are equiprobably 0 or 1, as being the upper digit of any 2-stack with a 1 at the bottom.

This shows that the 3-stacks $\begin{smallmatrix} 0 \\ x^1_i \\ 1 \end{smallmatrix}$ and $\begin{smallmatrix} 1 \\ x^1_i \\ 1 \end{smallmatrix}$ are equiprobable and, as $x^1_i$ is 0 or 1 with a probability $\frac{1}{2}$ the four possible 3-stacks $\begin{smallmatrix} 0 \\ 0 \\ 1 \end{smallmatrix}$ $\begin{smallmatrix} 0 \\ 1 \\ 1 \end{smallmatrix}$ $\begin{smallmatrix} 1 \\ 0 \\ 1 \end{smallmatrix}$ and $\begin{smallmatrix} 1 \\ 1 \\ 1 \end{smallmatrix}$ are equiprobable. This shows that the various elementary cylinders in $\{0,1,2,3,4\}^{\mathbb{Z}}$, as obtained by the coding process,have the probabilities $(\frac{1}{2}, \frac{1}{8}, \frac{1}{8}, \frac{1}{8}, \frac{1}{8})$. The fact that the measure is the product measure is a simple consequence of the independence of the digits 0 and 1 in the 2-stacks representation of $(\frac{1}{4}, \frac{1}{4}, \frac{1}{4}, \frac{1}{4})$.

Before ending this study of the Meshalkin isomorphism, let us note that, in this present case, the coding preserves almost surely the topology. This topology is defined by taking as a basis for the open sets the elementary cylinders and the a.e. continuity stems from the fact that the coding of an element just needs the knowledge of a finite number of neighboring 2-stacks. It is yet unknown if such a topological isomorphism exists for two arbitrary B. shift with the same entropy.

*Final remarks.*

(i) In this section, I only gave a sketchy introduction to the theory of B. shifts. The Ornstein proof and other theorems can be found in Ref. (37). It is possible[38] to define what is called a Bernouilli flow which may be a property of a system of ordinary differential equations. The classical examples of "ergodic flows", that is the Hadamard-geodesic flow on compact surfaces of negative curvatures[39], the Sinai-Billiard with convex scatterers on a torus[40], and more generally the Anosov flow[41] are Bernouilli flows. For the other more recent examples I refer to ref. (42). In number theory the natural extensions (to make them invertible) of the β-transform and the continued fraction expansion are Bernouilli.

(ii) I have not explained the general construction of the entropy for a dynamical system. In the sense of information theory, it was invented by Shannon[43], later on it was introduced in the theory of dynamical systems by Kolmogoroff and Sinai. (Their definition of the entropy requires the knowledge of invariant measures, it is called sometimes *metric* entropy).

Another sort of entropy (or *topological* entropy) of dynamical systems has been introduced by Adler et al.[44] : it needs continuity of the transform and compactness of the underlying space. It has been shown[45] that, when the two entropies (metric and topological) can be defined for the same dynamical system, the topological entropy is the sup of the metric entropies defined over all the invariant measures.

Very roughly speaking, one can say that the entropy measures the average rate (over the initial point) of divergence of two neighboring trajectories in a mixing system. Recently, it became an interesting tool for studying numerically simple dynamical systems[46]. Sinai has proven recently[47] that the topological entropy is an extensive quantity in a (particular) many body system, that is one may define a topological entropy "per particle". This is an important step toward the understanding of the manner in which a N-body dynamical system approaches the thermodynamic limit.

## 6. STRANGE ATTRACTORS, ITERATION OF MAPPINGS OF THE PLANE AND OF THE LINE, AND SO ON

In this last section, I shall approximately follow the following paths. First I shall give a brief account of a joint work with J.L. Ibanes[48] on the Lorenzz system. Then I shall explain the Hénon's simulation of the horseshoe mapping which is related with the Lorenz system. Finally I shall give some brief indications on the endomorphism of the line. At each step of this enumeration, I have gone farther from the original problem of studying a given system of ordinary differential equations, namely the Lorenz system. I hope to make it clear that even with drastic simplifications, this remains a difficult problem and many questions are yet unanswered.

*The Lorenz system and the horseshoe mapping*

The Lorenz system of ordinary differential equations(49) is now very popular among "non linear" physicists. It has been discussed by many authors for the range of parameters studied by Lorenz himself. Let us only recall that this is a system of 3 non linear coupled differential equations :

$$(*)\quad \begin{cases} \dfrac{dx}{dt} = \sigma(y-x) \\[2ex] \dfrac{dy}{dt} = -xz + rx - y \\[2ex] \dfrac{dz}{dt} = xy - bz \end{cases}$$

$\sigma$ , b and r are parameters. I studied, with J.L. Ibanes, this system by keeping fixed the value of b and $\sigma$ as in the original paper of Lorenz : $\sigma = 10$ and $b = 8/3$, and by varying the parameter r. We observed on the T.V. screen coupled with the analog computer a puzzling set of bifurcations around $r \simeq 220$. We then studied the same system on a digital computer in the same range of values of the parameters. When r is slightly above 220, say at 230, a pair of two stable limit cycles exist which are symmetric with respect to each other in the change $(x,y) \leftrightarrow (-x,-y)$ without having individually this symmetry however. Cutting then trajectories by a Poincaré plane, one gets a phase map with a pair of stable fixed points. When r is decreased, each of these fixed points of the Poincaré map becomes linearly unstable : a small fluctuation around the closed trajectory (which was stable at larger values of r) grows first and then is stabilized by non linearities. In the Poincaré plane, the evolution of small fluctuations around one of the fixed points may be decribed as follows : if $\delta\vec{P}_i$ [$\delta\vec{P}_i$ is a two component vector, (0,0) being the fixed point itself] is the (small) fluctuation around this fixed point at the $i^{th}$ step, after one turn it becomes $\delta\vec{P}_{i+1} = M\,\delta\vec{P}_i$, M being a constant $2 \times 2$ matrix. If the eigenvalues of M have a modulus less than 1, the fixed point is linearly stable. If the modulus of one of these eigenvalues becomes larger than 1 [as the divergence of the flow defined by (*) is negative , only one of the eigenvalues can have a modulus larger than 1], this is no longer so. In the present case, one of the eigenvalues goes through (-1) : starting near the fixed point, the successive iterates jump back and forth and diverge from it beyond the bifurcation. In the present case, due to non linear effects, a stable period 2 appears by bifurcation from the period 1, and the 2 points of this period collapse continuously on this fixed point at the instability threshold.

Looking at the closed trajectory in the 3d space of variables (x,y,z), one sees that, at the bifircation, this closed trajectory gives birth to a 2-loop trajectory when the fixed point becomes unstable :

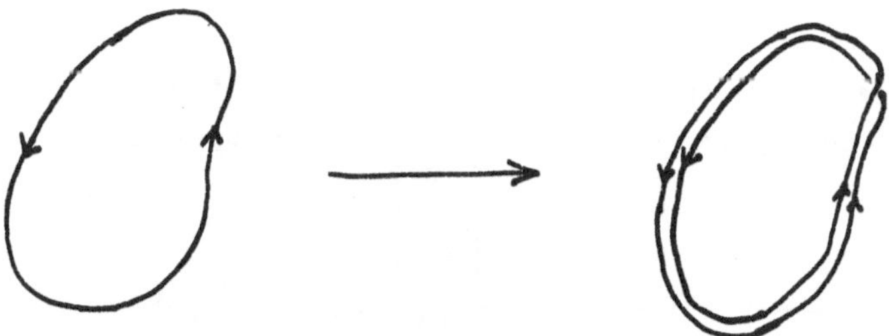

At the same time the period of the motion is multiplied by two. When the parameter r decreases a little more a new bifurcation occurs : the 2-loop limit cycle becomes again unstable and splits continuously into a 4-loop limit cycle. Then one observes a very striking phenomena : as r decreases a little more this splitting occurs again and again, and the initial period of the stable limit cycle is multiplied by $2,4,8,\ldots,2^n\ldots.$ Apparently this division of the frequency by two occurs an infinite number of times in a *finite* domain of variation of r. At the end of the process, the period is infinite, and one may guess that the system has got the mixing property (it is not clear if this is the strong or weak mixing, I suspect that this is weak mixing only, due to the existence of "quasi-periods" $T_o, 2T_o, \ldots, 2^nT_o \ldots.$).

Decreasing further r, the overall picture becomes very erratic : sometime a limit cycle exists, sometimes the point seems to move on a sort of surface. As this surface was apparently very different from the attractor described in the original paper of Lorenz, we studied it in detail.

This "surface" is rather close to the original stable limit cycle, it looks as closed ribbon, in particular there is a well defined hole in the middle. Accordingly it is possible to make a series of cuts, the last one being close to the first one, and one obtains the following series of pictures ; where time increases along the arrows.

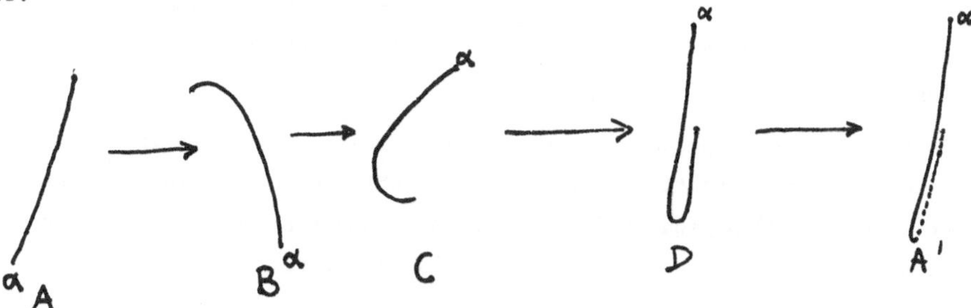


At the end of the process, the U-shaped section becomes so thin (its thickness is about $10^{-4}$ of the overall size) that one may neglect the fine structure of

A' = A at the accuracy of the calculation. This explains why the attractor is apparently a "surface" in the usual sense. But this is clearly not the case, as the two neighboring sheets of section D do not collapse in the evolution of D to A', as there is only one trajectory through each point [except through the fixed points of (*), of course, but they are outside the region under consideration ] . If one forgets the intermediate steps and considers the mapping from the initial section A into itself, as defined along the trajectory, it looks approximately as

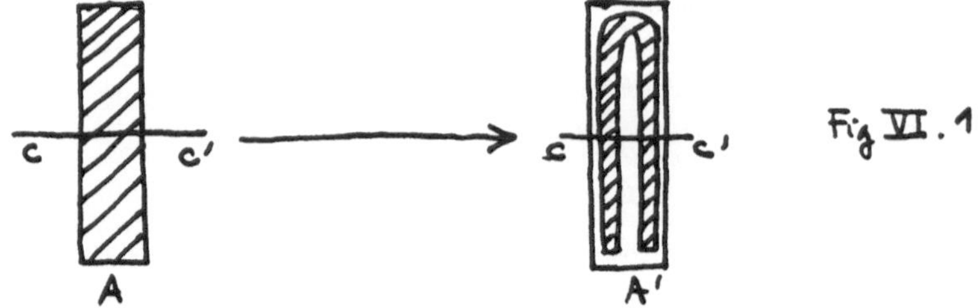

Fig VI.1

This sort of mapping is called the "horseshoe mapping" and has been studied, in particular, by Smale[11]. It is possible to show that a strange *object* (not attractor !) is stable by the mapping : consider the segment CC' ∩ A. After one application of the map the central part and two small parts on the border are deleted. This is precisely the method for generating the triadic Cantor set. After an infinite number of applications of the process, the cut of the remaining object by the line CC' is a topological Cantor set. But this does not exclude that inside this stable "object" a finite and stable period exists (which corresponds to a stable limit cycle for the original Lorenz systems).And this limit cycle exists actually for some values of the parameter .

After this (very) qualitative discussion, two questions come to one's mind :

(1) Is it possible to *prove* from the original equations that such stable objects exist with a Cantor like structure ?

This sort of question has been answered in a few explicit cases[50], but not for the Lorenz system however. In the present case, it does not seem to be beyond the possibilities : it is enough to prove that the topology of the A → A mapping is as described in Fig. 6.1. As one can see from the work of Levinson [50], it is enough (although rather uneasy practically) to handle a series of bounds for getting this topological insight.

(2) Is it possible to prove that such strange objects are strange "attractors" too,that is no simpler attracting structure, as stable periodic points exists inside this object? Here the situation is incredibly more complicated. I shall try to give a brief account of it for the case of the Hénon's transform.

M. Hénon has[55] modelled the above "horseshoe" transform by means of an explicit (and simple) mapping of the plane into itself (it is explicit, because one does not have to solve some ordinary differential equation to get $P_{i+1}$ from $P_i$). This is a quadratic mapping of the Cartesian plane into itself, defined as

$$x_{i+1} = 1 - a x_i^2 + y_i$$

$$y_{i+1} = b x_i$$

where a and b are parameters.

It is invertible (which makes it markedly different of the polynomial mapping studied by Stein and Ulam [51], although the fine layered structure of attractors appears in the class IV attractor of Stein and Ulam). It contracts the area by a constant factor if $|b| < 1$, as $dx_{i+1} \wedge dy_{i+1} = (-b)\, dx_i \wedge dy_i$. If $|b| < 1$ it is possible to find a quadrilater that maps into itself, roughly as shown in Fig. IV.1. Hénon has found values of the parameters for which the successive iterates of any point inside this quadrilateral move apparently erratically on an object with an infinitely sheeted structure.

There is another way for understanding the structure of this object. It is connected with a so called homoclinic point of Poincaré. The Hénon mapping has a pair of fixed points. Linearizing the map around a fixed point, one finds two directions defined by the tangent map (they are the eigenvector of the matrix with constant coeffciients obtained by linearizing the transform near the coordinates of the fixed point). For the values of a and b chosen by Hénon the fixed point is attracting along one of these directions and repulsive along the other one . By means of the Hadamard construction, it is possible to get along these directions two curves, or manifolds at a finite distance of the fixed point and intersecting of course, at the fixed point. These two manifolds are stable under the application of the Hénon map and of its inverse. The homoclinic point is a second intersection (beside the fixed point itself) of the two manifolds. This is *not* a fixed point. As the two manifolds are stable under the iteration, they must cross each other at all the iterates (and inverse iterates) of the homoclinic point, cutting each other again and again at many other points.

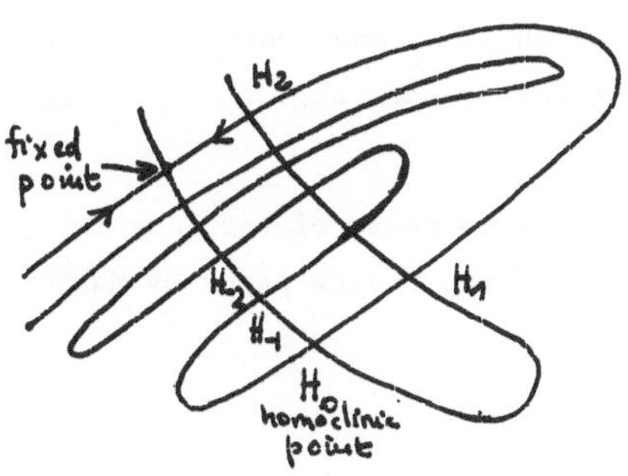

And the strange attractor of Hénon is nothing else but the attracting manifold (more exactly its closure) which is folded an infinite number of times, due to the existence of the homoclinic point.

Newhouse[52] has shown that under certain conditions, a wild attractor may exist for this sort of horseshoe transform : infinitely many stable periods may exist which are all attractive. Of course, their domain of attraction (or basin) can be very small. As these periods are certainly located on the strange object found by the infinite iteration of the mapping, it is likely very difficult to make experimentally a clear cut distinction between one of these "wild attractors" and a "strange attractor" without finite stable period.

One believes too that this strange attractor is not structurably stable. This means (approximately) that, in any open set of values of the parameter a topologically different attractor will exist, in other terms any very small perturbation in the equations will completely destroy its structure. This contrasts with the stable fixed point, for instance, which is structurably stable : if a mapping has a stable fixed point (or eventually, a stable period) any neighboring mapping has a stable fixed point too(this is true if one is not just at a bifurcation point). This idea of structural stability is very important indeed[53], but it is important to realize that a property might be non structurably stable, although one has a finite "chance" of finding it for some values of the parameter. This comes from the existence of topological Cantor sets with a finite (non zero) Lebesgue measure. Hermann[26], in his thesis and subsequent works has given an example of a dynamical property (the irrationality of the rotation number for a diffeomorphism of the circle) which is not structurally stable, although it is relevant in the sense of the measure in the parameter space.

The most simple dynamical system which is relevant here is obtained by putting $b = 0$ in the Hénon's transform. One gets the quadratic endomorphism of the line into itself :

$$x_{i+1} = 1 - a\,x_i^2 \quad .$$

This is a fair description of what happens in the original Lorenz system (although it is obviously not invertible) induced, when one considers the map of the linear coodinate along the sheet of the attractor, and neglects the 2-dimensional nature of the problem.

I just refer for this question of the endomorphism of the line to my joint work with B. Derrida and A. Gervois[10] and to the review paper by May[54].

## REFERENCES


1 - See for instance, Chap.3 in J.P. Hansen, I.R. McDonald "Theory of Simple Liquids", Academic Press (1976)

2 - S.D. Stoddard, J. Ford, Phys. Rev. A8, 1504 (1973) ;
- L. Galgani, A. Scotti, Revista del Nuovo Cimento II-2, 189 (1972)

3 - C. Froeschlé, J.P. Scheidecker, Phys. Rev. A12, 2137 (1975)

4 - D. Ruelle, F. Takens, Comm. Math. Phys. 20, 167 (1971) ; 23, 343 (1971)

5 - M. Hénon, Comm. Math. Phys. 50, 69 (1976) ;
- M. Hénon, Y. Pomeau in "Mathematical Problems in Turbulence Theory", Springer Verlag (1976)

6 - Ya.G. Sinai, "Theory of Dynamical Systems", Part I, Aarhus Univ.(Denmark)

7 - J.P. Gollub and H.L. Swinney, Phys. Rev. Lett. 35, 927 (1975)
- G. Ahlers, R.P. Behringer, Phys. Rev. Lett. 40, 712 (1978)
- A. Libchaber, J. Maurer, in preparation and communication at Euromech Colloquium Grenoble, Sept.1978
- P. Bergé, M. Dubois in "Les instabilités hydrodynamiques en convection forcée, libre et mixte", Lecture Notes in Physics 72, Springer Verlag (1978)

8 - P.R. Halmos, "Measure Theory", Van Norstrand, Reinhold C (1950)

9 - Section 16 in Ref.(8).

10 - B. Derrida, A. Gervois, Y. Pomeau, to be published in Annales de l'I.H.P.

11 - S. Smale, Bull. Am. Math. 73, 747 (1967)

12 - A. Renyi, Acta Math. Acad. Sci. Hung. 8, 477 (1957)

13 - W. Parry, Acta Math. Acad. Sci. Hung. 11, 401 (1960)

14 - The K.A.M. theorem is proved in Chap.III, vol.2 in S. Sternberg "Celestial Mechanics", W.A. Benjamin (1969)

15 - The elementary properties of the continued fraction expansion can be found in "Continued Fractions", by C.D. Olds, Random House (1963)

16 - A. Lasota, J.A. Yorke, Trans. Amer. Math. Soc. 186, 481 (1973)
- D. Ruelle, Comm. Math. Phys. 55, 471 (1977)

17 - "Diophantishes Approximation", Erg. Math. Grenzgcb. Band 4, Hoft 4, Julius Springer, Berlin (1936) by J.F. Koksma

18 - M. Keane, Isr. J. of Math. 26, 188 (1977)

19 - A.I. Khinchine, "Mathematical Foundation of Statistical Mechanics" Dover (1949)
- I.E. Farquhar, "Ergodic Theory in Statistical Mechanics", Interscience Pub. (1964)

20 - Book in Ref.(22), p.69 et seq.

21 - P.R. Halmos, "Lectures on Eergodic Theory", Chelsea Pub. Company, N.Y. (1956)

22 - Chap.2 in L. Kuipers and H. Niederreiter, "Uniform Distribution of Sequences" J. Wiley and Sons (1974)

23 - J. Hardy, Ph.D. Thesis, Univ. of Orsay (1978)

24 - Kuzmin, D.A.N. SSSR, Ser.A 375 (1928)

25 - H. Kesten, Acta Arith. 10, 183 (1964/65)

26 - M.R. Herman, C.R. Acad. Sci. Paris A283 (1976) ; Ph.D. Thesis, Orsay (1977) ; Bull. Soc. Math. de France 46, 181 (1976)

27 - H. Furstenberg, H. Kesten, Ann. Math. Statist. 31, 457 (1960)
V.I. Oseledec, Trans. Moscow Math. Soc. 19, 197 (1968)

28 - Simple frustrated systems : chains, strips and square, B. Derrida, J. Vaminnenus and Y. Pomeau, submitted to J. Phys. A

29 - D. Ruelle, Talk presented at the Conference on Bifurcation Theory and its Applications in Scientific Disciplines, October 31 - November 4 (1977) organized by the New York Acad. of Sciences
- D. Ruelle, preprint IHES/P/77/193

30 - Ref. 21 , p.26 et seq.

31 - G. Gallavotti, La Rivista del Nuovo Cimento 2, 133 (1972) ; Comm. Math. Phys. 27, 103 (1972)

32 - D.S. Ornstein in Advances in Math. 10, 49 (1973) gives an example of a K system that is not Bernouilli

33 - It is shown on Chap. X of the book of M. Smorodinsky (our Ref.(37)) that finite mixing Markov shift are Bernouilli

34 - Ya.G. Sinai, Dokl. Akad. Nauk SSSR 124, 768 (1959) ;
A.N. Kolmogoroff, Dokl. Akad. Nauk SSSR 119, 864 (1958)

35 - D.S. Ornstein, Advances in Math. 4, 337 (1970) and 5, 339 (1970)

36 - L.D. Meshalkin, Dokl. Akad. Nauk SSSR 128, 41 (1959)
- J. Blum, D. Hanson, Bull. Amer. Math. Soc. 69, 221 (1963)

37 - The proof of Ornstein's theorem and other properties of B. shift can be found
38 in M. Smorodinsky, "Ergodic Theory, Entropy", Lecture Notes in Math. 214, Springer Verlag (1971)
and J.R. Brown, "Ergodic Theory and Topological Dynamics", Academic Press (1976)

39 - D. Anosov, Proc. Steklov Inst. of Math. n°90 (1967)

40 - Katznelson has proved that ergodic automorphism of $T^n$ are Bernouilli shifts

41 - R. Azencott, C.R. Acad. Sci. Paris A270, 1105 (1970)

42 - M.C. Gutzwiller, J. Math. Phys. 18, 806 (1977)

43 - C. Shannon, Bell Syst. Tech. J. 27, 379 (1948)

44 - R.L. Adler, A.G. Konheim and M.H. McAndrew, Trans. Amer. Math. Soc. 114, 309 (1965)

45 - E.I. Dinaburg, Sov. Math. Dokl. 11, 13 (1970)

46 - G. Benettin, L. Galgani, J.M. Strelczyn, Phys. Rev. A14, 2338 (1976)

47 - Ya.G. Sinai, "On Measure Theoretical Entropy per Particle of the Dynamical System of Hard Spheres", n°172 in Non Linear Science Abstract, April 1978.

48 - J.L. Ibañes, Y. Pomeau, to appear in J. of Non Equilibrium Thermodynamics (1978)

49 - E.N. Lorenz, J. Atmo. Sciences 20, 130 (1963)

50 - N. Levinson, Ann. of Math. (2) 50, 127 (1949)

51 - P.R. Stein, S. Ulam, in "Studies in Math. Physics", A.O. Barut ed., D. Reidel Pub. Company (1973) and Rozprawy Matematyczne 39 (1964)

52 - S. Newhouse, Topology 12, 9 (1974)

53 - R. Thom, "Stabilité Structurelle et Morphogénèse", Reading Mass. (1973)

54 - R. May, Nature 261, 459 (1976)

# S E M I N A R S

and

# S H O R T   C O M M U N I C A T I O N S

# FOKKER-PLANCK DESCRIPTION OF A NAVIER-STOKES FLUID WITH INSTABILITY

M. DROZ, C.P. ENZ and A. MALASPINAS
Department of Theoretical Physics - University - 1211 Geneva 4

## 1. INTRODUCTION

The flow of a viscous incompressible fluid between two concentric cylinders, the inner of which rotates (Couette flow), becomes unstable at a sufficiently large angular velocity $\Omega$ . At the first instability, characterized by a critical value $\Omega_c$ of the angular velocity (Taylor instability), the flow forms a pattern of tori analogous to the rolls in the Bénard effect[1] . When appropriate fluctuations are taken into account a Fokker-Planck description may be applied to the Couette flow, along the lines pioneered by Graham in the Bénard problem[2,3].

Our aim was to make use of the Fokker-Planck formalism developed recently[4] in the problem of the Taylor instability. However, owing to the cylindrical geometry of the Couette flow this problem is complicated in a serious though inessential way by the need to work with Bessel functions[1]. For this reason we addressed ourselves to the essentially equivalent problem of a flow between two parallel plates at distance $h$ , one of which is moving with constant velocity $\alpha h$ , so that $\alpha$ plays the role of the angular velocity.

This geometry can be thought of as being the limit of infinite radii of the concentric cylinders in the Couette flow, their difference being kept at the constant value $h$ . In this limit it is possible to apply Graham's formalism[3] in essentially the same form as in the Bénard problem[5]. It is important, however, to retain an appropriate form of the centrifugal force, which is responsible for the Taylor instability. Since this form is not determined by the limit we simulate the destabilizing force by a linearized version of the centrifugal force similar to the Lorentz force.

## 2. NAVIER-STOKES EQUATIONS

We assume the velocity field $\vec{u}(\vec{r},t)$ to consist of a stationary flow in the x-direction :

$$\vec{u}_0 = (\alpha z, 0, 0) \quad ; \quad 0 \leqslant z \leqslant h \tag{1}$$

and a fluctuation $\vec{v}$ which at the instability develops into rolls with axis along the x-direction,

$$\vec{u} = \vec{u}_0 + \vec{v}(y,z,t) \ . \tag{2}$$

This field is assumed to satisfy the Navier-Stokes equations with the destabilizing force $-B\vec{v}$:

$$\dot{\vec{u}} + (\vec{u}\cdot\nabla)\,\vec{u} = -\nabla p + \nu\,\nabla^2\vec{v} - B\vec{v}, \tag{3}$$

where $p$ is the pressure (we put the density $\varrho = 1$), $\nu$ is the viscosity and $B$ is a constant matrix to be specified. Inserting (2), eliminating $p$ with the help of the transversality condition $\nabla\cdot\vec{u}=0$ and adding a random force $\vec{\xi}$, we may write eq. (3) as Langevin equations[4)]

$$\dot{v}_i = f_i^0 + f_i' + g_i + \xi_i \ . \tag{4}$$

Here[6)],

$$f_i^0 = -P_{i\ell}\ v_{\perp j}\nabla_j v_{\perp\ell} \tag{5}$$

is the non-dissipative mode-coupling force,

$$f_i' = \nu\,\nabla^2 v_{\perp i} \tag{6}$$

is the dissipative viscous force, and $v_{\perp i} = P_{ij}v_j$ where $P_{ij} = \delta_{ij} - \nabla^{-2}\nabla_i\nabla_j$ is the transverse projection, and repeated vector indices are to be summed over.

$$g_i = -P_{i\ell}\,B_{\ell j}v_{\perp j} - A_{ij}v_{\perp j} \tag{7}$$

contains the effect of the destabilizing force and of the drag due to the stationary flow (1). Here $A_{ij} = \nabla_j u_{0i}$, or

$$A = \begin{pmatrix} 0 & 0 & \alpha \\ 0 & 0 & 0 \\ 0 & 0 & 0 \end{pmatrix} \quad ; \quad B = \begin{pmatrix} 0 & 0 & \gamma \\ 0 & 0 & 0 \\ \beta & 0 & 0 \end{pmatrix} \tag{8}$$

and the form of $B$ is chosen to be such as to lead to an instability.

Eqs. (4) have to be supplemented by boundary conditions which express complete adherence of the fluid at the plates[1)] : $\vec{v} = 0$ for $z = 0$, $h$. Finally, the random forces are assumed to be of the usual Markovian white-noise type, characterized by a correlation or diffusion matrix $C_{ij}(\vec{r},\vec{r}\,')$[4)].

## 3. LINEAR STABILITY ANALYSIS

The linear deterministic part of eqs. (4) may be written as

$$-i\omega\vec{\Psi} = \Lambda\vec{\Psi} \tag{9}$$

where $\Lambda = \nu\nabla^2 - PB - A$ . This leads, for the Fourier components defined by

$$\vec{\Psi}(y,z) = \sum_{q_y,q_z} \vec{a}(q_y,q_z)\, e^{iq_y y + iq_z(z+h/2)} , \tag{10}$$

to the following explicit solution

$$\frac{a_1}{a_3} = \frac{\alpha+\gamma}{i\omega - \nu q^2} \tag{11}$$

with the condition

$$(i\omega - \nu q^2)^2 = \Omega^2 q_y^2 / q^2 . \tag{12}$$

Here $q^2 = q_y^2 + q_z^2$ , $\Omega^2 = (\alpha+\gamma)\beta$ and the component $a_2$ is determined by the transversality condition.

In the variable $z + h/2$ chosen in eq. (10) the $\Psi_i$ as functions of $z$ are eigenfunctions of parity. Here we are interested in the lowest instability, which is a soft, positive-parity mode[1) ;
$\omega = 0$ and $a_i(q_y,q_z) = a_i(q_y, -q_z)$ . The three roots $q^{(\lambda)}(q_y,\Omega)$ , $\lambda = 0, \pm$ , of eq. (12) give rise to three values $q_z^{(\lambda)}(q_y,\Omega)$ of $q_z$ , so that the 3-component of eq. (10) takes the form

$$\Psi_3(y,z) = \sum_{\lambda=0,\pm} A^{(\lambda)}(y) \cos q_z^{(\lambda)}(z+h/2) . \tag{13}$$

The boundary conditions

$$\Psi_1 = \Psi_3 = (\nabla_3\Psi_3) = 0 \quad ; \quad z = 0,h \tag{14}$$

then lead to a condition which determines $\Omega$ as a function of $q_y$ such that $\Omega$ has a minimum $\Omega_c = \sqrt{1708}\,\nu/h^2$ at $q_{yc}=3.117\ h^{-1}$, in complete agreement with the Bénard problem[1)].

## 4. THE FOKKER-PLANCK EQUATION

We shall assume here that the matrix B of eq. (8) is odd under time-reversal, as are $\vec{v}$ and A , which means that the destabilizing force is nondissipative, as is the centrifugal force. In this case the stationary dissipative probability current[4] may be made to vanish:

$$f_i'(\vec{r}) - \int (\delta C_{ji}(\vec{r}',\vec{r}) / \delta v_j(\vec{r}'))d^3r'$$
$$+ \int (\delta H / \delta v_j(\vec{r}'))\, C_{ji}(\vec{r}',\vec{r})d^3r' = 0 \ . \qquad (15)$$

These are the potential conditions[4], in which $H[\vec{v}]$ is defined by the stationary probability distribution $P^s[\vec{v}] = Z\exp(-H)$. In the case of a dissipative destabilizing force (which we do not consider here) the B-part of $g_i$ in eq. (7) has to be added to $f_i'$, and eq. (15) cannot be fulfilled in any obvious way. The solution of eq. (15) is (compare ref. 6)

$$H[\vec{v}] = \frac{1}{2T} \int d^3r(\vec{v} + \vec{w})^2 \qquad (16)$$

and

$$C_{ij}(\vec{r},\vec{r}') = - T\nu\nabla^2 P_{ij}\,\delta(\vec{r}-\vec{r}') \qquad (17)$$

where $\vec{w}(\vec{r})$ is independent of $\vec{v}$ and $T$ is the temperature ($k_B=1$). With eq. (15) the stationary Fokker-Planck equation reduces to a variational differential equation between $f_i^0 + g_i$ and $H$, which may be satisfied with an appropriate choice of $\vec{w}$ [7].

## 5. THE GREEN'S FUNCTIONS

Since Green's-function averages are to be taken with the stationary probability distribution $P^s = Z\exp(-H)$ and expression (16), the simplest form of the correlation function[4] is $\langle \bar{T}((v_i(\vec{r},t) + w_i(\vec{r}))(v_j(\vec{r}',t') + w_j(\vec{r}')))\rangle$. This amounts in the Fourier-transformed quantities to calculating

$$G_{i\vec{q},j-\vec{q}}(t) = \langle \bar{T}(v_{i\vec{q}}(t)\, v_{j-\vec{q}}(0))\rangle \qquad (18)$$

with $\vec{w} = 0$ in $H$. Since the soft-mode instability is determined by the linear time evolution (9) this instability must show up in the poles of the unperturbed correlation and response functions $G^0$ and $R^0$. Since $a_2$ may be expressed in terms of $a_1$ and $a_3$ the free time-dependence is determined by the 2x2 submatrix of $\Lambda$ with labels 1 and 3, whose eigenvalues are given by eq. (12) with $\omega = 0$ :

$$m_{\pm} = -\nu q^2 \pm \Omega_y \tag{19}$$

where $\Omega_y \equiv \Omega|q_y|/q$ . The calculation of the free correlation functions then proceeds as in ref. 4 and leads to

$$G^0_{i\vec{q},i-\vec{q}}(t) = T\cosh(\Omega_y t)\, e^{-\nu q^2|t|}; \; i = 1,3$$

$$G^0_{1\vec{q},3-\vec{q}}(t) = G^0_{3\vec{q},1-\vec{q}}(-t) = T\left(\frac{\Omega q}{\beta|q_y|}\,\Theta(t) + \frac{\beta|q_y|}{\Omega q}\,\Theta(-t)\right)\cdot$$

$$\cdot \sinh(\Omega_y|t|)\, e^{-\nu q^2|t|} \; . \tag{20}$$

With the help of the fluctuation-dissipation theorem $R^0(t) = \Theta(t)G^0(t)$, which is guaranteed by the potential conditions (15)[4,5], the Fourier-transformed free response functions are then easily obtained; e.g.

$$R^0_{ii}(q,\omega) = \frac{T}{2}\left(\frac{\nu q^2-\Omega_y}{i\omega-\nu q^2+\Omega_y} + \frac{\nu q^2+\Omega_y}{i\omega-\nu q^2-\Omega_y}\right)\; i = 1,3 \; . \tag{21}$$

Stability requires that the poles $\omega_{\pm}$ of $R^0$ are in the lower half $\omega$-plane, in other words, that $\nu^2 q^4 > \Omega_y^2$ . This implies that $\Omega^2 = (\alpha+\gamma)\beta > 0$ . The stability limit is reached at the soft modes $\omega = 0$ of eq. (12) and with a minimum value $\Omega_c$ at $q_{yc}$ .

## 6. THE CRITICAL REGION

The soft-mode behaviour near the bifurcation is reminiscent of a phase transition of the Ginzburg-Landau (GL) type. So one would expect that in appropriate variables the Langevin eqs. (4) would assume the form of time-dependent Ginzburg-Landau (TDGL) equations with fluctuations. This is not obvious, however, if one looks at the H-functional of eq. (16) that determines the stationary Fokker-Planck distribution $P^s$ . Indeed H is not of Ginzburg-Landau form but is just a free "Hamiltonian", the interaction being contained entirely in the mode-coupling force (5).

One is tempted to split off from H a term $H_{GL}$ which depends only on appropriate slow variables, and to factorize $\vec{v}$ into a "slow" and a "fast" term. This is the method used by Graham in the Bénard problem[3]. The idea is to examine how the variables scale in the vicinity of the bifurcation. Writing $\vec{q}-\vec{q}_c = \Delta\vec{q}$ with $\vec{q}_c=(0,q_{yc},q_{zc})$ and assuming $q_z$ to be locked by the boundary conditions, so that, to the order considered, $\Delta q_z = 0$ , we have

$$q^2-q_c^2 \cong (\Delta q_x)^2 + 2q_{yc}\Delta q_y$$

$$\Omega - \Omega_c \propto (\Delta q_y)^2 \qquad (22)$$

or, introducing a scaling parameter $\varepsilon$ ,

$$q_x = \varepsilon\kappa_x \;;\; q_y = q_{yc} + \varepsilon^2\kappa_y \;;\; q_z = q_{zc}$$

$$q^2 = q_c^2 + \varepsilon^2\kappa^2 \;;\; \omega = \varepsilon^4\sigma \qquad (23)$$

where $\kappa^2 = \kappa_x^2 + 2q_{yc}\kappa_y$ . Writing $(\vec{q}-\vec{q}_c)\cdot\vec{r} = \xi\kappa_x + \eta\kappa_y$ and $\omega t = \sigma\tau$ and assuming that we stay close to the stability limit $\Omega(q_y)$ , we obtain for the slow variables

$$\xi = \varepsilon x \;;\; \eta = \varepsilon^2 y \;;\; \tau = \varepsilon^4 t \, . \qquad (24)$$

It is now clear that we must also expand the velocity field,

$$\vec{v}(\vec{r},t) = \sum_n \varepsilon^n \vec{v}^{(n)}(\vec{r},t) \qquad (25)$$

and multiply the zeroth-order solution (10) by a slowly varying amplitude,

$$\vec{v}^{(0)}(\vec{r},t) = \varepsilon^s w(\xi,\eta,\tau)\vec{\Psi}(y,z) \, . \qquad (26)$$

Here $s > 0$ expresses the fact that the perturbation should vanish at the bifurcation but that it need not be specified further.

Equations (23), (24) and (26) show that the derivatives in the Langevin eqs. (4) are to be taken as if the "slow" and "fast" variables were independent :

$$\partial_x \rightarrow \varepsilon\partial_\xi \;;\; \partial_y \rightarrow \partial_y + \varepsilon^2\partial_\eta \;;\; \partial_z \rightarrow \partial_z \;;\; \partial_t \rightarrow \varepsilon^4\partial_\tau \, . \qquad (27)$$

Expanding eqs. (4) in powers of $\varepsilon$ one sees from (27) that the time derivative appears only at the order $\varepsilon^{4+s}$ . Hence it is at this order that the Langevin equation will appear as an equation for the slow amplitude $w(\xi,\eta,\tau)$ , while the lower orders serve to determine $\vec{v}^{(n)}$ with $n \leq 4$ . However, one has to make sure that the random force $\xi_i$ also appears at this order $\varepsilon^{4+s}$ . Now according to eqs. (17) and (24),

$$\langle \xi_i(\vec{r},t)\, \xi_j(\vec{r}',t')\rangle \propto \varepsilon^{7+s'}\, \delta(\xi - \xi')\ \ (\eta - \eta') \cdot$$
$$\cdot\ \delta(z-z') \cdot \delta(\tau - \tau')\ , \qquad (28)$$

assuming that the viscosity scales as $\varepsilon^{s'}$. Thus we must have $(7+s')/2 = 4+s$ , or $s' = 1+2s > 1$ .

The procedure described here is straightforward but quite tedious and is therefore left for detailed description elsewhere[7]. As in the case of the Bénard problem[3] the result is that $w(\xi, \eta, \tau)$ satisfies a TDGL equation with a random force. The associated stationary Fokker-Planck distribution will therefore indeed be proportional to $\exp(-H_{GL})$ and the soft response function will be dominated by the soft-mode poles of eq. (21).

REFERENCES

1) S. CHANDRASEKHAR, Hydrodynamic and hydromagnetic stability, (Clarendon, Oxford, 1961).

2) R. GRAHAM, Phys. Rev. Letters 31, 1479 (1973).

3) R. GRAHAM, Phys. Rev. A10, 1762 (1974).

4) C.P. ENZ, Physica 89A, 1 (1977).

5) R. GRAHAM, Private Communication.

6) C.P. ENZ, Physica A, to be published.

7) M. DROZ, C.P. ENZ and A. MALASPINAS, to be published.

# OPERATOR FORMALISM FOR CONTINUOUS MARKOV PROCESSES

L. GARRIDO and M. SAN MIGUEL
Dpto. Física Teórica, Facultad de Física
Diagonal 647, Barcelona-28, Spain

## 1. INTRODUCTION

We present here a brief account of recent work on the operator formulation of continuous Markov processes. The central point of this formalism is a set of operator equations of motion, known as Fokker-Planck dynamics, which represent a Heisenberg picture associated with the Fokker-Planck Equation (FPE) for that process. Our aim is to clarify the meaning , foundations and consequences of these equations. A first part is devoted to the introduction of the Fokker-Planck Dynamics and to the study of the corresponding thermodynamic propagators [1]. In the second part, underlying c-number equations that allow one to deduce Fokker-Planck dynamics by means of stochastic quantization are studied. Symmetry under time reversal is seen to be broken in this last step. This second part is the result of joint work with D. Lurie. [2]

## 2. FOKKER-PLANCK DYNAMICS

a) Equations of motion

In a great variety of physical problems one is faced with Langevin-type equations for a set of N gross variables in whose evolution we are interested.

$$\dot{q}_\mu = v_\mu \left(q_1(t), \ldots, q_N(t), t\right) + \xi_\mu(t) \tag{1}$$

where $\xi_\mu(t)$ is a stationary random force, assumed to be Gaussian, with zero mean, white-noise spectrum and a positive definite correlation matrix $D_{\mu\nu}$ independent of q:

$$E_\xi \left\{ \xi_\mu(t) \ \xi_\nu(t') \right\} = 2 \, D_{\mu\nu} \ \delta(t-t') \tag{2}$$

The quantities of interest are the correlation functions of the gross variables. A simple-minded way of attacking this problem is to solve eq. (1) formally for each realization of $\xi_\mu(t)$ so that $q_\mu(t)$ is a functional of $\xi_\mu(t)$ and of the initial conditions $q_u(0) \equiv q_u$:

$$q_\mu(t) = q_\mu\left(\left[\xi(t)\right], q, t\right) \tag{3}$$

The correlation function of $q_\mu(t)\, q_\nu(t')$ involves two kinds of averages ; an average over the realizations of $\xi_\mu(t)$ and a further average over the distribution of initial values. The main idea of the Fokker-Planck Dynamics that we present here is to introduce the first average into the dynamics. One is only left with the usual statistical-mechanical average over a probability distribution for the initial values, which defines an ensemble of equivalent systems with identical dynamics but starting from a different point in the q-space. To see how this can be achieved, let us consider eq. (1) for a definite realization of $\xi_\mu(t)$ and rewrite it as

$$\dot{q}_\mu(t) = \left[L_\xi\,(q(t), \hat{q}(t), t), q_\mu(t)\right] \tag{4}$$

$$L_\xi(q(t), \hat{q}(t), t) = \sum_\mu \left(v_\mu(q_1(t)\ldots, q_N(t);t)\, \hat{q}_\mu(t) + \xi_\mu(t)\hat{q}_\mu(t)\right) \tag{5}$$

where $\hat{q}_\mu(t)$ is an operator defined by the following equation of motion

$$\dot{\hat{q}}_\mu(t) = \left[L_\xi\,(q(t), \hat{q}(t), t),\ \hat{q}_\mu(t)\right] \tag{6}$$

with inital condition $\hat{q}_\mu(0) = \dfrac{\partial}{\partial q_\mu}$ (7)

so that it obeys $\left[\hat{q}_\mu(t), q_\nu(t)\right] = \delta_{\mu\nu}$ (8)

and can be interpreted as the derivative with respect to the value that $q_\mu(t)$ takes at the time t.

A statistical treatment of eq. (4) can be made by considering an ensemble of systems obeying eq. (4). The probability of finding the system at the point q at time t is given by the probability density $\rho(q,t)$, which obeys the continuity equation

$$\frac{\partial}{\partial t}\rho(q,t) = -\sum_\mu \frac{\partial}{\partial q}\left(\dot{q}_\mu\, \rho(q,t)\right) = \left(L_\xi^+\, \rho\right) \tag{9}$$

This equation represents a "Schrödinger picture" associated with (4),and we see that is governed by the adjoint operator $L^+$. Our interest is in the average over different realizations of $\xi_\mu(t)$. The solution to eq. (7) can be expressed as a functional of $\xi_\mu(t)$. It is by now well known [(3-7)] that if this solution is averaged over the realizations of $\xi(t)$,

$$P(q,t) = E_{\xi} \left\{ \rho\,(q,t) \right\} \tag{10}$$

it becomes the solution of the FPE associated with (1), i.e.,

$$\frac{\partial(P(q,t)}{\partial\, t} = - \sum_{\mu} \frac{\partial}{\partial q_{\mu}} (v_{\mu}\,(q,t)\; P(q,t)) + \sum_{\mu\nu} D_{\mu\nu} \frac{\partial^2}{\partial q_{\mu}\, \partial q_{\nu}} P(q,t)$$

$$= (L^{+}\; P(q,t)) \tag{11}$$

where the usual assumption $P(q,0) = \rho\,(q,0)$ has been made (see the lectures by Prof. Mazo and Prof. Santos in these proceedings).

By analogy of what happens in eqs. (4) and (9), when a single realization of $\xi_{\mu}(t)$ is considered, we now assume that the adjoint dynamics gives rise to a set of equations of motion in which the average over the realizations of $\xi_{\mu}(t)$ is already included. This assumption can be proved to be true [1] in every order of perturbation theory by means of a Wick theorem. In fact, this is a trivial statement when one considers averages of quantities depending only on one time, since (if $v_{\mu}$ does not depend explicitly on time).

$$\langle q_{\mu}\,(t)\rangle = \int d^n q P(q,t)\; q_{\mu} = \int d^n q (e^{L^{+}t}\; P(q,0))q_{\mu} =$$

$$= \int d^n q\;\; P(q,0) e^{Lt}\; q_{\mu}\; e^{-Lt} = \int d^n\; q\; \rho(q,0)\; q_{\mu}(t) \tag{12}$$

where $$\dot{q}_{\mu}\,(t) = \left[ L(q(t), \hat{q}(t)), q_{\mu}\,(t) \right] \tag{13}$$

and we see that in the last expression $q_{\mu}\,(t)$ is only averaged over the distribution of initial values $\rho(q,0)$. Rather than giving a general proof of the validity of our assumption we discuss here in a simple example the implications of these equations of motion.

The general equations of motion representing the "Heisenberg picture" associated with (11) (the so-called Fokker-Planck Dynamics) are [1,8]

$$\dot{q}_{\mu}\,(t) = \left[ L(q(t),\; \hat{q}(t),t),\; q_{\mu}(t) \right] = v_{\mu}\,(q(t),t) + \sum_{\nu} 2D_{\mu\nu}\;\; \hat{q}_{\nu}(t) \tag{14}$$

$$\dot{\hat{q}}_{\mu}\,(t) = \left[ L(q(t),\; \hat{q}(t),t),\; \hat{q}_{\mu}\,(t) \right] = -\sum_{\nu} \left[ \hat{q}_{\mu}(t),\;\; v_{\nu}(q(t),t) \right]\; \hat{q}_{\nu}(t) \tag{15}$$

$$\left[\hat{q}_\mu(t),\ q_\nu(t)\right] = \delta_{\mu\nu} \tag{16}$$

where $L(q(t),\hat{q}(t),t) = \sum_\mu v_\mu(q(t),t)\ \hat{q}_\mu(t) + \sum_{\mu\nu} D_{\mu\nu}\ \hat{q}_\mu(t)\ \hat{q}_\nu(t)$ (17)

Let us consider as a simple example the linear case

$$v_\mu = \lambda_\mu q_\mu \tag{18}$$

The solution of (14) + (15) reads in this case

$$q_\mu(t) = e^{\lambda_\mu t}\, q_\mu + \sum_\nu V_{\mu\nu}(t)\ \hat{q}_\nu \tag{19}$$

$$\hat{q}_\mu(t) = e^{-\lambda_\mu t}\hat{q}_\mu \tag{20}$$

where

$$V_{\mu\nu}(t-t) = \frac{2D_{\mu\nu}}{\lambda_\mu+\lambda_\nu}\left(e^{\lambda_\mu t} - e^{-\lambda_\nu t}\right) \tag{21}$$

From this solution, the non equal-time commutators can be evaluated:

$$\left[\hat{q}_\mu(t),\ \hat{q}_\nu(t')\right] = 0 \tag{22}$$

$$\left[\hat{q}_\mu(t),\ q_\nu(t')\right] = \delta_{\mu\nu}\ e^{-\lambda_\mu t}\ e^{\lambda_\nu t'} \tag{23}$$

$$\left[q_\mu(t),\ q_\nu(t')\right] = V_{\mu\nu}(t-t') \tag{24}$$

The whole effect of the stochastic force is characterized in this example by the second term on the right-hand side of (19), which is related by (24) to the commutator of the gross variables q(t) at different times. This lack of commutativity is the key implication of the equations of motion. What does it mean that classical variables do not commute? At any fixed time t, the system is completely specified by the values of the n gross variables $q_\mu(t)$ and, therefore, they can be chosen independently. This implies that no kinematical uncertainty exists, (kinematical uncertainty is inherent to Quantum Mechanics), and so the gross variables $q_\mu(t)$ should commute at equal times. This is indeed the case, since

$$V_{\mu\nu}(0) = 0 \tag{25}$$

Nevertheless, stochasticity introduces a dynamical uncertainty: This means physically that the probability density for the position of a Brownian particle is smeared out in the course of time. Mathematically, this is reflected by the commutation relation (24). The function $V_{\mu\nu}(t)$ is in fact the coefficient of the stochastic term in (19). In conclusion, the lack of commutativity accounts for the introduction of the stochastic average into the dynamics and can be thought of as a dynamical uncertainty relation. Equations (19) and (20) define a set of stochastic equations of motion from which physical results are obtained when their solutions are averaged over the probability distribution of initial values that defines the ensemble of systems under consideration.

In order to see the physical effects of the stochastic force we go on to study the correlation and response function that characterize our system.

b) Propagators

The correlation function for arbitrary times is defined by

$$G_{\mu\nu}(t,t') = \langle \bar{T}(q_\mu(t)q_\nu(t'))\rangle = \int d^nP(q,0)\bar{T}(q_\mu(t)q_\nu(t')) \tag{26}$$

where the brackets mean an average over the distribution of initial conditions and $\bar{T}$ is the time antiordering operator. The response function to an external force coupled to $q_\mu(t)$ through a function $\Gamma_{\mu\nu}(q(t),t)$ such that the original eq. (1) becomes

$$\dot{q}_\mu = v_\mu(q(t),t) + \sum_\nu \Gamma_{\mu\nu}(q(t),t)\, q_\nu(t) + \xi_\mu(t) \tag{27}$$

is

$$\bar{R}_{\mu\nu}(t,t') = \sum_\delta \langle \bar{T}(q_\mu(t)\,\Gamma_{\delta\nu}(q(t'),t')\,\hat{q}_\delta(t')) \tag{28}$$

For the simple example (18), the correlation function is found to be

$$G_{\mu\nu}(t,t') = g^{\circ}_{\mu\nu}(t,t') + U^{0}_{\mu\nu}(t,t') \tag{29}$$

where

$$g^{0}_{\mu\nu}(t,t') = e^{\lambda_\mu t} e^{\lambda_\nu t'} \langle q_\mu q_\nu \rangle \tag{30}$$

$$U^{0}_{\mu\nu}(t,t') = \Theta(t-t')e^{\lambda_\mu t} V_{\nu\mu}(t') + \Theta(t'-t)e^{\lambda_\nu t'} V_{\mu\nu}(t) =$$

$$= \frac{2D_{\mu\nu}}{\lambda_\mu+\lambda_\nu} e^{\lambda_\mu t} e^{\lambda_\nu t'} - \frac{2D_{\mu\nu}}{\lambda_\mu+\lambda_\nu}\left[\Theta(t-t')e^{\lambda_\mu(t-t')} + \Theta(t'-t)e^{\lambda_\nu(t'-t)}\right] \tag{31}$$

In these equations the effect of the stochastic force is clearly displayed. If it vanishes the correlation function is reduced to

$g^{0}_{\mu\nu}(t,t')$ and otherwise its whole effect is included in the so-called stochastic propagator $U^{0}_{\mu\nu}(t,t')$. Initial correlations appear only in $g^{0}_{\mu\nu}(t,t')$, while the stochastic dynamics are included in $U^{0}_{\mu\nu}(t,t')$ through the functions $V_{\mu\nu}$ related to the commutators (24).

It is instructive to see how initial correlations decay as the stationary solution is reached. Consider the equal-time correlation function $G_{\mu\nu}(t,t)$. Recalling that $[q_{\mu}(t),q_{\nu}(t)] = 0$, from (26) one has

$$U^{0}_{\mu\nu}(t,t) = e^{\lambda_{\nu}t}V_{\mu\nu}(t) = e^{\lambda_{\mu}t}\,V_{\nu\mu}(t) \tag{32}$$

$$G_{\mu\nu}(t,t) = g^{0}_{\mu\nu}(t,t) + \frac{2D_{\mu\nu}}{\lambda_{\mu}+\lambda_{\nu}}\left(e^{(\lambda_{\mu}+\lambda_{\nu})t}-1\right) \tag{33}$$

If Re $\lambda_{\mu} < 0$, $\mu = 1...N$, in the limit $t \to \infty$ one has

$$G_{\mu\nu}(t,t) = -\frac{2D_{\mu\nu}}{\lambda_{\mu}+\lambda_{\nu}} \tag{34}$$

which corresponds to the stationary state of the system[9)]. The function $g^{0}_{\mu\nu}(t,t)$ is not in general invariant under time translation, while in the second equality in (31) we have split $U^{0}_{\mu\nu}(t,t')$ into two parts : an invariant and a non-invariant part under time translation. In the limit considered above, only this second part of $U^{0}_{\mu\nu}(t,t')$ survives, and for the stationary state

$$G_{\mu\nu}(t-t') = -\frac{2D_{\mu\nu}}{\lambda_{\mu}+\lambda_{\nu}}\left[\Theta(t-t')\,e^{\lambda_{\mu}(t-t')} + \Theta(t'-t)\,e^{\lambda_{\nu}(t'-t)}\right] \tag{35}$$

If the system is initially already in its stationary state , $\langle q_{\mu}\, q_{\nu}\rangle$ in (30) must be replaced by $\frac{-2D_{\mu\nu}}{\lambda_{\mu}+\lambda_{\nu}}$ according to (34), and (35) is recovered.

This example shows the suitability of the formalism for the study of the detailed temporal evolution without knowledge of the solution of the time-dependent associated F.P.E., once the equations (14)-(15) have been solved. In several cases the problem can be reduced to the solution of the deterministic equations of motion. If these are known, the second term on the right-hand side of (14) can be introduced via an interaction picture.

The correlation and response function (26) to (28) can in general be expanded perturbatively in a diagrammatic series. In this expansion the unperturbed propagators correspond to the example considered here, so that the stochastic process is included exactly in first order. The validity of this expansion relies on the proof of the corresponding Wick theorem[1]. This theorem is proved in two steps. The first step is operational and includes stochastic effects through the equations of motion. In the second step the average over initial conditions is performed. If the distribution of initial values is not Gaussian, spurious diagrams appear which are often forgotten in functional formulations[10]. It can be seen that the whole effect of the stochastic force in the diagrammatic expansion as compared with the deterministic case[11] is the replacement of $g^{o}_{\mu\nu}(t,t')$ by $G^{o}_{\mu\nu}(t,t')$, that is, the inclusion of $U^{o}_{\mu\nu}(t,t')$ .

In conclusion, Fokker-Planck dynamics represents an interesting alternative to the problem of solving Fokker-Planck equations. In this formalism systematic diagrammatic expansions are based on a well established, if formal basis. The effect of the stochastic force can be treated exactly or alternatively introduced systematically after the deterministic equation has been solved. Relaxation of initial correlations can also be studied in a simple manner. On the other hand the operator equations of motion (14)-(15) have revealed themselves to be of great utiliy in the proof of general relations such as the fluctuation dissipation theorem[12].

## 3. OPERATOR FORMALISM AS A STOCHASTIC QUANTIZATION OF c-NUMBER HAMILTON EQUATIONS

While in the first part of this seminar we have sketched the meaning, consequences and possible relevance of Fokker-Planck dynamics, we now give a somehow deeper foundation of the formalism in terms of a c-number Lagrangian.

a) Stochastic Quantization

In the path-integral formulation of the FPE[7,13,14], a probability density for the occurrence of different paths is expressed in terms of a Lagrangian $\mathcal{L}(q,\dot{q})$ . Depending on the precise definition of the path integral, different Lagrangians can be considered[7,15]. If one requires that the variational principle applied to that Lagrangian give via the Euler-Lagrange equation a differentiable path such that the paths close to it are the ones that occur with maximum probability, the Lagrangian to be chosen[16] is the one considered by Graham[14].

$$\mathcal{L} = \sum_{\mu\nu} \left( \frac{1}{4} D^{-1}_{\mu\nu} \; (\dot{q}_\mu - v_\mu)(\dot{q}_\nu - v_\nu) + \frac{1}{2} \frac{\partial v_\mu}{\partial q_\mu} \right) \tag{36}$$

This Lagrangian defines a classical-mechanics problem for the set of variables $q_\mu$ and the stochastic problem consists in allowing fluctuations around the most probable path that makes the action $\int \mathcal{L}(q,\dot{q})d\tau$ stationary. The stochastic quantization of the Euler-Lagrange equations associated with (36) is a procedure consisting in the systematic introduction of fluctuations. The word quantization has nothing to do here with Planck's constant, but is used because of the parallelism existing with quantum mechanics : Fluctuations can be introduced by means of the path integral (Feynman quantum mechanics) or by imposing commutation relations (Dirac's prescription to obtain quantum mechanics). This second procedure gives rise to the Fokker-Planck dynamics described before.

The derivation[2)] of Fokker-Planck dynamics from the path-integral formalism involves some technical details outside the scope of this seminar. The final recipe is obtained from the c-number Hamiltonian associated with $\mathcal{L}$ :

$$\mathcal{H}(q,p)=\sum_\mu \dot{q}_\mu p_\mu - \mathcal{L}(q,\dot{q})= \sum_{\mu\nu} D_{\mu\nu} p_\mu p_\nu + \sum_\mu v_\mu p_\mu - \sum_\mu \frac{1}{2}\frac{\partial v_\mu}{\partial q_\mu} \quad (37)$$

$$p_\mu = \frac{\partial \mathcal{L}}{\partial \dot{q}_\mu} = \sum_\nu \frac{1}{2} D^{-1}_{\mu\nu}(\dot{q}_\nu - v_\nu) \quad (38)$$

as follows : To $p_\mu(t)$ corresponds the operator $\hat{q}_\mu(t)$, and to the c-number Hamiltonian corresponds the operator

$$\frac{1}{2}\lim_{\delta\to 0} T\left\{\mathcal{H}(q(t),\hat{q}(t+\delta)) + \mathcal{H}(q(t),\hat{q}(t-\delta))\right\} = -L(q(t),q(t),t) \quad (39)$$

where $T$ is the time-ordering operator. From this correspondence one is led to equations(14)-(16) defining the Fokker-Planck dynamics.

The physical difference from the parallel derivation of the Schrödinger-Heisenberg formulation of quantum mechanics from the Feynman formulation consists in the fact that we are here considering real transition probabilities and not complex probability amplitudes. This difference is reflected in the inequivalence between the forward and backward FPE for the transition probability, while the parallel equations in quantum mechanics are related by complex conjugation (even if a non Hermitian Hamiltonian is considered).

b) Time reversal properties

We consider three different levels of description of a system. The first level is the one in which the system is described by the positions and momenta of the constituent particles, that is, a microscopic description. On a second level, gross variables are already included, and the system is described by the Euler-Lagrange equations, associated with the Lagrangian (36), for these gross variables. We are thus considering the most probable path in configuration space, which does not coincide in general with the mean path with respect to the fluctuations. On a third level, fluctuations are taken into account. Mathematically, this amounts to a stochastic quantization, either by a path integral or by an operator formalism.

The Fokker-Planck equation is manifestly noninvariant under time reversal. Since the microscopic equations are invariant under time reversal, we may ask : on which level of description, and how, is time-reversal invariance broken? Since a nonequilibrium-thermodynamics interpretation of $\mathcal{L}$ has been proposed[14] we believe that a careful answer to the above question should be the first step towards significant progress in this direction.

For the gross variables $q_\mu$ , the time-reversal operation T is defined by

$$\begin{aligned} T \;:\; & t \rightarrow -t \\ & q_\mu \rightarrow q_\mu \\ & v^R \rightarrow -v^R \\ & v^I \rightarrow v^I \\ & D_{\mu\nu} \rightarrow D_{\mu\nu} \end{aligned} \tag{40}$$

where even variables have been assumed and where the drift $v_\mu = v^R_\mu + v^I_\mu$ has been split into a reversible (odd) and an irreversible (even) part. Thus, two sources of irreversibility are present - the irreversible drift $v^I_\mu$ and the diffusion matrix $D_{\mu\nu}$ .

The Lagrangian (36) yields the Euler-Lagrange equations

$$\ddot{q}_\mu - \sum_\nu \frac{\partial v_\mu}{\partial q_\nu}\,\dot{q}_\nu + \sum_{\rho\sigma\nu} D_{\mu\sigma}\, D^{-1}_{\rho\nu}\, \frac{\partial v_\rho}{\partial q_\sigma}\,(\dot{q}_\nu - v_\nu) - \sum_{\sigma\rho} D_{\mu\sigma}\, \frac{\partial^2 v_\rho}{\partial q_\sigma \partial q_\rho} = 0 \tag{41}$$

Microscopic reversibility implies detailed balance, which in turn is equivalent to the Graham-Haken potential conditions[17]

$$v^I_\mu = \sum_\nu -D_{\mu\nu}\,\frac{\partial \phi_{st}}{\partial q_\nu} \tag{42}$$

$$\sum_\mu \frac{\partial v^R_\mu}{\partial q_\mu} = \sum_\mu v^R_\mu\,\frac{\partial \phi_{st}}{\partial q_\mu} \tag{43}$$

where

$$\phi_{st} = -\log P_{st}(q) \tag{44}$$

and $P_{st}(q)$ is the stationary solution of eq. (11). If these potential conditions are imposed it is easy to check that the Euler-Lagrange equations (41) which define the most probable path are invariant under time reversal. Thus, detailed balance guarantees that the symmetry under time reversal is still present at the second level of description we have considered.

What happens when fluctuations are included? If they are included by a path integral, what plays a crucial role is not the Euler-Lagrange equations but the Lagrangian itself. The Lagrangian (36) is not invariant under time reversal, so that the two directions of time evolution are differently weighted by $\mathcal{L}$ in the path integral. Nevertheless, the following Lagrangian

$$\mathcal{L}^{inv}(q,\dot{q}) = \sum_{\mu\nu}(\tfrac{1}{4}D^{-1}_{\mu\nu} \quad (\dot{q}_\mu - v^R_\mu)(\dot{q}_\nu - v^R_\nu) + v^I_\mu v^I_\mu \quad + \tfrac{1}{2}\frac{\partial v^I_\mu}{\partial q_\mu}) \tag{45}$$

is invariant under time reversal, and if the potential conditions (42), (43) are assumed it is easy to see that

$$\mathcal{L}(q,\dot{q}) - \mathcal{L}^{inv}(q,\dot{q}) = \frac{1}{2}\ \frac{d\phi_{st}}{dt} \tag{46}$$

Since the difference between $\mathcal{L}$ and $\mathcal{L}^{inv}$ is a total derivative with respect to time, both Lagrangians yield the same Euler-Lagrange equations and irreversibility is related to the term $\frac{1}{2}\frac{d\phi_{st}}{dt}$, whose physical interpretation remains for us as an open question.

Let us see how the breaking of time-reversal symmetry appears when fluctuations are introduced by the Fokker-Planck dynamics. We first need to know how $\hat{q}_\mu(t)$ transforms under time reversal. If it is assumed that if transforms as $p_\mu(t)$ does in the time-reversal invariant c-number Hamilton equations equivalent to (41), i.e., if

$$\hat{q}_\mu \xrightarrow{T} -\hat{q}_\mu - \sum_\nu D^{-1}_{\mu\nu} \quad v^I_\nu \tag{47}$$

then equation (14) is invariant under $T$ but the commutation relation (16) is not, so that over all invariance of the formalism under time reversal breaks down. On the other hand, the transformation

$$\hat{q}_\mu \xrightarrow{T} \hat{q}_\mu \tag{48}$$

which follows from the identification of $\hat{q}_\mu(t)$ as $\frac{\partial}{\partial q_\mu(t)}$, leaves the commutation relation (16) invariant, but then the equation of motion (14) is not invariant. Therefore, time-reversal symmetry breaks down in Fokker-Planck dynamics owing to the inconsistency of the simultaneous requirement of invariance for both the equations of motion and the equal-time commutation relations (16).

As a conclusion we state that the inclusion of fluctuations at the third level of description through the stochastic quantization of the Euler-Lagrange equation breaks time-reversal invariance.

Up to now we have been dealing with a constant $D_{\mu\nu}$ . If q-dependent diffusion matrix $D_{\mu\nu}(q)$ represents a Euclidean metric, the problem can be reduced to the constant diffusion case[2]. Nevertheless it should be pointed out that the Lagrangian[14]

$$\mathcal{L} = \sum_{\mu\nu} \left( \frac{1}{4} \; D^{-1}_{\mu\nu} \; (\dot{q}_\mu - W_\mu)(\dot{q}_\nu - W_\nu) + \frac{1}{2} \sqrt{D} \frac{\partial}{\partial q_\mu} \frac{W_\mu}{\sqrt{D}} \right) \tag{49}$$

$$W_\mu = V_\mu - \sum_\nu \sqrt{D} \frac{\partial}{\partial q_\nu} \frac{D_{\mu\nu}}{\sqrt{D}} \tag{50}$$

$$D = \| D_{\mu\nu} \| \tag{51}$$

has the same properties mentioned for the constant diffusion case : The associated Euler-Lagrange equations are invariant under time reversal if detailed balance is assumed, and

$$\mathcal{L}(q,\dot{q}) - \mathcal{L}^{inv}(q,\dot{q}) = \frac{1}{2} \frac{d}{dt} (\phi_{st} - \ln \sqrt{D}) \tag{52}$$

Finally, we should like to make some comments on the so-called stochastic formulation of quantum mechanics[18,19]. An important point of this formulation is the derivation of a FPE for the quantum mechanical probability density $p(\bar{x},t) = \Psi(\bar{x},t)\,\Psi^*(\bar{x},t)$ . The point we want to clarify is how quantum mechanics, which is a time-reversal invariant theory, can be based on a FPE for which this symmetry is absent. In this formulation of quantum mechanics one writes

$$\Psi = e^{R+iS} \tag{53}$$

and the related FPE reads

$$\frac{\partial p}{\partial t} = - \bar{\nabla} (\bar{c} p) + D_0 \bar{\nabla}^2 p \tag{54}$$

where the drift $\bar{c}$ is written as

$$\bar{c} = \bar{v} + \bar{u} \tag{55}$$

$$\bar{v} = 2D_0 \bar{\nabla} S \tag{56}$$

$$\bar{u} = 2D_0 \bar{\nabla} R \tag{57}$$

and

$$D_0 = \frac{\hbar}{2m} \tag{58}$$

Under time reversal, $\bar{v} \to -\bar{v}$, $\bar{u} \to \bar{u}$ , so that $\bar{u}$ corresponds to what we have called $v^I$ , the irreversible drift, and $\bar{v}$ is the reversible drift $v^R$ . Equation (57) implies that

$$\vec{u} = D_0 \vec{\nabla} \log P \tag{59}$$

which compared with (42) means an extension of the potential conditions to every time. This unphysical extension of the detailed-balance condition guarantees that

$$- \vec{\nabla}(\vec{u}P) + D_0 \vec{\nabla}^2 P = 0 \tag{60}$$

Thus, the two sources of irreversibility in the FPE cancel each other identically and the FPE (54) reduces to

$$\frac{\partial P}{\partial t} = - \vec{\nabla} (\vec{v}P) \tag{61}$$

which is nothing but the continuity equation of quantum mechanics.

Therefore, the stochastic formulation of quantum mechanics relies on an unphysical extension of the detailed-balance condition that allows one to write a time-reversal invariant FPE by the simultaneous addition and substraction of the same term to the continuity equation of quantum mechanics. We may conclude the stochastic formulation of quantum mechanics lacks physical meaning.

ACKNOWLEDGEMENTS

We are grateful to D. Roekaerts for drawing our attention to ref. 16.

REFERENCES


1) L. GARRIDO and M. SAN MIGUEL, Progr. Theor. Phys. 59, 40 (1978).

2) L. GARRIDO, D. LURIE and M. SAN MIGUEL, Phys. Lett. A (to appear) and unpublished work.

3) S.F. EDWARDS and W.D. McCOMB, J. Phys. A2, 157 (1969).

4) N.G. VAN KAMPEN, Phys. Rep. 24C (1976).

5) R. GRAHAM, Z. Phys. B26, 397 (1977).

6) M. SAN MIGUEL, Ph. D. Thesis, Barcelona University (1978).

7) F. LANGOUCHE, D. ROEKAERTS, F. TIRAPEGUI, Preprints KUL-TF-77/023 (1977), KUL-TF-78/007 (1978), KUL-TF-78/015 (1978).

8) C.P. ENZ, Physica A89, 1 (1977).

9) M.C. WANG and G.E. UHLENBECK, Rev. Mod. Phys. 17, 323 (1945).

10) P.C. MARTIN, E.D. SIGGIAmand H.A. ROSE, Phys. Rev. A8, 423 (1973).

11) C.P. ENZ and L. GARRIDO, Phys. Rev. A14, 1258 (1978).

12) L. GARRIDO and M. SAN MIGUEL, Progr. Theor. Phys. 59, 55 (1978).

13) H. HAKEN, Z. Phys. B24, 321 (1976).

14) R. GRAHAM, Z. Phys. B26, 281 (1977), (see also lectures by R. GRAHAM in these proceedings).

15) H. LESCHKE and M. SCHMUTZ, Z. Phys. B27, 85 (1977).

16) D. DÜRR and A. BACH, Comm. Math. Phys. (to appear).

17) R. GRAHAM and H. HAKEN, Z. Phys. 243, 289 (1971).

18) E. NELSON, Phys. Rev. 150, 1079 (1966).

19) C. DE LA PEÑA-AUERBACH, Jour. Math. Phys. 10, 1620 (1969).

# APPLICATIONS OF PATH INTEGRALS

EUGENE P. GROSS

Martin Fisher School of Physics,

Brandeis University, Waltham, Mass., U.S.A.

## 1. INTRODUCTION

In the present seminar we summarize some methods and results that have been obtained using path integrals in practical applications. The cases considered are electrons in random potentials, the polymer excluded-volume problem, and the polaron problem. We are interested in assessing approximation schemes of two types. First, there are conventional Hamiltonian techniques that can also be expressed in path integral language. Path integrals then offer only improvement in computational convenience. Second, there are methods that are specific to the path integral viewpoint. It appears to be difficult or even impossible to obtain equivalent results with conventional formalisms.

We start with the one-dimensional potential problem for a particle of unit mass with Hamiltonian ($\hbar = 1$).

$$H = \frac{p^2}{2} + V(x) \tag{1.1}$$

The initial value problem is solved when one knows the eigenfunctions and eigenvalues and constructs the space-time propagator

$$K(x_2 x_1 / t) = \sum_n \psi_n(x_2)\, \psi_n^*(x_1)\, e^{-iE_n t} \tag{1.2}$$

We will usually be interested in the associated density matrix

$$\rho(x_2 x_1 | \beta) = \sum_n \psi_n(x_2)\, \psi_n^*(x_1)\, e^{-\beta E_n} \tag{1.3}$$

which obeys the equation

$$\left\{ \frac{\partial}{\partial u} - \tfrac{1}{2} \frac{\partial^2}{\partial x_2^2} + V(x_2) \right\} \rho(x_2 x_1 / u) = \delta(x_2 - x_1)\, \delta(u)$$

$$\text{and} \quad \rho(x_2 x_1 / u) = 0 \,, \quad u < 0 \,. \tag{1.4}$$

For a free particle the explicit form is

$$\rho(x_2 x_1 | \beta) = \frac{1}{\sqrt{2\pi\beta}}\, e^{-(x_2 - x_1)^2 / 2\beta} \,. \tag{1.5}$$

The path integral representation is

$$\rho(x_2 x_1 | \beta) = \int_{x(0) = x}^{x(\beta) = x_2} Dx\; e^{-\frac{1}{2}\int_0^\beta \dot{x}^2 du}\; e^{-\int_0^\beta V(x(u)) du} \tag{1.6}$$

---

Research supported by a grant from the N.S.F.

We want to find approximations for $\rho(x_2x_1/\beta)$ when the exact answer cannot be found. Much of what I say about the potential problem is contained in the book by Feynman and Hibbs[1] and in Feynman's lecture notes on statistical mechanics[2]. I want to focus attention and comment on certain points that are important in our later considerations.

i) Perturbation-Variational Treatments

For real actions one may write

$$\rho(x_2x_1/\beta) = \int_{x(0)=x_1}^{x(\beta)=x_2} Dx\, e^{-S_0}\, e^{(S_0-S)} \tag{1.7}$$

Introduce the weight function

$$\begin{aligned} w_0(x_2x_1/\beta) &= e^{-S_0}/\rho_0(x_2x_1/\beta) \\ \rho_0(x_2x_1/\beta) &= \int_{x_1}^{x_2} Dx\; e^{-S_0} . \end{aligned} \tag{1.8}$$

Using the convexity of the exponential, one obtains the Feynman variational principle

$$\begin{aligned} \rho(x_2x_1/\beta) &= \rho_0(x_2x_1/\beta)\; \{\langle e^{S_0-S}\rangle\, w_0(x_2x_1/\beta)\} \\ &\geq \rho_0(x_2x_1/\beta)\; \exp\{\langle S_0-S\rangle\, w_0(x_2x_1/\beta)\}. \end{aligned} \tag{1.9}$$

Here

$$\{\langle A\rangle w_0(x_1x_2/\beta)\} \equiv \int_{x_1}^{x_2} Dx\; w_0(x_2x_1/\beta) A \;. \tag{1.10}$$

The variational principle is usually used to fix some parameters in a trial action $S_0$. To obtain a systematic perturbation expansion based on $S_0$, one then makes a cumulant expansion of $\langle e^{S_0-S}\rangle$.

It does not seem to be appreciated that there exists an infinite set of similar variational principles, each of which is more accurate (but more complicated) than the previous one. The idea rests on simple matrix multiplication together with

$$e^{-\beta H} = e^{-\beta H/N} \ldots\ldots e^{-\beta H/N} \quad \text{with } N \text{ factors} .$$

For example for $N = 2$

$$\begin{aligned} \langle x_2| e^{-\beta H}|x_1\rangle &= \int \langle x_2|e^{-\beta H/2}|\xi\rangle\, d\xi\, \langle \xi|e^{-\beta H/2}|x_1\rangle \\ \rho(x_2x_1/\beta) &= \int \rho(x_2\xi/\beta/2)\, d\xi\, \rho(\xi x_1/\beta/2) . \end{aligned} \tag{1.11}$$

The Feynman principle is applied to each factor separately

$$\rho(x_2x_1|\beta) \geq \int d\xi\, \rho_0(x_2\xi/\beta/2)\rho_0(\xi x_1/\beta/2)\exp\big[\{\langle S_0-S\rangle w_0(x_2\xi/\beta/2)\} + \{\langle S_0-S\rangle w_0(\xi x_1/\beta/2)\}\big] \quad (1.12)$$

For the special case that the action $S_0$ is for a potential $V_0(x)$, the Feynman principle involves the exponential of

$$\{\langle S_0-S\rangle w_0(x_2x_1|\beta)\} = \frac{1}{\rho_0(x_2x_1|\beta)}\int dy \int_0^\beta du\, \rho_0(x_2y/\beta-u)(V_0(y) - V(y))\,\rho_0(y\, x_1/u). \quad (1.13)$$

The extended form uses

$$\{\langle S_0-S\rangle w_0(x_2\xi/\beta/2)\} = \frac{1}{\rho_0(x_2\xi/\beta/2)}\int dy\int_0^{\beta/2} du\, \rho_0(x_2y/\tfrac{\beta}{2}-u)(V_0(y) - V(y))\,\rho_0(y\xi/u). \quad (1.14)$$

From the eigenfunction expansion we have

$$\int \rho(x\xi/u)d\xi\, \rho(\xi x_1/u_1) = \rho(xx_1/u+u_1). \quad (1.15)$$

The Feynman estimate is obtained by taking $\dfrac{\rho_0(x_2\xi/\beta/2)\rho_0(\xi x_1/\beta/2)}{\rho_0(x_2x_1/\beta)}$ as a weight function and applying the convexity inequality to the integration. Thus the new inequality is stronger than the Feynman inequality.

ii) Specifically Path Integral Approximations

Feynman notes that the classical limit of the partition function is obtained by approximating the diagonal element of the density matrix by

$$\rho(x_1x_1|\beta) \simeq \int_{x_1}^{x_1} Dx\, e^{-\frac{1}{2}\int_0^\beta \dot{x}^2 du}\, e^{-\beta V(x_1)} = \frac{1}{\sqrt{2\pi\beta}}\, e^{-\beta V(x_1)} \quad (1.16)$$

In $\int_0^\beta V(x(u))du$, $V(x(u))$ is replaced by its value at the starting point. The argument is that for $\beta \ll 1$ the kinetic energy term restricts the particle to small excursions. This suggests that a good trial action is

$$S_0 = \tfrac{1}{2}\int_0^\beta \dot{x}^2 du + \beta\, w(\bar{x})$$

where

$$\bar{x} = \frac{1}{\beta}\int_0^\beta x(u_1)du_1 \quad (1.17)$$

is the mean position for a given path. $w(x)$ is to be chosen using the variational principle.

This is an intrinsically path integral type of approximation. The paths are grouped according to the value of the mean position. A weight is assigned to all paths with a given mean position. (They differ only in their kinetic energy contribution.) The result is

$$\rho(y\, y|\beta) \geq \int d\xi \; e^{-\beta w(\xi\, y)} \oint_y D_w x \; \delta(\bar{x}-\xi)$$
$$D_w x \equiv Dx \; e^{-\frac{1}{2}\int_0^\beta \dot{x}^2 du} \tag{1.18}$$

$$w(\xi\, y) = \oint_y D_w x \left[\int_0^\beta V(x(u))du\right] \delta(\bar{x}-\xi) \Big/ \oint_y D_w x \; \delta(\bar{x}-\xi)\,. \tag{1.19}$$

The two path integrals occurring can be done explicitly. This led Feynman to a novel treatment of the quantum correlations to the classical partition function.

One can also use trial actions of a type

$$\beta w_1(\bar{x}) + \beta w_2\left[\overline{x^2} - (\bar{x})^2\right]. \tag{1.20}$$

In more than one dimension one can partially classify paths according to their shapes. There are a number of obvious generalizations which allow explicit evaluation of the path integral. For example, expansion of $V(x(u))$ about the mean position leads to the trial action $\beta w_1(\bar{x}) + w_3(\bar{x}) \int_0^\beta (x(u)-\bar{x})^2 du$. These ideas do not appear to have been studied seriously.

iii) WKB with Path Integrals

There is a vast and growing literature based on applying the method of steepest descent together with a treatment of fluctuations in the Gaussian approximation. This can be done either for the propagator $K(x_2 x_1 / t)$ or for the density matrix $\rho(x_2 x_1 | \beta)$. In general one works in the complex $t$ plane to include quantum tunnelling effects[3].

## 2. MULTITIME PATH INTEGRALS

Our main concern is with approximation methods to deal with the more difficult multitime path integrals that arise in practical applications. We start by listing some cases.

We examine path integrals of the type

$$\int_{x(0)=x_1}^{x(\beta)=x_2} Dx \; e^{-\frac{1}{2}\int_0^\beta \dot{x}^2 du} \; e^{I} \quad . \tag{2.1}$$

One type of action has

$$I \;=\; \epsilon \int_0^\beta \int_0^\beta W(x(u) - x(u'))du\, du', \quad W > 0 \quad . \tag{2.2}$$

For $\epsilon = 1$, $I$ is positive and there is an enhanced contribution when $I$ is large. This case corresponds to the averaged partition function for an electron moving in a random potential subject to Gaussian statistics. Usually $W(x)$ is large for $x$ near zero. The electron prefers paths that return to earlier positions provided the kinetic energy cost is not too great. This is particularly true for large $\beta$ and leads to a tendency to form localized states.

The case $\epsilon = -1$ corresponds to the excluded volume problem for a polymer chain. $\beta$ becomes the length of the chain and $\vec{x}(u)$ the actual position of a point at a distance $u$ along the chain. The tendency of chain units with different values of $u$ to avoid each other finds expression in the fact that $e^{I} \to 0$ for configurations where two points touch. This path integral describes only a 'Gaussian equivalent' chain with the kinetic energy term providing the 'backbone' of the chain. The representation is adequate for the study of asymptotic properties. The path integral counts the number of configurations for a chain of length $\beta$ with endpoints at $\vec{x}_1$ and $\vec{x}_2$.

Another type of action is

$$I \;=\; \int_0^\beta \int_0^\beta W(x(u_1)-x(u_2);\; u_1-u_2)du_1\, du_2 \quad . \tag{2.3}$$

It has an explicit time dependence. This arises from the inertial properties of the field oscillators in a particle-scalar field Hamiltonian with linear coupling. The path integral is the reduced density matrix after elimination of the field oscillators[1]. This is a simple version of the problems of quantum electrodynamics which originally led Feynman to path integrals.

Finally, there is the Edwards-Gulyaev action[4]

$$I \;=\; \frac{N}{\Omega} \int \left[\exp(-\int_0^\beta v(\vec{x}(u) - \vec{R})du) - 1\right] d\vec{R} \quad . \tag{2.4}$$

It arises from the Hamiltonian

$$H = \frac{p^2}{2} + \sum_{i=1}^{N} v(\vec{x} - \vec{R}_i) \tag{2.5}$$

The electron interacts with N potentials located at positions $\vec{R}_1, \ldots, \vec{R}_N$ in a volume $\Omega$. One wants the averaged density matrix with the weight

$$w(\vec{R}_1, \ldots, \vec{R}_N) = \prod_{i=1}^{N} \frac{d\vec{R}_i}{\Omega} \quad . \tag{2.6}$$

This corresponds to Poisson statistics and the Gaussian limit is ontained by keeping only the first three terms in the expansion of $e^{-\int v(\vec{x}-\vec{R})du}$.

It should be noted that $\beta$ is not a physical temperature for the random potential problem. The quantity of interest is the average density of states $n(E)$. $Z(\beta) = \int \rho(x\ x/\beta)dx$ is the Laplace transform of this quantity

$$Z(\beta) = \int_{-\infty}^{+\infty} e^{-\beta E}\, n(E)dE \tag{2.7}$$

$$n(E) = \frac{1}{2\pi i}\int_{c-i\infty}^{c+i\infty} Z(\beta)\, e^{\beta E}\, d\beta \ .$$

The two sided transform is used since there is a tail' in the density of states for $E < 0$. $Z(\beta)$ is only the simplest path integral of interest. One is also interested in averaged current-current time correlation functions in order to compute the conductivity. This involves a double path integral. While the extra complication is significant it is not central to the kind of question discussed here.

All of the cases described take advantage of one virtue of path integrals that has been stressed by Feynman. It is that the averaging or integration over 'other' degrees of freedom can <u>sometimes</u> be done first. Of course this only transfers the difficulty to the problem of evaluating multitime path integrals. We will describe some of the main approaches to the problem.

A key feature of all of the actions listed is translation invariance. I is unchanged under the rigid displacement $\vec{x}(u) \rightarrow \vec{x}(u)+\vec{a}$ for any $\vec{a}$. Should this invariance be maintained for acceptable trial actions? Or, can one argue[5,6] that since paths start at $x_1$ and end at $x_2$, the translation symmetry is broken, and one is free to use trial actions without the invariance property. The answer is not obvious. If one thinks of the polymer problem for very large $\beta$, one notes that very long sections of the chain can be moved rigidly with negligible change in the value of the action. We will argue that experience with particular examples shows that one should use translation invariant trial actions and that failure to do so leads to spurious discontinuities for physical quantities.

## 3. TRIAL ACTIONS FOR 2-TIME PATH INTEGRALS

In the present section we describe the most common theories used to analyse our path integrals. For concreteness we talk about evaluating the average partition function for the one dimensional problem of an electron subject to Gaussian white noise. The partition function is given[8)] as the functional integral

$$\langle Z \rangle = \int \delta V \; Z_V \; e^{-\frac{1}{2}\iint V(x) W^{-1}(x-y) V(y) dx\, dy} \tag{3.1}$$

with autocorrelation function

$$\langle V(x)V(y) \rangle = \frac{1}{\gamma}\,\delta(x-y) \quad .$$

$V(x)$ will be represented by a discrete, real orthonormal set $\phi_n(x)$. Choose units with $\hbar = m = 1$ and measure lengths in terms of the thermal de Broglie wavelength $\sqrt{\beta}$ . Then the averaged partition function is

$$\frac{\langle Z \rangle}{L} = \frac{1}{\sqrt{\beta}} \int \prod_n dq_n \; \frac{e^{-q_n^2/2}}{\sqrt{2\pi}} \int_0 Dx \; e^{-\frac{1}{2}\int_0^1 \dot{x}^2 du} \; e^{-\lambda \sum_{n=0}^{\infty} q_n \int_0^1 \phi_n(x(u))du} \tag{3.2}$$

$\langle Z \rangle \sqrt{\beta}/L$ is a function of the single parameter $\lambda$ :

$$\lambda = \sqrt{\gamma}\;\beta^{3/4} \quad . \tag{3.3}$$

The factor L can be extracted because of the translational invariance (T.I.) of the average partition function. We call this the 'field' form. The two-time form results from doing the random average first. It is

$$\frac{\langle Z \rangle}{L} = \frac{1}{\sqrt{\beta}} \int_0 Dx \; e^{-\frac{1}{2}\int_0^1 \dot{x}^2 du} \; e^{(\lambda^2/2)\int_0^1\int_0^1 \delta(x(u)-x(u'))du\, du'} . \tag{3.4}$$

We now list some of the trial actions that have been used.

(i) Cumulant Perturbation Theory

Here $S_0$ is just the free particle action. The theory comes in two variants. In the field form one applies the Feynman principle and a cumulant development. This is done for each field configuration, and the random average is taken last. In the two time form one cumulant development is applied directly to equation (3.4). Since all of the path integrals are pure numbers, the quantity $\frac{\langle Z\rangle}{L}\cdot\sqrt{2\pi\beta}$ is just the exponential of a power series in $\lambda^2$.

(ii) Non T-I Self-Consistent Field(4,5,6,7)

$$S_0 = \tfrac{1}{2}\int_0^1 \dot{x}^2 du + \int_0^1 U(x(u),u)du \qquad (3.5)$$

The self-consistent field $U(x,u)$ (which can be allowed to be explicitly time dependent) is chosen with the aid of the variational principle. This approach also comes in the field and two-time variants.

(iii) Collective Variable Theory (T.I. Self-Consistent Field)

This theory is a path integral version and extension of the deep-trap theory of Halperin and Lax[10)] and of Zittartz and Langer[8)]. It is essentially the same approach as we used in recent collective variable treatments of the quantum field theory of extended objects[11)].

(iv) Generalized Quadratic Actions

$$S_0 = \tfrac{1}{2}\int_0^1 \dot{x}^2 du + \int_0^1\int_0^1 \left[x(u) - x(u')\right]^2 F(u - u')du\,du' \qquad (3.6)$$

Here $F(u)$ is to be chosen in accordance with the variation principle. This is the same type of action that Feynman used in his spectacularly successful theory of the ground state energy and effective mass of the polaron. He obtained accurate results over the entire range of coupling constants that have never been achieved by conventional Hamiltonian methods. Other methods yield a more precise description over limited coupling constant regions but do not cover the entire range in a unified way. The Feynman polaron approach showed for the first time that two time trial actions can lead to powerful new approximation schemes that are outside of conventional formalisms.

For the averaged partition function of the random potential problem, one can obtain an excellent theory with a much simpler one parameter quadratic action(12,13). This is

$$S_0 = \tfrac{1}{2}\int_0^1 \dot{x}^2 du + \frac{w^2}{4}\int_0^1\int_0^1 \left[x(u) - x(u')\right]^2 du\,du' \qquad (3.7)$$

$$= \tfrac{1}{2}\int_0^1 \dot{x}^2 du + \frac{w^2}{4}\int_0^1 (x(u) - \bar{x})^2 du$$

$$\bar{x} = \int_0^1 x(u')du' .$$

In fact it is only necessary to fix $\omega$ by its optimal value in the low temperature region ( $\beta \gg 1$, $\lambda \gg 1$) as $\omega = \frac{1}{2\pi}\lambda^4$. This trial action is sufficiently simple that the second cumulant correction can be examined in detail[13] and shown to give only small corrections.

(v) T-I Average Path Actions

We consider

$$S_0 = \tfrac{1}{2}\int_0^1 \dot{x}^2 du + \int_0^1 U(x(u) - \bar{x})du \tag{3.8}$$

$$\bar{x} = \int_0^1 x(u_1)du_1 \quad .$$

This is a new type of trial action which combines some of the desirable features of some of the other theories. It contains the simple quadratic action as a special case. It makes possible a T-I self-consistent field theory without the use of collective variables. Clearly one can write down generalizations with explicit time dependence. However the simple action exhibited here is only on the edge of solubility. To work with it one must have a good approximation for the propagator of a particle in a general potential $U(x)$ and an additional constant electric field, for all values of the electric field.

## 4. ASSESSMENT OF THE TRIAL ACTIONS

We now summarize the results of using the trial actions of the previous section when applied to the one-dimensional Gaussian white noise case. Here there exists a number of exact results that were found with entirely different, intrinsically one-dimensional methods. It should be noted that the actions we have described do not involve dimensionality in any way other than in the relatively trivial space integrals. (Such considerations do, however, show that pure white noise leads to collapse in more than one dimension, so that one must have a finite correlation length for the noise.)

The exact results of interest here were obtained a long time ago by Frisch and Lloyd[14] and by Halperin[15]. They refer to the density of states $n(E)$, which can be exhibited as a quadrature and has a smooth behavior in the entire range $-\infty < E < +\infty$. There is no closed form for the Laplace transform, but the limiting expressions for $\lambda \ll 1$ and $\lambda \gg 1$ can be easily obtained. In $\lambda \ll 1$ the results agree with cumulant perturbation theory. The limit $\lambda \gg 1$ contains the deep trap contributions which give a negative energy tail. The $\lambda \gg 1$ limit, which we denote as the 'strong coupling limit', has the expansion

$$\frac{\langle Z\rangle}{L}\sqrt{2\pi\beta} = \exp\left[b_1\lambda^4 + \delta\log\lambda + b_2\lambda^0 + \text{inverse powers}\right] \quad (4.1)$$

$$= \lambda^\delta \exp\left[b_1\lambda^4 + b_2\lambda^0 + \ldots\right] .$$

The coefficients $b_1$ and $b_2$ are known and the prefactor is $\lambda^4$ or $\delta = 4$. Let us discuss the different trial actions in sequence.

(i) Cumulant Perturbation Theory

There is little to be said about cumulant perturbation theory as applied to the two time form. It agrees with the expansion of the exact theory term by term, but is only useful for $\lambda \ll 1$. However an interesting point arises in the comparison of the two time and field forms. In the two time form the simple variational bound is

$$\frac{\langle Z\rangle}{L} \geq \frac{1}{\sqrt{2\pi\beta}}\, e^{\lambda^2 J} \quad (4.2)$$

where
$$J = \int dx \int_0^1 du_1 \int_0^{u_1} du_2\ \rho_0(0x/1-u_1)\frac{1}{\sqrt{u_1-u_2}}\rho_0(x0/u_2) \quad (4.3)$$

$$\rho_0(x\,0/u) = (e^{-x^2/2u})/\sqrt{2\pi u} .$$

The exponential contains all the terms of order $\lambda^2$.

On the other hand the variational bound applied to the field form leads to an expression with $J$ replaced by $J_0$

$$J_0 = \pi\int dx\,\Lambda^2(x) \quad (4.4)$$

$$\Lambda(x) = \int_0^1 \rho_0(0x/1-u)\,\rho_0(x0/u)\,du .$$

This is not the correct coefficient. It is necessary to obtain the second cumulant correction before doing the random average to pick up the complete contribution of order $\lambda^2$.

(ii) Non T.I. Self-Consistent Field Theories

These theories yield the correct value of the coefficient $b_1$ corresponding to deep trap formation. The coefficient $\delta$ has the value 2 in contrast to the correct value $\delta = 4$. The coefficient $b_2$ is 'slightly' wrong. The coefficient $\delta$ is wrong 'forever' in the sense that a cumulant perturbation theory aiming at a systematic strong coupling expansion never corrects the logarithmic term. The subsequent terms in the expansion are then inaccurate.

The simple problem of a pair of particles bound harmonically illustrates the nature of the difficulty. The exact solution is given in terms of the center of mass motion and relative motion. However, if the interaction $\frac{\omega^2}{2}(x-y)^2$ is treated by using the coupling $-\omega^2 xy$ as a perturbation the unperturbed propagator corresponds to two particles bound to the origin. In any finite order of perturbation theory we do not obtain a satisfactory description of the center of mass motion.

With the field form, using a time independent potential obeying equation (1.4), the variational bound is

$$\frac{\langle Z\rangle}{L} \geq \rho(00|1)\exp\left[\frac{\lambda^2}{2}\int\Gamma^2(x)dx + \int U(x)\,\Gamma(x)dx\right] \qquad (4.5)$$

with

$$\Gamma(x) = \frac{1}{\rho(00|1)}\int_0^1 \rho(0x/1-u)\,\rho(x0/u)du \quad . \qquad (4.6)$$

Functional variation leads to the self-consistent potential $U(x) = -\lambda^2\,\Gamma(x)$. It is difficult to analyse the coupled equations in the intermediate coupling case. Approximate treatments have been given by Edwards[5)] and by Freed[6)] for the three-dimensional case in attacking the mobility edge problem.

For strong coupling the theory is simple. There is a discrete eigenvalue $E_0$ that dominates, and $U(x)$ tends to $-\lambda^2\psi_0^2(x)$. We then have the well known and exactly soluble non-linear eigenvalue problem.

$$\left\{-\tfrac{1}{2}\frac{\partial^2}{\partial x^2} - \frac{\lambda^2}{2}\psi_0^2(x)\right\}\psi_0(x) = E_0\,\psi_0 \quad . \qquad (4.7)$$

Thus there is a new length $a$, such that $E_0 \sim 1/a^2$ and $\psi_0(x) = \frac{1}{\sqrt{a}}\,\psi_0(x/a)$. We have $a = \lambda^{-2}$. The partition function tends to $\frac{\langle Z\rangle}{L} \rightarrow \frac{1}{a}\,\psi^2(0)\,\exp\left[-E_0 - \frac{\lambda^4}{2}\int E_0^4\,dy\right]$ . (4.8)

This gives the correct dominant term, but the prefactor is $\lambda^2$, i.e. $\delta = 2$. There is only one bound state and it becomes more extended as one moves to weak coupling.

The density matrix corresponding to the potential $-\lambda^4\psi_0^2(x/a)$ can be used in the variational bound to obtain an estimate of $\langle Z\rangle$ over the entire coupling range. This is indeed a smooth function of $\lambda$, but at some critical value of $\lambda \sim 1$, the free particle action yields a better bound. The cost in kinetic energy needed to form a bound state is not paid back by the gain from the potential energy.

Thus $\partial Z/\partial\lambda$ is discontinuous at this value of $\lambda$. The same feature is shared by the general self-consistent solution. The exact solution for the density of states shows that this discontinuity is spurious.

The same self.consistent field analysis may be made with the two time action. The results for the dominant terms and the prefactor are the same. However, in the two time version, the first step includes terms of order $\lambda^0$ that can only be obtained at the second cumulant level when one uses the field form. The two time form allows us to treat the small wavelength fluctuations of the random potential in a more economical way.

### (iii) Collective Variable Self-Consistent Field Theory

The collective variable theory of Halperin and Lax[10] and Zittartz and Langer[8] can be done in a very direct way that exhibits its virtues and limitations. Assume that the Gaussian random field $V(x) = \sum_{n=0}^{\infty} q_n \phi_n(x)$ can be described by

$$V(x) = \sum_{n\neq 1}^{\infty} Q_n \phi_n(x-R) \quad . \tag{4.9}$$

The translation coordinate $R$ replaces $Q_1$. Here the function $\phi_1(x)$ is chosen in a special way related to the shifting of $\phi_0(x)$. Let

$$\phi_1(x) = -\frac{\partial\phi_0}{\partial x} \Big/ \sqrt{\int \left(\frac{\partial\phi_0}{\partial z}\right)^2 dz} \quad . \tag{4.10}$$

We ignore for the moment the fact that it is not legitimate to transform from the $q_n$ to the $Q_m, R$ variables. Proceeding formally we find a Jacobian that is independent of $R$.

$$J = \left| \sum_{n\neq 1} Q_n \int \phi_n \frac{\partial\phi_1}{\partial x} dx \right| . \tag{4.11}$$

The orthonormal set can be chosen so that there are only two terms

$$J = \left| Q_0 \int \phi_0 \frac{\partial\phi_1}{\partial x} dx + Q_2 \int \phi_2 \frac{\partial\phi_1}{\partial x} dx \right| . \tag{4.12}$$

The averaged partition function is

$$\frac{\langle Z\rangle}{L}\sqrt{\beta} = \int \frac{dR}{\sqrt{2\pi}} \prod_{n\neq 1} \frac{dQ_n}{\sqrt{2\pi}} e^{-\sum_{n\neq 1} Q_n^2/2} \; J \int_{-R} Dx(u) e^{-\frac{1}{2}\int_0^1 \dot{x}^2 du}$$
$$e^{-\lambda \sum_{n\neq 1} Q_n \int_0^1 \phi_n(x(u))du} . \tag{4.13}$$

The field calculation is performed by writing $Q_C = \bar{Q}_0 + (Q_0 - \bar{Q}_0)$ and using the trial action

$$S_0 = \tfrac{1}{2}\int_0^1 \dot{x}^2 du - \lambda\, \bar{Q}_0 \int_0^1 \phi_0(x(u))du \quad . \qquad (4.14)$$

For strong coupling, we make the same choice of $\phi_0$ as in the non T-I theory. Then the contributions from $n \neq 0$ modes vanish. The variational principle leads to $Q_0$ proportional to $\lambda^2$. The Jacobian is simply $|\bar{Q}_0 \int \phi_0 \frac{\partial \phi 1}{\partial x} dx|$ to order $\lambda^0$. The dominant exponential is the same. The only difference is the prefactor which is now

$$J \sim |\bar{Q}_0 . \tfrac{1}{a}| \sim \lambda^4 \quad . \qquad (4.15)$$

This is in agreement with the exact result. Zittarz and Langer calculated the $\lambda^0$ term by going to the second cumulant approximation. The coefficients $b_1$, $\delta$, $b_2$ are all in agreement with the exact theory. They used a different formalism and studied the normal mode spectrum. In accordance with our earlier remarks it is easier to first do the average over $Q_n$ for $n \neq 0$, obtaining a two time action. This handles the shortwave fluctuations directly.

The collective variable theory is satisfactory in the deep trap, strong coupling regime and can be applied with confidence in the three-dimensional case. What is wrong with it? The transformation only has a meaning for configurations where $Q_0$ is $\sim \bar{Q}_0$ and is large compared to the $Q_n$. It can be used to obtain asymptotic series for integrals of functions that pick out this region. Halperin and Lax already noted that the theory fails to describe situations where there are nearby traps. The collective variable theory is a one trap theory which is T-I with a trap depth that decreases and a range that increases as one moves to weaker coupling. It can be extended to treat many well separated traps, but fails in the region where there are shallow overlapping traps. It is interesting that the same situation was encountered earlier in polaron theory, where the strong coupling theory of Bogolyubov and Tyablikov[16)] is the analogue of the Halperin-Lax theory.

(iv) Quadratic Actions

We first discuss the simple quadratic action. Details are contained in a paper by the author[13)] and in an earlier paper by Samathiyakanit[12)], using the variation principle, one finds an explicit expression for $\langle Z \rangle$ as a function of $\lambda$ in terms of a single parameter $\omega$. The analysis of the $\lambda >> 1$ limit leads us to choose $\omega = \lambda^4/2\pi$ in this regime. But the same value leads to a smooth $\langle Z \rangle$

for all $\lambda$ , and the result is better than perturbation theory in the weak coupling limit. The way in which the correct prefactor appears is interesting. The term $\rho(00/1)$ in the simple variation principle is

$$\rho(00/1) = \frac{1}{\sqrt{2\pi}} \frac{\omega/2}{\sinh(\omega/2)} \tag{4.16}$$

which has the $\lambda^4$ factor for $\omega >> 1$. Actually only the $\delta$ coefficient is correct. The dominant $b_1$ coefficient is appropriate to a harmonic approximation to the bound state and $b_2$ is considerably in error. However in the paper cited, I carried out a calculation of the second cumulant correction. This drives $b_1$ and $b_2$ to within a few per cent of the correct values.

The simple quadratic action thus yields a very successful theory for the averaged partition function over the entire range of coupling strengths. Physically it is a single trap interpolation from strong to weak coupling. $\langle Z \rangle$ is completely smooth and the bound is everywhere better than the free action.

I have analysed the behavior of $\langle Z \rangle$ when one uses the most general quadratic action with a time delay function $F(u-u')$. This improves the theory but the corrections are unimportant. However this more general action may be needed in the conductivity problem, but there are no definitive results at this time.

It is also not clear physically what the general quadratic action means in terms of the multitrap picture. The mathematical equations for this action were already obtained by Des Cloizeaux[17)] (except for appropriate sign changes) in an analysis of the polymer excluded volume problem. The use of quadratic actions is more dubious in the repulsive case. For example the simple action without time delay represents an oscillator with imaginary frequency which blows the chain apart. To save the situation one lets $F(u-u')$ fall off to zero as $|u-u'|$ approaches the chain length $\beta$ (in ordinary units). An exponential decay or finite range leads however to a free flight chain. Des Cloizeaux examines inverse power decays and makes a very careful analysis of the general equations. He does obtain nonBrownian chain behavior but the critical exponents are not satisfactory.

### (v) T-I Mean Path Actions

The mean path action is expected to give one correct leading term $b_1$ if the potential $U(x)$ is chosen to be the non T-I self-consistent field. It should also be usable for the polymer problem. We only outline how the correct prefactor emerges and do not present the complete calculation of $\langle Z \rangle$ .

Consider the path integral

$$\rho(00/1) = \oint_0 D_W x \, e^{-\int_0^1 U(x(u)-\bar{x})du} = \oint_0 D_W x \int_{-\infty}^{+\infty} e^{-\int_0^1 U(x-\xi)du} \delta(\bar{x}-\xi)d\xi. \tag{4.17}$$

Interchange the order of integrations and write

$$\rho(00/1) = \int_{-\infty}^{+\infty} d\xi R(\xi) \tag{4.18}$$
$$R(\xi) = \oint_{-\xi} D_W y \, e^{-\int_0^1 U(y(u))du} \delta(\bar{y}) .$$

If $\xi$ is large, $R(\xi)$ tends to a Gaussian. The contributing paths start $-\xi$ and end there but must pass to a point of order $+\xi$ since the mean position is required to be zero. These paths have high velocity and spend a short time in the region where the potential is large. For small values of $\xi$ we use the representation

$$R(\xi) = \frac{1}{2\pi} \int d\alpha \oint_{-\xi} D_W y \, e^{-\int_0^1 U(y(u))du} \, e^{i\alpha\int_0^1 y(u)du} . \tag{4.19}$$

For the propagator $K(00/t)$ this represents a particle in a potential $U(y)$ and also in a uniform field of strength $\alpha$. We have to integrate over all $\alpha$. When $U(y)$ is a deep trap a cumulant analysis can be applied to the factor involving $\alpha$. For strong coupling $\langle\int_0^1 y(u)du\rangle \to 0$ and

$$R(\xi) \to \frac{1}{\sqrt{2\pi}} \oint_{-\xi} D_W y \, e^{-\int_0^1 U(y(u))du} \left(\left\langle\left(\int_0^1 y(u)du\right)^2\right\rangle\right)^{-\frac{1}{2}} . \tag{4.20}$$

We obtain the correct prefactor since

$$\left\langle\left(\int_0^1 y(u)du\right)^2\right\rangle \to 2 \sum_{n\neq 0} \frac{|\int \psi_n^* \, x \, \psi_0 dx|^2}{E_n - E_0} \sim a^4 . \tag{4.21}$$

REFERENCES


1) R.P. FEYNMAN and A.R. HIBBS, Quantum Mechanics and Path Integrals, McGraw Hill, New York (1965).

2) R.P. FEYNMAN, Statistical Mechanics, W.A. Benjamin, New York (1972).

3) K.F. FREED, J. Chem. Phys. 56, 692 (1972).
D.W. McLAUGHLIN, J. Math. Phys. 13, 1099 (1972).

4) S.F. EDWARDS and Y.B. GULYAEV, Proc. Phys. Soc. (London) 83, 495 (1964).

5) S.F. EDWARDS, J. Non Cryst. Solids 4, 417 (1970).
R.A. ABRAM and S.F. EDWARDS, J. Phys. C: Solid State Phys. 5, 1183 (1972).

6) K.F. FREED, Phys. Rev. B5, 4802 (1972).

7) R. FRIEDBERG and J.M. LUTTINGER, Phys. Rev. B12, 4460 (1975).

8) J. ZITTARTZ and J.S. LANGER, Phys. Rev. 148, 741 (1966).

9) R. JONES and T. LUKES, Proc. Roy. Soc. A309, 457 (1969).

10) B.I. HALPERIN and M. LAX, Phys. Rev. 148, 722 (1966); 153, 802 (1967).

11) J.L. GERVAIS, A. JEVICKI and B. SAKITA, Phys. Rev. D12, 1038 (1975).

12) V. SAMATHIYAKANIT, J. Phys. C 7, 2849 (1974).

13) E.P. GROSS, J. Statist. Phys. 17, 265 (1977).

14) H.L. FRISCH and S.P. LLOYD, Phys. Rev. 120, 1175 (1960).

15) B.I. HALPERIN, Phys. Rev. 139, A104 (1965); Adv. Chem. Phys. 13, 123 (1967).

16) N.N. BOGOLYUBOV and S.V. TYABLIKOV, Zh. Eksp. Teor. Fiz. 19, 256 (1949). Cf. also E.P. GROSS, Ann. Phys. (N.Y.) 99, 1 (1976).

17) J. DES CLOIZEAU, Journal de Physique, 31, 715 (1970).

# FUNCTIONAL INTEGRAL METHODS FOR RANDOM FIELDS

F. LANGOUCHE, D. ROEKAERTS and E. TIRAPEGUI
Instituut voor Theoretische Fysika, Universiteit Leuven, België

We treat here macrosystems in terms of slow variables obeying Langevin equations. For the problems we want to discuss the number of slow variables plays no role so we shall consider only one, the generalization to a finite or infinite number being straightforward. Most of the work reported is contained in refs. 1-4.

Let the Langevin equation be

$$\dot{q}(t) + A(q(t)) = -\frac{c}{2}\, b(q(t))\frac{\partial b(q(t))}{\partial q(t)} + b(q(t))f(t) \tag{1}$$

where $A(q)$ and $b(q)$ are functions of $q$, and the white noise $f(t)$ has the correlation $\{f(t)f(t')\} = c\delta(t-t')$. This means that the average of a functional $F[f]$ of $f(t)$ is computed by the functional integral

$$\{F[f]\} = \frac{\int Df(\tau)\, F[f]\, \exp\left[-\frac{1}{2c}\int_{t_0}^{\infty} d\tau\, f(\tau)^2\right]}{\int Df(\tau)\, \exp\left[-\frac{1}{2c}\int_{t_0}^{\infty} d\tau\, f(\tau)^2\right]} \quad . \tag{2}$$

Let $q_f(t;Q_0,t_0)$ be the solution of (1) such that $q_f(t_0;Q_0,t_0)=Q_0$ for $f(t)$ fixed; then the probability density $P_f(Q,t;Q_0,t_0) = \delta(Q-q_f(t;\, Q_0,t_0)$ satisfies a differential equation easily obtained using (1), and taking the average of this equation over $f(t)$ using (2) one obtains the Fokker-Planck equation[1,2] $(D(q) \equiv b(q)^2)$

$$\dot{P}(Q,t) = \frac{\partial}{\partial Q}\left(A(Q) + \frac{c}{2}\frac{\partial}{\partial Q}D(Q)\right)P(Q,t) \tag{3}$$

No ambiguities arise in doing this when one replaces the $\delta$-function in $\{f(t)\, f(t')\}$ by a symmetric function $\Delta_\eta(t-t') \underset{\eta\to 0}{\longrightarrow} \delta(t-t')$, and this is the physical way of doing it[1]. If one keeps $\eta$ finite one would obtain instead of (3) an equation containing an infinite series[5] since the process $q(t)$ will be non-Markovian. The fact that for $\eta=0$ it is Markovian is easily checked with this technique[1]. One has then a process $q(t)$ whose conditional probability density is the fundamental solution $P(Q,t;Q_0,t_0)$ of (3) satisfying $P(Q,t_0;Q_0,t_0) = \delta(Q-Q_0)$. The joint probability density is $W_n(q_1,t_1;\, q_2,t_2;\ldots;q_n,t_n) = P(q_{\pi(1)},t_{\pi(1)};q_{\pi(2)},t_{\pi(2)})\cdot P(q_{\pi(2)},t_{\pi(2)};\, q_{\pi(3)},t_{\pi(3)})\ldots P(q_{\pi(n)},t_{\pi(n)};\, Q_0,t_0)$, where $\pi$ is the permutation such that $t_{\pi(1)} > t_{\pi(2)} > \ldots > t_{\pi(n)} > t_0$, and in terms of it the average $\{q_f(t_1;Q_0,t_0)\ldots q_f(t_n;Q_0,t_0)\}$ over $f$ of the product of the solutions $q_f$ (the correlation function), which we denote by $\langle q(t_1)\ldots q(t_n)\rangle$, is given by

$$\langle q(t_1)\cdots q(t_m)\rangle = \int \prod_{i=1}^{m} dq_i \, W_m(q_1,t_1;\ldots;q_m,t_m)\, q_1 q_2 \cdots q_m \, . \tag{4}$$

Our aim here is to give functional integral representations for $P(Q,t;Q_0,t_0)$, which is the basic quantity of the formalism. Nothing changes if we write $A(Q,t)$ and $D(Q,t)$, thus allowing for an explicit time dependence (the process $q(t)$ is of course not stationary in this case). We shall briefly recall now the operator formalism we have introduced in ref. 1. Consider operators $\hat{q}_s$ and $\hat{p}_s$, $[\hat{q}_s,\hat{p}_s] = i$, and in the usual notation of quantum mechanics let $|Q\rangle$ be the ket defined by $\hat{q}_s\,|Q\rangle = Q|Q\rangle$, and $\langle Q|$ the conjugate bra with the normalization $\langle Q'|Q\rangle = \delta(Q'-Q)$. Then we can write the solution $P(Q,t;Q_0,t_0)$ of (3) as

$$P(Q,t;Q_0,t_0) = \langle Q|U(t,t_0)|Q_0\rangle = {}^{L}\langle Q,t|Q_0,t_0\rangle^{R} \tag{5}$$

where $U(t',t)$ satisfies the differential equation

$$i\,\frac{\partial U(t',t)}{\partial t'} = \hat{H}(\hat{p}_s,\hat{q}_s,t)\,U(t',t) \quad , \quad U(t,t)=1 , \tag{6}$$

with

$$\hat{H}(\hat{p}_s,\hat{q}_s,t) = -i\,\frac{c}{2}\,\hat{p}_s^2\,D(\hat{q}_s,t) - \hat{p}_s\,A(\hat{q}_s,t) \, . \tag{7}$$

We put $U(t) \equiv U(t,0)$ and then ${}^{L}\langle Q,t| = \langle Q|U(t)$ , $|Q_0,t_0\rangle^{R} = U^{-1}(t_0)\,|Q_0\rangle$ . The Heisenberg operators $\hat{q}(t)$ and $\hat{p}(t)$ are defined by

$$\hat{q}(t) = U^{-1}(t)\,\hat{q}_s\,U(t) \quad , \quad \hat{p}(t) = U^{-1}(t)\,\hat{p}_s\,U(t) \tag{8}$$

The bra $\langle L,t| = \int dQ\,{}^{L}\langle Q,t| = \int dQ\,\langle Q|\,U^{-1}(t)$ is independent of $t$ (we call it $\langle L|$ ). One then shows that (4) can be written as

$$\langle q(t_1)\cdots q(t_m)\rangle = \langle L|\,T\,\hat{q}(t_1)\ldots\hat{q}(t_m)|Q_0,t_0\rangle^{R} \tag{9}$$

where $T$ is the chronological product and we call the right-hand side of (9) the Green's functions of the theory. The Green's functions containing $\hat{p}(t)$ are easily interpreted as linear response functions with respect to a force $\xi(t)$ added to the right-hand side of (1).

Averages over the initial condition $Q_0$ at $t_0$ are easily done; they just change the vector $|Q_0,t_0\rangle^R$ in (9) to $|b,t_0\rangle = U^{-1}(t_0)\int dQ_0 b(Q_0)\,|Q_0\rangle$ if $b(Q_0)$ ( $\int dQ_0 b(Q_0) = 1$) is the initial distribution at $t_0$ . From (5) one has $\int dQP(Q,t;Q_0,t_0) = \langle L|Q_0\rangle = 1$ and writing $U(t,t_0) = U(t,t')\,U(t',t_0) = \int U(t,t')|q\rangle\, dq\, \langle q|U(t',t)$ one obtains the Chapman-Kolmogorov relation. Operator formalisms developed with similar purposes in mind can be found in refs. 6-11 .

Let us derive now the functional-integral representations for $P(Q,t;Q_0,t_0)$ . In order to simplify the presentation we confine ourselves to $A(q,t)$ and $D(q,t)$ with no explicit time dependence, since this point is not relevant for our discussion. The representations are of two types :

$$P(Q,t;Q_0,t_0) = \int_{q(t_0)=Q_0}^{q(t)=Q} Dp\,Dq\; \exp i\int_{t_0}^{t} d\tau\,\big(p(\tau)\dot{q}(\tau) - H^{\gamma}(p(\tau),q(\tau))\big) \qquad (10)$$

$$P(Q,t;Q_0,t_0) = \int_{q(t_0)=Q_0}^{q(t)=Q} D\!\left(\frac{q(\tau)}{\sqrt{D(q(\tau))}}\right) \exp\int_{t_0}^{t} d\tau\; L^{\gamma}(q(\tau),\dot{q}(\tau)) \;, \qquad (11)$$

which we call, respectively, the Hamiltonian and the Lagrangian type. The letter $\gamma$ stands for the discretization prescription necessary for the definition of the functional integrals, and we note that $H^{\gamma}$ and $L^{\gamma}$ depend on it (this is the origin of the different representations one finds in the literature). $L^{\gamma}(q,\dot{q})$ is usually referred to as the Onsager-Machlup function (we shall call it the Lagrangian from now on). If one is familiar with quantum field theory, where functional integrals are popular, one recognizes immediately in (10) and (11) the path integrals associated with the field theory with Green's functions given by (9). There is a voluminous literature on the representations (10) and (11) ; some representative works are refs. 12 - 15 . In ref. 15 Leschke and Schmutz gave, independently of our work and using a different technique, a systematic and careful treatment of a special kind of prescriptions (the ones we call $\gamma_1(\alpha)$ in ref. 4 and their relation to the ordering of operators in $\hat{H}(\hat{p}_s,\hat{q}_s)$ in the Hamiltonian-type representation (10). Their result and conclusions coincide with burs. In order to get some feeling and to motivate the definition of a discretization $\gamma$ that we shall introduce to establish (10) and (11), let us do a simple calculation in the case of constant diffusion ($D(q)= 1$) . Doing a commutation one one can write $\hat{H}$ as ($\alpha$ is a real number)

$$\hat{H}(\hat{p}_s,\hat{q}_s) = -i\frac{c}{2}\hat{p}_s^2 - (1-\alpha)\hat{p}_s A(\hat{q}_s) - \alpha A(\hat{q}_s)\hat{p}_s + i\alpha\frac{\partial A(\hat{q}_s)}{\partial \hat{q}_s} \tag{12}$$

One wants to compute $\langle Q|U(t,t_0)|Q_0\rangle$. Dividing the interval $t-t_0$ into $(n+1)$ parts of length $\varepsilon$, $t-t_0 = (n+1)\varepsilon$, and putting $t_j = t_0 + j\varepsilon$, $t_{n+1} = t$, $q_0 = Q_0$, $q_{n+1} = Q$, one has

$$\begin{aligned}\langle Q|U(t,t_0)|Q_0\rangle &= \int \prod_{i=1}^{n} dq_i \prod_{j=1}^{n+1} \langle q_j|U(t_j,t_{j-1})|q_{j-1}\rangle \\ &= \int \prod_{i=1}^{n} dq_i \prod_{j=1}^{n+1} \langle q_j|1 - i\varepsilon\hat{H}|q_{j-1}\rangle ,\end{aligned} \tag{13}$$

where we have put $U(t_j,t_{j-1}) = 1 - i\varepsilon\hat{H}(\hat{p}_s,\hat{q}_s)$. The fact that only terms up to order are needed is well-known, and this also applies to the non-constant diffusion case; we have discussed and proved this delicate point in detail in ref. 4. If we put from now on $A_{j,j-1} = \langle q_j|U(t_j,t_{j-1})|q_{j-1}\rangle$, a short calculation up to order $\varepsilon$ gives

$$A_{j,j-1} = \int \frac{dp}{2\pi} \exp i\varepsilon\Big[p\frac{q_j-q_{j-1}}{\varepsilon} + i\frac{c}{2}p^2 + p\big((1-\alpha)A(q_{j-1}) + \alpha A(q_j)\big) - i\alpha\frac{\partial A}{\partial q_{j-1}}\Big] \tag{14}$$

Replacing $p$ by $p_j$ in (14) and putting $A_{j,j-1}$ back in (13), one obtains

$$\langle Q|U(t,t_0)|Q_0\rangle = \int \prod_{j=1}^{n+1}\frac{dp_j}{2\pi}\prod_{i=1}^{n} dq_i \exp i\varepsilon\sum_{j=1}^{n+1}\Big[p_j\frac{q_j-q_{j-1}}{\varepsilon} + i\frac{c}{2}p_j^2 + p_j\big((1-\alpha)A(q_{j-1}) + \alpha A(q_j)\big) - i\alpha\frac{\partial A}{\partial q_{j-1}}\Big] \tag{15}$$

The integration over $p_j$ is a Gaussian, and doing it one has

$$\langle Q|U(t,t_0)|Q_0\rangle = \int \prod_{i=1}^{n} dq_i \prod_{j=1}^{n+1}\frac{1}{\sqrt{2\pi\varepsilon c}} \exp \varepsilon\sum_{j=1}^{n+1}\Big[-\frac{1}{2c}\Big(\frac{q_j-q_{j-1}}{\varepsilon} + (1-\alpha)A(q_{j-1}) + \alpha A(q_j)\Big)^2 + \alpha\frac{\partial A}{\partial q_{j-1}}\Big] \tag{16}$$

One can now take the limit $\varepsilon \to 0$ ($n \to \infty$) and notice that the sums in the argument of the exponentials can be written formally as integrals. Consequently, still in a formal way, one may write (15) and (16) as path integrals (the limits $q(t_0) = Q_0$ and $q(t) = Q$ are there, since $q_0 = Q_0$, $q_{n+1} = Q$)

$$\langle Q|U(t,t_0)|Q_0\rangle = \int_{\substack{\gamma_2(\alpha)\\ q(t_0)=Q_0}}^{q(t)=Q} Dp\,Dq \exp i\int_{t_0}^{t} d\tau\Big[p(\tau)\dot{q}(\tau) + i\frac{c}{2}p^2(\tau) + p(\tau)A(q(\tau)) - i\alpha\frac{\partial A(q(\tau))}{\partial q(\tau)}\Big] \tag{17}$$

$$\langle Q|U(t,t_0)|Q_0\rangle = \int_{\gamma_2(\alpha)\, q(t_0)=Q_0}^{q(t)=Q} D(q(\tau)) \exp \int_{t_0}^{t} d\tau \left[ -\frac{1}{2c}\left(\dot{q}(\tau) + A(q(\tau))\right)^2 + \alpha \frac{\partial A(q(\tau))}{\partial q(\tau)} \right] \quad (18)$$

The expression (17) and (18) are the desired representations (compare with (10) and (11))and are to be interpreted essentially as short-hand notations for (15) and (16), and that is the meaning of the subscript $\gamma_2(\alpha)$ attached to the path integral: it is the discretization prescription that is specified by (15) and (16) and that one must know in order to give a meaning to (17) and (18). Let us illustrate this point with a simple and exact calculation corresponding to the case $A(q) = \mu q$, $D(q) = 1$ (Brownian motion with friction). Putting $c = \mu = 1$, $t_0 = 0$, one obtains for (16) (the factor $\exp(\alpha t)$ comes from the term $\alpha \frac{\partial A}{\partial q} = \alpha$ )

$$\langle Q|U(t,t_0)|Q_0\rangle = \exp(\alpha t) \int \prod_{i=1}^{m} dq_i \prod_{j=1}^{m+1} \frac{1}{\sqrt{2\pi\varepsilon}} \exp\left[ -\frac{\varepsilon}{2} \sum_{j=1}^{m+1} \left( \frac{q_j - q_{j-1}}{\varepsilon} + (1-\alpha) q_{j-1} + \alpha q_j \right)^2 \right] \quad (19)$$

The integral in the right-hand side of (19) can be calculated (see ref. 1, section IV)and has the value

$$\frac{1}{\sqrt{\pi(1-\exp(-2t))}} \exp\left[ -\frac{(Q - Q_0 \exp(-t))^2}{1-\exp(-2t)} \right] \cdot \exp(-\alpha t) \quad (20)$$

Putting this into (19) one can see that the $\alpha$-dependence cancels, as it should (there is no $\alpha$-dependence in $\langle Q|U(t,t_0)|Q_0\rangle$ ), and one obtains the known result. But suppose now that one had discretized the Lagrangian, putting

$$-\frac{1}{2}(\dot{q}+q)^2 + \alpha \longrightarrow -\frac{1}{2}\left( \frac{q_j - q_{j-1}}{\varepsilon} + q_{j-1} \right)^2 + \alpha \quad . \quad (21)$$

Then the value of $\langle Q|U(t,t_0)|Q_0\rangle$ would be wrong by a factor $\exp(\alpha t)$. What this simple example shows is that the value of the functional integral depends on the discretization, which in this case is parametrized by $\alpha$ , and consequently if one wants to represent a fixed quantity, here $\langle Q|U(t,t_0)|Q_0\rangle$ , by a functional integral, then when one changes the discretization the integrand must change. In our case this means, of course, that the functions $H^{\gamma}(p,q)$ in (10) and

$L^{\gamma}(q,\dot{q})$ in (11) will be prescription-dependent, as the notation indicates: this is the origin of the different Lagrangians $L^{\gamma}(q,\dot{q})$ one finds in the literature. We see then that in order to give meaningful functional-integral representations we have to give a general definition of discretization. In the Hamiltonian type situation we have to define

$$I_1 = \int_{\gamma\ q(t_0)=Q_0}^{q(t)=Q} Dp\, Dq \ \exp i \int_{t_0}^{t} d\tau \left(p\dot{q} - H(p,q)\right) \tag{22}$$

for an $H(p,q)$ quadratic in $p$, of the form $H=p^2D_1(q) + pD_2(q) + D_3(q)$. The discretization is defined when one is given functions $D_i^{\gamma}(q',q)$, $i = 1,2,3$, such that $D_i^{\gamma}(q',q) \xrightarrow[q'\to q]{} D_i(q)$, and then the functional integral $I_1$ is defined as the limit when $\varepsilon \to 0 (n\to\infty)$ of the multidimensional integral

$$I_1 = \lim_{n\to\infty} \int \prod_{j=1}^{n+1} \frac{dp_j}{2\pi} \prod_{i=1}^{n} dq_i \ \exp i\,\varepsilon \sum_{j=1}^{n+1} \left[p_j \frac{q_j - q_{j-1}}{\varepsilon} - F^{\gamma}(p_j, q_j, q_{j-1})\right] \tag{23}$$

with $F^{\gamma}(p,q',q) = p^2D_1^{\gamma}(q',q)+pD_2^{\gamma}(q',q)+D_3^{\gamma}(q',q)$. In the Lagrangian-type case we shall see that we need more generally to define the functional integral

$$I_2 = \int_{\gamma\ q(t_0)=Q_0}^{q(t)=Q} D\left(\frac{q(\tau)}{\sqrt{D(q(\tau))}}\right) \exp \int_{t_0}^{t} d\tau \ L(q(\tau), \dot{q}(\tau)) \tag{24}$$

for $L(q,\dot{q})$ of the form $L=\dot{q}^2B_1(q) + \dot{q}B_2(q) + B_3(q)$ (in our case we have $B_1(q) = -(2D(q))^{-1}$). The discretization $\gamma$ is defined when one is given functions $D^{\gamma}(q',q)$, $B_i^{\gamma}(q',q)$, $i = 1,2,3$, such that $D^{\gamma}(q',q) \xrightarrow[q'\to q]{} D(q)$, $B_i^{\gamma}(q',q) \xrightarrow[q'\to q]{} B_i(q)$, and then $I_2$ is defined as the limit when $n \to \infty$ of the n-dimensional integral

$$I_2 = \lim_{n\to\infty} \int \prod_{i=1}^{n} dq_i \prod_{j=1}^{n+1} \frac{1}{\sqrt{2\pi\varepsilon D^{\gamma}(q_j,q_{j-1})}} \exp \varepsilon \sum_{j=1}^{n+1} E^{\gamma}\left(\frac{q_j - q_{j-1}}{\varepsilon}, q_j, q_{j-1}\right), \tag{25}$$

with

$$E^{\gamma} = \left(\frac{q_j - q_{j-1}}{\varepsilon}\right)^2 B_1^{\gamma}(q_j, q_{j-1}) + \left(\frac{q_j - q_{j-1}}{\varepsilon}\right) B_2^{\gamma}(q_j,q_{j-1}) + B_3^{\gamma}(q_j, q_{j-1}). \tag{26}$$

With these definitions, formulas (10) and (11) are unambiguous.

The prescription $\gamma_2(\alpha)$ of formulas (17) and (18) can be characterized in words by saying that functions of $q$ have to be discretized as $f(q(\tau)) \to (1-\alpha) f(q_{j-1}) + \alpha f(q_j)$. What does one obtain for this discretization prescription in the case of non-constant diffusion? Doing commutators in both terms of $\hat{H}$ given by (7) one can write $\hat{H}$ in the form $(c = 1)$

$$\hat{H} = -\frac{i}{2}\left[(1-\alpha)\hat{p}_s^2 D(\hat{q}_s) + \alpha D(\hat{q}_s)\hat{p}_s^2\right] - (1-\alpha)\hat{p}_s\left(A(\hat{q}_s) + \alpha D'(\hat{q}_s)\right) \tag{27}$$
$$- \alpha\left(A(\hat{q}_s) + \alpha D'(\hat{q}_s)\right)\hat{p}_s + i\alpha\left(A'(\hat{q}_s) + (\alpha - \tfrac{1}{2}) D''(\hat{q}_s)\right)$$

where the prime denotes differentiation with respect to $q$. The same calculation as before[4)] gives now

$$A_{j,j-1} = \int \frac{dp}{2\pi} \exp\left[ip(q_j - q_{j-1}) - \frac{\varepsilon}{2} p^2 D_{j,j-1} + i\varepsilon p B_{j,j-1} + \varepsilon C_{j,j-1}\right] \tag{28}$$

with

$$D_{j,j-1} = (1-\alpha) D(q_{j-1}) + \alpha D(q_j), \tag{29}$$

$$B_{j,j-1} = (1-\alpha)\left(A(q_{j-1}) + \alpha D'(q_{j-1})\right) + \alpha\left(A(q_j) + \alpha D'(q_j)\right), \tag{29'}$$

$$C_{j,j-1} = \alpha\left(A'(q_{j-1}) + (\alpha - \tfrac{1}{2}) D''(q_{j-1})\right). \tag{29''}$$

After integration over $p$ this gives representation (11) for $\gamma_2(\alpha)$, with the Lagrangian

$$L^{\gamma_2(\alpha)}(q,\dot{q}) = -\frac{1}{2D(q)}\left(\dot{q} + A(q) + \alpha D'(q)\right)^2 + \alpha\left(A'(q) + (\alpha - \tfrac{1}{2}) D''(q)\right) \tag{30}$$

The functions $D^{\gamma_2(\alpha)}$, $B_i^{\gamma_2(\alpha)}$, are determined from (28) and (29) and have the values $D^{\gamma_2(\alpha)}(q_j, q_{j-1}) = D_{j,j-1}$, $B_1^{\gamma_2(\alpha)} (2D_{j,j-1})^{-1}$ $B_2^{\gamma_2(\alpha)} = -B_{j,j-1}/D_{j,j-1}$, $B_3^{\gamma_2(\alpha)} = -(B_{j,j-1}^2/2D_{j,j-1}) + C_{j,j-1}$.

We shall now discuss briefly a Lagrangian that is formally covariant under point transformations (using the normal rules of calculus), proposed by Graham (ref. 13, second paper). As we have already explained from our point of view, a path-integral representation is meaningless when the discretization is not specified. Then in order to discuss the covariant Lagrangian we have first to see if it corresponds to some discretization . For this we introduce an invariant probability density $S(Q,t;Q_0,t_0) = b(Q)P(Q,t;Q_0,t_0)$ , $b(Q) = \sqrt{D(Q)}$ , which obeys

$$\dot{S} = b(Q)\frac{\partial}{\partial Q}\left(\frac{h(Q)}{b(Q)} + \frac{1}{2}\, b(Q)\frac{\partial}{\partial Q}\right)S \tag{31}$$

with $h(Q) = A(Q) + \frac{1}{2}\, b(Q)\,\frac{\partial b(Q)}{\partial Q}$ . S can be written as $S(Q,t;Q_0,t_0) = b(Q)\,\langle Q|U_1(t,t_0)|Q_0\rangle$ with

$$i\,\frac{\partial U_1(t',t)}{\partial t'} = \hat{H}_1(\hat{p}_s,\hat{q}_s)\,U_1(t',t) \quad , \; U_1(t,t) = 1, \tag{32}$$

with $(\hat{b} = b(\hat{q}_s),\ \hat{h} = h(\hat{q}_s))$

$$\hat{H}_1 = -\frac{i}{2}\,\hat{b}\,\hat{p}_s\,\hat{b}\,\hat{p}_s - \hat{b}\left((1-\alpha)\,\hat{p}_s\,\frac{1}{\hat{b}}\,\hat{h} - \alpha\,\frac{1}{\hat{b}}\,\hat{h}\,\hat{p}_s\right) + i\alpha\,\frac{\partial}{\partial \hat{q}_s}\,\frac{1}{\hat{b}}\,\hat{h}\,. \tag{33}$$

One has as before $\langle Q|U_1(t,t_0)|Q_0\rangle = \int \prod_{i=1}^{n} dq_i \prod_{j=1}^{n+1} A^{(s)}_{j,j-1}$ and one obtains for $A^{(s)}_{j,j-1}$ expression (28), but now with $(q^{(\alpha)}_{j-1} \equiv q_{j-1} + \alpha(q_j - q_{j-1}))$

$$D^{(s)}_{j,j-1} = b(q_j)\left[2\,b(q^{1/2}_{j-1}) - \frac{1}{2}\left(b(q_j) + b(q_{j-1})\right)\right] , \tag{34}$$

$$B^{(s)}_{j,j-1} = b(q_j)\left[(1-\alpha)\,\frac{h(q_{j-1})}{b(q_{j-1})} + \alpha\,\frac{h(q_j)}{b(q_j)}\right] , \tag{34'}$$

$$C^{(s)}_{j,j-1} = \alpha\, b(q_{j-1})\,\frac{\partial}{\partial q_{j-1}}\,\frac{h(q_{j-1})}{b(q_{j-1})} \quad . \tag{34''}$$

Integrating (28) over p gives then

$$A^{(s)}_{j,j-1} = \frac{1}{\sqrt{2\pi\varepsilon\, D^{(s)}_{j,j-1}}}\,\exp\left[-\frac{\varepsilon}{2D^{(s)}_{j,j-1}}\left(\frac{q_j - q_{j-1}}{\varepsilon} + B^{(s)}_{j,j-1}\right)^2 + \varepsilon\, C^{(s)}_{j,j-1}\right]. \tag{35}$$

One now puts (35) back into $\int \prod dq_i \prod A^{(s)}_{j,j-1}$ and since $P=(b(Q_0)/b(Q))S$ one has to incorporate this factor in the measure to obtain a Lagrangian type representation. This is done by writing

$$\frac{b(Q_0)}{b(Q)} = \sum_{j=1}^{n+1} \frac{b(q_{j-1})}{b(q_j)} \tag{36}$$

which then changes $D^{(s)}_{j,j-1}$ to $\frac{b(q_j)^2}{b(q_{j-1})^2} D^{(s)}_{j,j-1}$ in (35); then the discretization, that we call $\gamma_3(\alpha)$ is characterized by $D^{\gamma_3(\alpha)}(q_j,q_{j-1}) = \frac{b(q_j)^2}{b(q_{j-1})^2} D^{(s)}_{j,j-1}$, and the $B_i^{\gamma_3(\alpha)}(q',q)$ can be read from (35). This gives finally the Lagrangian

$$L^{\gamma_3(\alpha)}(q,\dot{q}) = -\frac{1}{2D(q)}\left(\dot{q} + A(q) + \frac{1}{4}\frac{\partial D}{\partial q}\right)^2 + \alpha\sqrt{D}\frac{\partial}{\partial q}\frac{1}{\sqrt{D}}\left(A + \frac{1}{4}\frac{\partial D}{\partial q}\right) \tag{37}$$

which for $\alpha = \frac{1}{2}$ is the result of Graham (it is in fact formally a scalar for any ). We remark that there is no Hamiltonian-type representation corresponding to $\gamma_3(\alpha)$ because of the inclusion of the factor (36) in the measure. This has as a consequence that the function $(2D(q))^{-1}$ in (37) has to be discretized in a different way from the same function $D(q)$ in the measure, and consequently one cannot go back from $(b(q_{j-1})/b(q_j))\, A^{(S)}_{j,j-1}$ as given by (35) to the analogue of (38). In order to do this one has to put $b(Q_0)/b(Q)$ in the exponential, thus defining a new discretization $\gamma_4(\alpha)$ and a new Lagrangian $L^{\gamma_4(\alpha)} = L^{\gamma_3(\alpha)} - \frac{1}{2}\dot{q}\frac{D'}{D}$ which was proposed in the first paper of ref. 13 (the technique to prove this is just integration by parts, which we have shown to be valid in the functional integral when one discretizes the corresponding term in the midpoint[33]). One has $D^{\gamma_3(\alpha)} = D^{(s)}_{j,j-1}$, $B_i^{\gamma_4(\alpha)} = B_i^{\gamma_3(\alpha)}$ [4]. This last remark is pertinent for perturbation theory that is done starting from the Hamiltonian-type representation and for which one must then use the function $H^{\gamma_4(\alpha)}(p,q)$. We have also shown in ref. 4 how one can obtain the different Lagrangians proposed in ref. 12, together with their discretizations.

Before discussing further $L^{\gamma_3(\alpha)}(q,\dot{q})$ and other ways of deriving it, following ref. 4, we must comment briefly on the general covariance of the path integral. Under a general coordinate transformation $\bar{q} = \bar{q}(q)$ the probability density will transform as $\bar{P}(\bar{q},t)d\bar{q} = P(q,t)dq$, and if $P(q,t)$ obeys (3) then $\bar{P}$ obeys the same equation with $A(q) \to \bar{A}(\bar{q}) = \bar{q}'A(q) - \frac{1}{2} D(q)\bar{q}''$, $\bar{D}(\bar{q}) = (\bar{q}')^2 D/q)$ (the primes denote derivatives with respect to $q$). Putting $\bar{Q} = \bar{q}(Q)$, $\bar{Q}_0 = \bar{q}(Q_0)$, one has $\bar{P}(\bar{Q},t;\bar{Q}_0,t_0) = (\bar{q}'(Q))^{-1}P(Q,t;Q_0,t_0)$. We introduce $L_1^\gamma(\dot{q},A(q), D(q)) = L^\gamma(\dot{q},q)$; $L_1$ exhibits explicitly the dependence on $A(q)$ and $D(q)$. One must have then, for any discretization $\gamma$,

$$\int_{\gamma\,\bar{q}(t_0)=\bar{Q}_0}^{\bar{q}(t)=\bar{Q}} \mathcal{D}\left(\frac{\bar{q}(\tau)}{\sqrt{\bar{D}(\bar{q})}}\right) \exp\int_{t_0}^{t} d\tau\, L_1^\gamma(\dot{\bar{q}}, \bar{A}(\bar{q}), \bar{D}(\bar{q})) = \left(\frac{d\bar{q}}{dq}\bigg|_Q\right)^{-1} \int_{\gamma\, q(t_0)=Q_0}^{q(t)=Q} \mathcal{D}\left(\frac{q(\tau)}{\sqrt{D(q)}}\right) \exp\int_{t_0}^{t} d\tau\, L_1^\gamma(\dot{q}, A(q), D(q)) \quad . \tag{39}$$

It has been argued that (39) is only valid for $\gamma_3(\alpha)$. We shall see that it is valid for arbitrary $\gamma$. In order to prove this we explain first how to go from one discretization to another. We start with a discretization $\gamma$ and we shall see how to go to $\gamma_2(0)$ which corresponds to discretization of all functions of $q$ in the prepoint $q_{j-1}$. One has $(\Delta_j \equiv q_j - q_{j-1})$

$$I = \int \prod_{i=1}^{n} dq_i \prod_{j=1}^{n+1} \frac{1}{\sqrt{2\pi\varepsilon D^\gamma(q_j,q_{j-1})}} \exp \varepsilon \sum_{j=1}^{n+1} E^\gamma\left(\frac{\Delta_j}{\varepsilon}, q_j, q_{j-1}\right) \quad . \tag{40}$$

Since when $q' \to q$ one has $D^\gamma(q',q) \to D(q)$, $B_i^\gamma(q',q) \to B_i(q)$, one can expand everything around $q_{j-1}$ in (40). One obtains then an expression of the form

$$I = \int \prod_{i=1}^{n} dq_i \prod_{j=1}^{n+1} \frac{1}{\sqrt{2\pi\varepsilon D(q_{j-1})}} \exp\left[-\sum_{j=1}^{n+1} \frac{\Delta_j^2}{2\varepsilon D(q_{j-1})}\right]\cdot\left(1 + f(\Delta_j, \varepsilon, q_{j-1})\right) \tag{41}$$

Owing to the exponential factor in the n-dimensional integral, the following replacements are allowed[16)]

$$\Delta_j^2 \doteq \varepsilon D(q_{j-1}) \quad , \quad \Delta_j^3 \doteq 3\varepsilon D(q_{j-1})\, \Delta_j \quad , \tag{42}$$

$$\Delta_j^4 \doteq 3\varepsilon^2 D(q_{j-1})^2 \quad , \quad \Delta_j^6 \doteq 15\varepsilon^3 D(q_{j-1})^3 \quad . \tag{42}$$

The notation $\doteq$ was introduced by de Witt[17)] and means equality as far as replacement in the n-dimensional integral is concerned. The relations (42) show that $\Delta_j^2$ is effectively of order $O(\varepsilon)$ and this tells us the terms that are needed in $f(\Delta_j, \varepsilon, q_{j-1})$ in (41). Using (42) in (41) to eliminate $\Delta_j^n$, $n \geqslant 2$, one obtains, as one must, the $\gamma_2(0)$ result; that is,

$$I = \int \prod_{i=1}^{m} dq_i \prod_{j=1}^{m+1} \frac{1}{\sqrt{2\pi\varepsilon D(q_{j-1})}} \exp - \varepsilon \sum_{j=1}^{m+1} \frac{1}{2D(q_{j-1})} \left( \frac{q_j - q_{j-1}}{\varepsilon} + A(q_{j-1}) \right)^2 \tag{43}$$

corresponding to the Lagrangian

$$L^{\gamma_2(0)} = -\frac{1}{2D(q)} \left( \dot{q} + A(q) \right)^2 . \tag{44}$$

We see then that we need only to prove (39) for the $\gamma_2(0)$ prescription, and we have to do this of course using the definition of the path integral that involves discretization. We write then in the $\gamma_2(0)$ prescription $(\bar{q}(t_j) = \bar{q}_j = \bar{q}(q(t_j)) = \bar{q}(q_j))$

$$\bar{I} = \int \prod_{i=1}^{m} d\bar{q}_i \prod_{j=1}^{m+1} \frac{1}{\sqrt{2\pi\varepsilon \bar{D}(\bar{q}_{j-1})}} \exp \varepsilon \sum_{j=1}^{m+1} E^{\gamma_2(0)}\left( \frac{\bar{\Delta}_j}{\varepsilon}, \bar{q}_j, \bar{q}_{j-1} \right) \tag{45}$$

Using $\bar{D}(\bar{q}_j) = (\bar{q}'(q_j))^2 D(q_j)$ one obtains

$$\bar{I} = \left( \left.\frac{d\bar{q}}{dq}\right|_{Q_0} \right)^{-1} \int \prod_{i=1}^{m} \left.\frac{d\bar{q}}{dq}\right|_{q_i} dq_i \left.\frac{d\bar{q}}{dq}\right|_{q_{m+1}} \prod_{j=1}^{m+1} \frac{1}{\left.\frac{d\bar{q}}{dq}\right|_{q_{j-1}} \sqrt{2\pi\varepsilon D(q_{j-1})}} \exp \varepsilon \sum_{j=1}^{m+1} E_j^{\gamma_2(0)} \tag{46}$$

The $E_j^{\gamma_2(0)}$ in (46) is written in the $\bar{q}$ coordinate, that is

$$E_j^{\gamma_2(0)}\left( \frac{\bar{\Delta}_j}{\varepsilon}, \bar{q}_j, \bar{q}_{j-1} \right) = -\frac{\bar{\Delta}_j^2}{2\varepsilon \bar{D}(\bar{q}_{j-1})} - \frac{\bar{\Delta}_j \bar{A}(\bar{q}_{j-1})}{\bar{D}(\bar{q}_{j-1})} - \frac{\varepsilon}{2} \frac{\bar{A}(\bar{q}_{j-1})^2}{\bar{D}(\bar{q}_{j-1})} . \tag{47}$$

The quantity $\bar{\Delta}_j$ transforms as $(\bar{q}'_j = \bar{q}'(q_j))$

$$\bar{\Delta}_j = \bar{q}'_{j-1} \Delta_j + \frac{1}{2} \bar{q}''_{j-1} \Delta_j^2 + \frac{1}{6} \bar{q}'''_{j-1} \Delta_j^3 + O(\Delta_j^4) \tag{48}$$

and one can easily convince oneself that one needs all the terms written in (48) in order to be correct up to order $\varepsilon$ (remember $\Delta_j^2 = O(\varepsilon)$) . The continuation is clear; one has to develop everything in (46) around $q_{j-1}$ , use (48) for $\bar{\Delta}_j$ , and also the known transformation properties of $\bar{A}(\bar{q})$ and $\bar{D}(\bar{q})$ . After some lengthy calculations and after using of relations (42) to eliminate $\Delta_j^n, n \geq 2$, one obtains, as we have shown in ref. 4, that

$$\bar{I} = \left( \frac{d\bar{q}}{dq}\Big|_Q \right)^{-1} I \tag{49}$$

with I given by (43), which is the desired result.

We are now prepared, after the technical details we have summarized, to understand some methods of derivation of the formally covariant Lagrangian $L^{\gamma_3(\alpha)}$ . When one has only one variable there is no curvature and one can always find a coordinate transformation such that D(q) becomes constant. We can then obtain a Lagrangian for non-constant diffusion starting from D constant and doing a coordinate transformation, and if one uses the formal rules of calculus one easily checks that one obtains $L^{\gamma_3(\alpha)}$ as should be the case since it is a scalar (see our discussion in ref. 4; this procedure was used in ref. 14). But by proceeding in this way one is not transforming correctly, since, for instance, in (48) one will only keep the first term, while we have seen that the other ones should be kept also. But one can easily convince oneself that the difference between the correct transformation and the transformation according to the formal rules of calculus consists in the terms that are dropped in the derivation of $L^{\gamma_3(\alpha)}$ in ref. 13 using the argument of restriction to differentiable paths, and consequently the result should be the formally transformed version of the constant-diffusion Lagrangian, which is indeed $L^{\gamma_3(\alpha)}$ .

Let us comment now briefly on perturbation theory. We have shown in ref. 1 that for the $\gamma_2(\alpha)$ discretization the $\alpha$-dependence cancels,as it must, and that the rule in perturbation theory is to put the equal-time propagator $\langle p(t)q(t) \rangle = i\alpha$ . This independence of the discretization is completely general in perturbation theory, but of course the simple rule we have mentioned is not a good one for arbitrary $\gamma$, in fact it is a good rule for the $\gamma_1(\alpha)$ discretization of Leschke and Schmutz only[15] , which coincides with $\gamma_2(\alpha)$ for constant diffusion. It should be clear that the simplest thing to do is to perform the perturbation expansion with the $\gamma_2(0)$ discretization, that is with $H_{\gamma_2}^{(0)}(p,q)$ obtained from the Lagrangian (44). But if, for some reason, one wants to use another $H_\gamma$, then the rules for the cancellations can become very complicated and to derive them it is necessary to know the discretization. It is of course just the mathematical consistency of the formalism that gives the rules for the cancellations, and not the physics as is sometimes stated. This problem of different but equivalent perturbation expansions has been treated recently, with the same conclusions, using only operator methods, in ref. 18. However, if one wants to apply nonperturbative methods we think that one must previously reduce the problem to a prescription-independent functional integral (no result independent of the discretization is known apart from in perturbation theory). An illustration of this is provided by our steepest-descents calculation in ref. 2, where we transform the functional integral to a prescription-independent one by a suitable factorization. The factorization can be clearly generalized to $N > 1$ variables if the drift $A_i(q) = \frac{\partial V}{\partial q_i}$ , and this case is realized for instance if the non-dissipative part of the drift vanishes and the dissipative part obeys the potential conditions[19] (this can be checked using the formulation of ref. 7).

Let us end this brief review of functional-integral techniques by stating our conclusions :

a) there is no unique Lagrangian,
b) this is not relevant for perturbation theory, which is prescription-independent,
c) if one wants to use methods other than perturbation theory one must previously make the reduction, where possible, to prescription-independent path integrals.

We have not considered the question of the most probable path. Considering this is a different point of view, and we think that the solution of this problem (in the sense of finding a Lagrangian whose Euler-Lagrange equation gives the most probable path) should be the Lagrangian proposed by Graham[13]. This can easily be shown when there is no curvature, and when there is curvature the only doubt is in the coefficient of the scalar curvature.

REFERENCES


1) F. LANGOUCHE, D. ROEKAERTS and E. TIRAPEGUI, preprint KUL-TF-77/023.
2) F. LANGOUCHE, D. ROEKAERTS and E. TIRAPEGUI, preprint KUL-TF-78/003.
3) F. LANGOUCHE, D. ROEKAERTS and E. TIRAPEGUI, preprint KUL-TF-78/007.
4) F. LANGOUCHE, D. ROEKAERTS and E. TIRAPEGUI, preprint KUL-TF-78/015.
5) F. LANGOUCHE D. ROEKAERTS and E. TIRAPEGUI, in preparation.
6) C.P. ENZ and L.M. GARRIDO, Phys. Rev. A14, 1258 (1976).
7) C.P. ENZ, Physica 89A, 1 (1977).
8) P.C. MARTIN, E.D. SIGGIA and H.A. ROSE, Phys. Rev. A8, 423 (1973).
9) B. JOUVET, Quantum Aspects of Stochastic Fields, Orsay preprint, France (1977).
10) L. GARRIDO and M. SAN MIGUEL, Prog. Theor. Phys. 59, 40 (1978) and 59, 55 (1978).
11) P. PHYTHIAN, J. Phys. A8, 1423 (1975).
12) H. DEKKER, Physica 87A, 419 (1977).
H. DEKKER, On the Fourier Series Analysis of the Functional Integral for General Diffusion Processes, preprint TNO Den Haag (1978).
13) R. GRAHAM, Springer Tracts on Modern Physics 66 (1973);
Z. Physik B26, 281 (1977) and B26, 397 (1977).
14) W. HORSTHEMKE and A. BACH, Z. Phys. B22, 189 (1975)
15) H. LESCHKE and M. SCHMUTZ, Z. Phys. B27, 85 (1977).
16) D.W. Mc LAUGHLIN and L.S. SCHULMAN, J. Math. Phys. 12, 2520 (1971).
17) B.S. De WITT, Rev. Mod. Phys. 29, 377 (1957).
18) H. LESCHKE, A.C. HIRSCHFELD and T. SUZUKI, Canonical perturbation theory for nonlinear systems, preprint Institut für Physik, Universität Dortmund, March (1978).
19) L. GARRIDO and M. SAN MIGUEL (Private Communication).

## KINETIC EQUATIONS FROM HAMILTONIAN DYNAMICS : THE MARKOVIAN LIMIT

H. SPOHN

Fachbereich Physik - Universität München - Germany

## 1. INTRODUCTION

The development of kinetic theory in the nineteenth century posed the problem of how this description is related to the full microscopic dynamics. After some thought it is clear that the two theories necessarily give different answers to the same physical question. However, in some limiting situations, such as, e.g., low density or weak coupling, the solutions of the kinetic equation may approximate well certain solutions of Hamilton's equation of motion for the corresponding mechanical system. In recent years, some of the physical arguments have been put on a rigorous mathematical basis. Fairly diverse (quantum as well as classical) model systems and various limits have been studied. The unifying principle seems to be that the limiting procedure eliminates certain memory (non-Markovian) effects. The goal of this lecture is to discuss, non-technically, as a paradigm, the Rayleigh gas, i.e., a Brownian particle immersed in an ideal gas. We then indicate rigorous results about other model systems. (This list is not claimed to be complete.)

## 2. THE RAYLEIGH GAS

The Rayleigh gas consists of a single Brownian particle in an ideal gas interacting via a pair potential $V$. We enclose the total system in a box $\Lambda$. If we assume that the gas is in thermal equilibrium, then kinetically the motion of the Brownian particle may be described by the linear Boltzmann equation[1)]. This is just the nonlinear Boltzmann equation with $f(v_*')$ and $f(v_*)$ replaced by the Maxwellian distribution. After some manipulations[1)], the linear Boltzmann equation can be brought to the form

$$\frac{\partial}{\partial t} f(q,p,t) = - p\cdot \frac{\partial}{\partial q} f(q,p,t) + \int dp' \left[ k(p|p')\, f(q,p',t) - k(p'/p)\, f(q,p,t) \right] \qquad (2.1)$$

with specular reflection at the boundary of $\Lambda$. $q,p$ stands for the position and momentum of the Brownian particle and $f(q,p,t)$ is its probability density at time $t$. Equation (2.1) is the forward equation of a Markov process denoted by $\{q(q^o,p^o,t)\ ,\ p(q^o,p^o,t)\}$ (sometimes called a random-flight or transport process) : The Brownian particle, which starts at $q^o,p^o$ , moves under the combination of free flow and of random jumps in the momentum as governed by the collision term. $\int dp' k(p'|p) = \nu(p)$ is the jump rate, which regulates how often the particle will undergo a collision. In a collision, if the particle has momentum $p$ , then $\int_A dp' k(p'|p)\nu(p)^{-1}$ is the probability for the momentum to jump into the set $A$ .

Microscopically, the mechanical system is specified by having $N$ gas molecules of mass $M$ with an interaction potential $V_R$ of range $R$ between the Brownian particle and the gas molecules. Therefore, the Hamiltonian function is

$$H = \frac{p^2}{2} + \sum_{j=1}^{N} \frac{p_j^2}{2M} + \sum_{j=1}^{N} V_R(q_j - q)\ ; \tag{2.2}$$

$(q_1,p_1,\ldots,q_N,p_N)$ stands for the positions and momenta of the gas particles. We assume that at time $t=0$ the gas is in thermal equilibrium at a temperature $\beta^{-1}$ uncorrelated with the Brownian particle, which starts at $q^o$ with momentum $p^o$ . If $(q_1^o,p_1^o,\ldots,\ q_N^o,p_N^o)$ are the initial data of the gas particles, then solving Hamilton's equation of motion with specular reflection at the boundary of $\Lambda$ gives the position and momentum of the Brownian particle at a later time as $\{q^{(R,N)}(q^o,p^o,q_1^o,p_1^o,\ldots,\ q_N^o,p_N^o,t)\ ,\ p^{(R,N)}(q^o,p^o,q_1^o,p_1^o,\ldots,\ q_N^o,p_N^o,t)\}$. We have explicitly indicated $R$ and $N$ , since we regard them as parameters still at our disposal. Since the initial data are distributed according to the Maxwellian distribution $\{\prod_{j=1}^{N} \exp[-\beta(p_j^o)^2/2M]\}/Z$ we may suppose that $\{q^{(R,N)}(q^o,p^o,t),p^{(R,N)}(q^o,p^o,t)\}$ is a stochastic process starting at $q^o,p^o$ . This process is <u>non-Markovian</u>. If the Brownian particle collides with the same gas molecule again, because of the dynamical laws both collisions are correlated. This constitutes a memory effect.

The problem raised at the beginning can now be made precise. In what sense is the solution $f(q,p,t)$ of the linear Boltzmann equation (2.1) with the initial data $f(q,p)$ a good approximation to the reduced dynamics

$$f^{(R,N)}(q,p,t) = \int dq_1 dp_1,\ldots,\ dq_N dp_N\ h(q,p,q_1,p_1,\ldots,\ q_N,p_N,t)\ \ ? \tag{2.3}$$

Here $h(t)$ is the time-evolved probability density corresponding to the initial data $h(q,p,q_1,p_1,\ldots,q_N,p_N,0) = f(q,p)\ \{\prod_{j=1}^{N} \exp[-\beta\, p_j^2/2M]\}/Z$.

More ambitiously, one might ask in what sense the stochastic process $\{q(q^o,p^o,t),\ p(q^o,p^o,t)\}$ approximates the process $\{q^{(R,N)}(q_0,p_0,t),\ p^{(R,N)}(q_0,p_0,t)\}$.

For these questions to be meaningful we already use two "rules" implicitly :

1) Only a partial description of the microscopic system is considered. (For the Rayleigh gas the partial description corresponds to considering only a subsystem, whereas for an interacting N-particle system the partial description usually corresponds to considering only the one-particle correlation function.)
2) We restrict the class of initial states. (If the gas is in some wild nonequilibrium state, there is no hope of finding a simple kinetic description of the motion of the Brownian particle.)
   We still have to ensure that the non-Markovian effects are negligible. This leads to:
3) The Markovian limit. (Mathematically, this is where all the work goes.) Note that for the same system one may take different Markovian limits and, accordingly, one ends up with different kinetic equations.

For the Rayleigh gas the idea is to make the gas molecules smaller and smaller. This will decrease the probability of recollisions. To retain a nontrivial effect, we have at the same time to increase the number of gas molecules. How should we scale?

i) We assume that

$$V_R(q) = V(\tfrac{q}{R}) ,$$

where $V$ is a given central, twice continuously differentiable potential of range 1. This scaling leaves the (normalized) differentiable cross-section invariant. The total cross-section decreases as $R^2$ ($R^{d-1}$ in $d$ dimensions).

ii) We let $R \to 0$, $N \to \infty$, such that $NR^2 = \rho$ with given $0 < \rho < \infty$ ($NR^{d-1} = \rho$ in $d$ dimensions). This means that the mean free path $\lambda \approx |\Lambda|/NR^2$ is kept constant. Note that although the number density of the gas becomes infinite, the volume density, $\sim NR^3$, goes to zero. So, we really consider the low-density limit. i) with ii) is called the Boltzmann-Grad limit.

Theorem[2]: Under the assumptions i) and ii), the process $\{q^{(R,N)}(q^o,p^o,t),\ p^{(R,N)}(q^o,p^o,t)\}$ converges to the process $\{q(q^o,p^o,t),\ p(q^o,p^o,t)\}$.

By convergence of the process we mean that all multi-time correlation functions of $\{q^{(R,N)}(q^o,p^o,t),\ p^{(R,N)}(q^o,p^o,t)\}$ converge to the corresponding ones of $\{q(q^o,p^o,t),\ p(q^o,p^o,t)\}$ (all the finite-dimensional distributions converge weakly; cf. the lectures by van Kampen). In particular, the reduced dynamics $f^{(R,N)}(t)$ converges pointwise to the solution $f(t)$ of the linear Boltzmann equation. We cannot show that the convergence is uniform in $t$ (and physically, presumably, it should not be so).

By an additional argument, one shows that random variables such as $\int dq^o dp^o\ f(q^o,p^o)\ g(q^{(R,N)}(q^o,p^o,t),\ p^{(R,N)}(q^o,p^o,t))$ do not fluctuate in the Boltzmann-Grad limit. Up to a small error, for a "typical" initial configuration of the gas particles, such an expectation value will be the same as that computed from the linear Boltzmann equation.

## 3. SURVEY OF OTHER RESULTS

The models studied fall into two broad classes. Either they are of the form "system plus reservoir" like the Rayleigh gas. This leads in the Markovian limit to a linear kinetic equation for the time evolution of the system. Or they are interacting N-particle systems. In this case one is led to a kinetic equation with quadratic nonlinearity for the one-particle correlation function. Then, of course, the models may be classified as classical and quantum mechanical, and by what kind of limit is taken.

Roughly one might say that at the moment one understands the Boltzmann-Grad limit classically and the weak-coupling limit quantum mechanically. Furthermore, although not completely written out, the mean-field limit can be taken. There are many open problems of a similar type, such as, e.g., the Brownian particle in the limit of small mass ratio (cf. Mazo's lectures), or the quantum-mechanical Boltzmann equation.

<u>Classical systems</u> : For the Lorentz gas (static scatterers) and for a tagged particle in a real fluid (the gas particles also interact via the potential $V_R$, here positive) one can prove results similar to the ones for the Rayleigh gas (convergence of the process)[2,3,4]. There are some one-dimensional results[5-10]. For an N-particle system of hard spheres, Lanford has proved that in the Boltzmann-Grad limit an initially uncorrelated state stays uncorrelated and the the one-particle correlation function satisfies the Boltzmann equation[11].

The main technical restriction is the convergence up to only one-fifth of the mean free time. These results have been extended to positive interaction potentials[12)]. Braun and Hepp[13)] showed that for an N-particle system in the mean-field limit, $V_\lambda(q) = \lambda V(q)$, $\lambda \to 0$, $N \to \infty$, $N\lambda = \rho$, again an initially uncorrelated state stays uncorrelated, and that the one-particle correlation function satisfies the Vlasov equation. They also study fluctuations. For harmonic sys-systems the weak-coupling limit has been investigated[14)].

Quantum mechanical systems : Davies has investigated the weak-coupling limit for the Ford-Kac-Mazur model $V_\lambda(q) = \lambda V(q)$, $\lambda \to 0$, $t \to \infty$, $\lambda^2 t = \tau$ [15)], for the Wigner-Weisskopf atom[16)], and for a general system (with a system Hamiltonian with a pure point spectrum) weakly coupled to a thermal reservoir[17)]. The theory has been extended to time-dependent Hamiltonians[18)] and to various other cases[19,20)]. In all cases one obtains a quantum dynamical semigroup in the limit. (This is a quantum generalization of the Pauli master equation.) In three and more dimensions the weak-coupling limit has been shown to exist for the spatially homogeneous polaron up to a finite time[21,22)]. The resulting kinetic equation is the linear Boltzmann equation. Gorini et al. investigated an N-level system coupled to an electromagnetic field in the vacuum in the singular-coupling limit[23,24)]. This limit was introduced by Hepp and Lieb[25)] and corresponds to an instantaneous interaction. By appropriately adjusting the coupling parameters one obtains all possible quantum dynamical semigroups in the limit. The mathematical theory developed in connection with the treatment of open quantum systems may be found in ref. 26.

REFERENCES


1) C. CERCIGNANI, The Boltzmann Equation, Elsevier, New York (1976).
2) H. SPOHN, unpublished notes.
3) G. GALLAVOTTI, Phys, Rev. 185, 308 (1969).
4) H. SPOHN, Comm. Math. Phys., to be published.
5) F. SPITZER, J. Math. Mech. 18, 973 (1969).
6) D.W. JESPEN, J. Math. Phys. 6, 405 (1965).
7) T.E. HARRIS, J. Appl. Prob. 2, 322 (1965).
8) W. SZATZSCHNEIDER, in Lecture Notes in Mathematics, Vol. 472, Springer, Berlin (1975).
9) P. MAJOR and D. SZASZ, preprint (1977)
10) R. HOLLEY, Trans AMS 144, 523 (1969).
11) O.E. LANFORD, in Lecture Notes in Physics, Vol. 38, Springer, Berlin (1974).
12) F. KING, Ph. D. Thesis, Dept. of Math., Univ. of Cal., Berkeley, (1975).
13) W. BRAUN and K. HEPP, Comm. Math. Phys. 57, 101 (1978).
14) H. SPOHN and J.L. LEBOWITZ, Comm. Math. Phys. 54, 97 (1977).
15) E.B. DAVIES, Comm. Math. Phys. 33, 171 (1973).
16) E.B. DAVIES, J. Math. Phys. 15, 2036 (1974).
17) E.B. DAVIES, Comm. Math. Phys. 39, 91 (1974).
18) E.B. DAVIES and H. SPOHN, preprint (1978).
19) E.B. DAVIES, Math. Ann. 219, 147 (1976).
20) E.B. DAVIES and J.P. ECKMANN, Helv. Phys. Acta 48, 731 (1975).
21) P. MARTIN and G.G. EMCH, Helv. Phys. Acta,48, 59 (1975).
22) H. SPOHN, J. Stat. Phys. 17, 385 (1977).
23) V. GORINI and A. KORSAKOWSKI, J. Math. Phys. 17, 1298 (1976).
24) V. GORINI and A. FRIGERIO, J. Math. Phys. 17, 1708 (1976).
25) K. HEPP and E.H. LIEB, Helv. Phys. Acta 46, 573 (1973).
26) E.B. DAVIES, Quantum Theory of Open Systems, Academic Press, London (1976).

BOLTZMANN BEHAVIOUR OF A SPATIALLY INHOMOGENEOUS GAS

J. BIEL

Facultad de Ciencias, Universidad Autónoma, Bellaterra, Barcelona

and

J. MARRO

Depart. de Física Teórica, Universidad de Barcelona, Barcelona-28.

We present a formalism for the study of spatially nonuniform gases[1] far from equilibrium which generalizes previous work on spatially uniform systems[2]. Starting from the familiar BBGKY hierarchy we obtain a new hierarchy of equations for the reduced distribution functions which gives their rate of change at any given order in the mean density of the gas as a sum of a finite number of terms. This is accomplished by means of a projection operator which removes the relative coordinates. By means of a Fourier transformation we introduce a wave vector (which plays the role of an inhomogeneity parameter) corresponding to the dependence of the reduced n-particle function on the position of the centre of mass of the set of $n$ particles.

The Boltzmann integro-differential equation and molecular chaos at all times can be derived in the framework of this formalism by introducing the following hypotheses:

(a) The interparticle potential is repulsive and finite-ranged.

(b) The thermodynamic limit is taken in the usual sense.

(c) The system is initially weakly nonuniform; that is, the distances over which the properties of the system change significantly at $t = 0$ are large compared with the range of the interparticle potential.

(d) The spatial correlations are finite in the initial state; this leads to the initial molecular chaos.

(e) The particle density of the system is low and the times which we consider are large in the sense

$$D = \frac{N}{V} \rightarrow 0, \quad t \rightarrow \infty \;, \quad Dt \text{ finite}$$

Assuming that a well-defined function of the relative distance of two particles, $r$, has a proper limit when $r \rightarrow \infty$, the Boltzmann equation follows exactly from Hamiltonian dynamics.

---

Becario de la Fundación Juan March.

REFERENCES

1) J. BIEL and J. MARRO, "Statistical approach to the kinetics of nonuniform systems", preprint.
2) P. MAZUR and J. BIEL, Physica 32, 1633 (1966).

## AN EXAMPLE OF PROCESSES IN PHYSICS THAT ARE NONLOCAL IN TIME : THE INTERDEPENDENCE OF ROTATIONAL PROCESSES IN THE DYNAMICS OF SIMPLE FLUIDS

A. GERSCHEL
Lab. de Physico-Chimie des Rayonnements, Université, Paris-Sud, France

The time-domain analysis of experimental absorption spectra in liquids displays important specific characteristics of the rotational dynamics. Among these, the interdependence between the short-time and the long-time features of the total angular path are displayed in the pattern of the memory functions of first and second order. A tentative interpretation of these features is possible in the case when the time dependence of the generalized friction originates in the autocorrelation of the angular velocities.

REFERENCES

1) A. GERSCHEL, Comm. on Physics 1, 111 (1976).
2) A. GERSCHEL, in Proceedings of the Faraday Symposium no. 11, The Chemical Society, London, 1977.

# STOCHASTIC THEORY FOR HYDRODYNAMICAL SYSTEMS

CH. VAN DEN BROECK
V.U.B., Brussels
and
L. BRENIG
U.L.B. Brussels.

We derive nonlinear fluctuating kinetic equations for a one- and two-component Boltzmann gas, from a limiting form of a master equation where the motion and collisions of the particles are described by stochastic displacement and collision terms[1]. The linear fluctuating equations[2], as obtained from equilibrium thermodynamics, are easily recovered. The importance of the nonlinear fluctuating equations is their applicability to nonequilibrium hydrodynamic instabilities, where fluctuations and consequently nonlinearities become important. As a simple illustration for a two-component system, we derive the fluctuating diffusion and friction equations for Brownian motion from its fluctuating kinetic equation. Finally, as an answer to a question raised by Kac[3] concerning the possibility of a stochastic description of an inhomogeneous Boltzmann gas, we prove that in our master equation the stochastic displacement and collision term are complementary instead of contradictory. It is illustrated that the effects of collisions are transferred from the stochastic collision term to the displacement term as we change the scaling of the initial master equation (for similar ideas, see Mori[4]).

REFERENCES

1) L. BRENIG, W. HORSTHEMPKE, M. MALEK-MANSOUR, Phys. Lett., 59A, 341 (1976).

2) FOX and UHLENBECK, Phys. Fluids 13, 2881 (1970).

3) LOGAN and KAC, Phys. Rev. A13, No. 1 (1976).

4) H. MORI, Prog. Theor. Phys. 33, 423 (1965).

# FLUCTUATION SPECTRA NEAR THE THRESHOLD OF A CURRENT INSTABILITY

M. BÜTTIKER and H. THOMAS
University of Basel, Klingelbergstr. 82, 4056 Basel.

We discuss the fluctuation spectra of the uniform stationary-current state of a model in which Bragg scattering of hot electrons leads to bulk negative differential conductivity[1]. The system is described by a set of nonlinear Langevin equations for the transport fields, which are coupled via the current density and the charge density to the Maxwell fields. We find two types of slow long-wavelength modes; flux conservation leads to hydrodynamic $\underset{\sim}{B}$-field diffusion modes[2] and charge conservation leads to dielectric-relaxation modes[3]. The dielectric-relaxation modes that become soft at the threshold are strongly coupled to the hydrodynamic modes. This coupling has interesting consequences in the resulting fluctuation spectra. Corresponding to the two different types of modes, one finds that the long-time decay of the correlation functions is governed by a power law and an exponential law showing the critical slowing down.

REFERENCES

1) M. BÜTTIKER and H. THOMAS, Phys. Rev. Lett. 38, 78 (1977); and Solid State Electr. 21, 95 (1978).

2) J.D. JACKSON, Classical Electrodynamics, Wiley, New York.

3) E. PYTTE and H. THOMAS, Phys. Rev. 179, 431 (1969).

# STOCHASTIC ELECTRODYNAMICS: EXAMPLE OF A NONLINEAR STOCHASTIC DIFFERENTIAL EQUATION WITH A NON WHITE NOISE STOCHASTIC FORCE

P. CLAVERIE
University of Paris VI, France.

Stochastic electrodynamics (SED) studies the (classical) motion of (charged) particles subjected to three kinds of forces: the usual deterministic force $\vec{F}(\vec{q})$, the radiation damping force $(2e^2/3c^3)\times(d^3\vec{q}/dt^3)$, and a stochastic force $\vec{f}(t)$ due to a stochastic electromagnetic field with zero mean value and a spectral density $\rho(\omega) = (2\hbar/3c^3)\,|\omega|^3$. SED has been introduced as a possible classical model for simulating some aspects of the quantum behaviour of charged particles (see ref. 1 and references therein). Since $\vec{f}(t)$ is <u>not</u> white noise, the phase-space process $\{\vec{q}, \vec{p}\}$ is <u>not</u> Markovian. Nevertheless, for small damping and stochastic forces, approximate Fokker-Planck type equations may be found for the one-time probability density $W(\vec{q},\vec{p};t)$. Two such different equations, respectively denoted "Lax 5 - Khas'minskiĭ" and "Lax 6 - van Kampen" may be considered[1]. But (except for the case of the harmonic oscillator) it is practically impossible to evaluate analytically the diffusion coefficients of these Fokker-Planck type equations, because these coefficients involve integrals over time $\theta$ of the correlation function of $f(t)$ multiplied by some elements of the Jacobian matrix $\dfrac{D(\vec{q},\vec{p})}{D(\vec{q}^{-\theta}, \vec{p}^{-\theta})}$ associated with the deterministic evolution (from time $-\theta$ to time $0$)[1]. But, according to a procedure devised by Haken (ref. 4, section XI,c.2), we may reduce the problem to a space whose coordinates will be the "relevant" constants of motion of the deterministic undamped problem. (Then, remarkable recombinations involving complicated derivatives $\left(\dfrac{\partial\vec{q}}{\partial\vec{p}^{-\theta}}, \dfrac{\partial\vec{p}}{\partial\vec{p}^{-\theta}}, \ldots\right)$ occur, leading to much simpler expressions for the coefficients of the Fokker-Planck equation in the "reduced" space[3]. Thus, the passage to this reduced space involves several simultaneous advantages:

1) the number of variables is reduced;
2) the coefficients of the reduced Fokker-Planck equation are much easier to calculate;
3) we get the same reduced coefficients whether we start from the Lax 5 - Khas'minskiĭ or Lax 6 - van Kampen Fokker-Planck type equations. This is an important result, since it ensures that these two equations, although different, are actually consistent.

## REFERENCES

1) P. CLAVERIE and S. DINER, "Stochastic Electrodynamics and Quantum

Theory", Intern. J. Quantum Chem. 13 (1978) (in the press).

2) H. HAKEN, Rev. Mod. Phys. 47, 67 (1975).

3) P. CLAVERIE, L. DE LA PEÑA-AUERBACH and S. DINER, Langevin equation with small arbitrary damping and stochastic forces: the Fokker-Planck equation in a reduced space of constants of motion, to be published.

## BROWNIAN MOTION IN PERSISTENT FLUCTUATIONS (WITHOUT THERMAL EQUILIBRIUM)

R. GRAPPIN

Observatoire de Meudon, DAF, 92190 Meudon, France.

We know that the energy of a frictionless test particle subjected to a random, time-dependent, $\delta$-correlated force $f(t)$ undergoes Brownian motion and its mean energy increases linearly with time. This is the case for

$$\frac{dv}{dt} = f(t), \quad \text{as well as for} \quad \frac{dv}{dt} + \omega^2 x = f(t), \tag{1}$$

where $\omega$ is a constant. We are interested here in fluctuations of the parameter, as in

$$\frac{dv}{dt} + \omega^2(t)x = f(t) \tag{2}$$

or

$$\frac{dv}{dt} + \omega^2(t)\sin x = f(t) , \tag{3}$$

where $\omega^2(t)$ is a random function of time. These equations occur in trapping problems. After analysing the validity of the usual diffusion description, we give a study of eq. (2), as well as preliminary results for eq. (3), which show both the quantitative and the qualitative difference between the Brownian motion of a trapped particle and that of a free particle.

REFERENCES

1) R. GRAPPIN, Physica (1977).

2) R. GRAPPIN and J. HEYVAERTS, submitted to Phys. Rev.

# COVARIANT WIGNER-FUNCTION APPROACH TO RELATIVISTIC QUANTUM STATISTICS

R. HAKIM
Groupe d Astrophysique Relativiste, Observatoire, Paris-Meudon, France

As a simple example, the techniques of the covariant Wigner function are illustrated on the model of spin-$\frac{1}{2}$ baryons interacting via scalar mesons (the so-called "scalar plasma"). A BBGKY hierarchy is derived and truncated at the lowest order (i.e., the Hartree approximation). The case of thermal equilibrium is studied and interesting phase transitions are found. The excitation spectrum of the quasi-mesons is derived quite easily by methods similar to those used in conventional plasma physics. More-general approximations can also be obtained. These techniques have been applied to baryonic matter, quark matter and QED plasmas with astrophysical problems in view (strong magnetic fields, white dwarfs, neutron stars, primeval Universe).

## REFERENCES

1) R. DOMINGUEZ TENREIRO and R. HAKIM, Phys. Rev. D15, 1435 (1977).
2) R. HAKIM, Riv. Nuovo Cim. (July 1978).

# STOCHASTIC MODEL OF CURRENT-COUPLED CHANNELS IN NERVE MEMBRANES

I. v.d.HEYDT, N. v.d.HEYDT and G. OBERMAIR
Institut Physik II - Universität - D 8400 Regensburg - W. Germany

The strong voltage dependence of nerve membrane conductances, which vanish at high voltages, results from permeability changes of discrete, highly selective ion channels influenced by the local electric field. This well-known fact is taken into account here by the simplest possible model assumptions : each channel (label i) has only two states, open (+) or closed (-), the transitions occur stochastically by the interaction with the electrolyte of constant temperture, and the transition rates are Boltzmann factors with activation energies linearly dependent on the instantaneous voltage $V_i$ across the channel: $F^{(\pm)} = F_0^{(\pm)} \mp \alpha V_i^{(\pm)}$ . The channel voltages $V_i$ are determined by the voltage $V$ between the electrodes and by the conductance states of all other channels. They turn out to depend only on the number of open channels in many cases, because of a long-range current interaction between the channels. The electrolyte conductivity divided by the membrane conductance per unit area gives the range; for nerve membranes it is larger than 100 channel distances. In such a mean-field case a single-step master equation for the probability that n channels are open can be derived. From its rigorous stationary solution the stationary mean conductance $g$ between the electrodes (divided by its maximum) is calculated in the thermodynamic limit as a function of the electrode voltage $V$ and the temperature $T$ . The inverse function $V(g,T)$ can be given analytically :

$$V(g,T) = \frac{1}{\alpha(1+B)} \frac{1+A}{1+A-Ag} \left\{ k_B T \ln \left[ (1+A)(\frac{1}{g} - 1) \right] + F_0^{+} - F_0^{-} \right\} ,$$

where $A$ and $B$ are some resistance ratios. It agrees well with the measurements of the Na-conductance of nerve membranes, and explains for the first time the observed dependences on the electrolyte composition and on the density and conductance of the channels, for instance, in the different species: frog, sea worm and squid, or under different experimental conditions. The stationary current fluctuations are discussed. The mean-field character of the current interaction reduces the current noise of the membrane to a minimum compared to other possible mechanisms, e.g., with independent channels. This might be one of the means by which a signal transmission with a signal-to-noise ratio as high as possible is attained in nerves.

REFERENCES

1) Proceedings of the 27th International Congress of Physiol. Sci., Paris, 1977, abstract 2356.
2) Verhandlungen der Deutschen Physikalischen Gesellschaft 1, 1978, abstract TH 56.

## DERIVATION OF STOCHASTIC TRANSPORT EQUATIONS FOR SYSTEMS OF INTERACTING BROWNIAN PARTICLES

W. HESS,

University of Konstanz, Germany.

Concentrated colloidal systems and solutions of chain polymers can be described as systems of interacting Brownian particles. Using the inverse of the friction coefficient as an expansion parameter we derive from the Langevin equations for the momenta of the particles Langevin equations for their coordinates. As a by-product, this procedure gives a new derivation of Kirkwood's generalized diffusion equation. With a local-equilibrium approximation for the configurational distribution function a mode-mode coupling equation is derived for the concentration fluctuations. For the interaction-free case the relation to the results of the usual theory of Brownian motion is established.

REFERENCE

W. HESS and R. KLEIN, Physica A, to be published.

# THERMODYNAMIC INSTABILITY, OSCILLATIONS, AND MODIFICATION OF NERVE SYNAPSES

K. KAUFMANN,
Max-Planck-Institut für Biophysikalische Chemie, D-34 Göttingen.

Owing to their logistic position in nerve systems, modifiable synapses play a crucial role in nerve-network theories of "memory"[1]. The biochemical substrate for such modification has to be kinetically unstable upon stimulation. Such an instability was found for acetylcholine, which induces the excitation in the specialized nerve cells of the electric organ of the ray Torpedo Marmorata, leading to oscillations with a period in the order of 10 sec, upon stimulation[2].

Starting from the Glansdorff-Prigogine stability criteria, the kinetic system underlying this physiological modifiability was in-investigated[3]. Analysis of the frequency dispersion with stimulation intensity allows us to specify the enzyme ATPase as the possible kinetic cause of the instability. This leads to the prediction that oscillations of the concentration of ATP (adenosine triphosphate), an energy-rich substrate for acetylcholine synthesis, should occur[3].

Recently, these have been observed[4]. The cross-correlation with the acetylcholine oscillations, computed from the experimental data, indicates that the ATP oscillations precede those of the acetylcholine[5]. This supports the idea that external energy flow and physiologically stimulated ATP metabolism have a causal role in the "short-term-memory" of this nerve synapse. Long-term memory is not investigated; it apparently requires an instability in the metabolism of the proteins.

REFERENCES

1) One example is reported in H. HAKEN: Synergetics, Springer Verlag 1973 and 1977, and in articles by Z. COWAN.

2) Y. DUNANT ET AL., Nature 252, 485 (1974).

3) K. KAUFMANN, Naturwissenschaften 64, 371 (1977).

4) M. ISRAËL ET AL., J. Neurochem. 28, 1259 (1977).

5) K. KAUFMANN, to be published.

# DISCUSSION OF A RANDOM-WALK MODEL WITH CORRELATED JUMPS

J.W. HAUS and K.W. KEHR

Institut für Festkörperforschung der Kernforschungsanlage Jülich, 5170 Jülich, West Germany

In this seminar the results of a generalization of the continuous-time random walk (CTRW) theory of Montroll and Weiss to include correlations over two jumps are reported. A simple one-dimensional example is discussed and a homogeneous second-order differential equation is presented. This CTRW theory is shown to be equivalent to a set of extended Markovian master equations[1]. This description also produces a frequency-dependent diffusion coefficient of the form

$$D(\omega) = D_{\infty} + \frac{\Delta D}{\omega^2 + \Gamma^2} ,$$

where the coefficients are determined by the phenomenological rates in the extended CTRW theory. We briefly allude to the generalization to Bravais and non-Bravais lattices in higher dimensions.

## REFERENCE

1) J.W. HAUS and K.W. KEHR, Solid State Commun., to be published.

# DISCUSSION OF A QUANTUM THEORY OF DIFFUSION INCLUDING MEMORY EFFECTS

KAZUO KITAHARA

Department of Physics, University of Tokyo, Bunkyo-ku, Tokyo, Japan

and

J.W. HAUS

Institut für Festkörperforschung der Kernforschungsanlage Jülich, 5170 Jülich, West Germany.

We introduce a phenomenological Hamiltonian for a particle in a medium with site-diagonal and off-diagonal disorder, namely

$$\mathcal{H}(t) = \sum \epsilon |n\rangle\langle n| + J \sum_{\langle n,m\rangle} |n\rangle\langle m| + \delta\epsilon_n(t)|n\rangle\langle n| + \delta J_{nm}(t)\,|n\rangle\langle m| \;.$$

The last two terms contain stochastic variables $\delta\epsilon_n(t)$ and $\delta J_{nm}(t)$. These variables are assumed to be Gaussian-correlated, but the averages include memory:

$$\langle \delta\epsilon_n(t)\,\delta\epsilon_n(t')\rangle = \frac{\gamma_0}{\tau_c}\, e^{-|t-t'|/\tau_c}$$

$$\langle \delta J_{nm}(t)\,\delta J_{nm}(t')\rangle = \frac{\gamma_1}{\tau_c}\, e^{-|t-t'|/\tau_c}\,(\delta_{nn'}\delta_{mm'} + \delta_{nm'}\delta_{n'm})\,\delta_{n,m\pm1} \;.$$

We develop a master equation which is correct in second-order perturbation theory. When $\tau_c \to 0$ (white-noise limit) the results are exact and agree with published results[1]. We develop an expansion in $\tau_c$ and calculate the decay of the initial velocity, the mean-square position and diffusion coefficient to seoond order in $\tau_c$ when $\gamma_1 = 0$, and to first order in $\tau_c$ with off-diagonal disorder ($\gamma_1 \neq 0$). The results for the diffusion coefficient in both cases are

$$D = \frac{2J^2}{\gamma_0}(1 + \tau_c{}^2)$$

and

$$D = 4\gamma_1 + \frac{2J^2}{\gamma_0 + 2\gamma_1} - \frac{8J^2\tau_c\gamma_1}{\gamma_0 + 2\gamma_1}$$

respectively.

## REFERENCES

1) The results for the diffusion coefficient, we believe, were first presented by: H. HAKEN and P. REINEKER, Z. Physik 249, 253 (1972).

# THE DYNAMO EFFECT IN MAGNETHYDRODYNAMIC TURBULENCE

J. LÉORAT

Observatoire de Meudon et Université Paris VII - France

The origin of cosmic (planetary, solar or galactic) magnetic fields remains a major problem of astrophysics in spite of many different approaches. It is possible to study non-linear homogeneous and isotropic MHD turbulence by means of the eddy-damping quasi-normal Markovianized approximation. Above a critical magnetic Reynolds number (about 29) the "dynamo effect" is obtained: a stationary magnetic energy spectrum is obtained in the turbulence with only kinetic energy injection; initially the most unstable scales are about ten times smaller than the energy scale. If helicity is also injected, there is no stationary spectrum as energy is transferred towards greater and greater scales.

## REFERENCES

1) J. LEORAT, U. FRISCH, A. PRIQUET and A. MAZURE, J. Fluid Mechanics 68, 769 (1975)

2) J. LEORAT, U. FRISCH and A. PRIQUET, J. Fluid Mech. 77, 321 (1976).

3) Determination of the critical magnetic Reynolds number in MHD turbulence (preprint, 1978).

# DYNAMICAL CORRELATIONS IN NONLINEAR CHEMICAL REACTIONS

R. SCHRANNER and S. GROSSMANN
Max-Planck-Institut für Biophysikalische Chemie, Göttingen - Germany

Projection-operator techniques are applied to stochastic master equations to arrive at a continued-fraction expansion for dynamical correlations. This provides a systematic way of investigating the dynamics of systems far from thermal equilibrium that show unstable behaviour and enhanced fluctuations.

The method is used to describe two simple nonlinear chemical reaction systems showing analogies to first- and second-order phase transitions, respectively. Analytical as well as numerical results are obtained for the relaxation and memory properties near the instabilities, showing the advantages as well as the limitations of the approach.

REFERENCES

1) S. GROSSMANN and R. SCHRANNER, Z. Phys. B (1978) in press.
2) S. GROSSMANN, Phys. Rev. A17, 1123 (1978).

# A COOPERATIVE EFFECT WITH SIGNIFICANCE IN BIOLOGICAL EVOLUTION

J. WAGENSBERG
Facultad de Física, Universidad de Barcelona, Spain

We have obtained thermograms (time evolution of the heat dissipation rate) of the growth histories of several wild marine bacteria isolated in the sea in different environmental conditions. The thermogram of fig. 1 has the typical profile of a bacterium coming from an ambient rich in oxygen, a Flavobacterium isolated at 20 m depth. If we look closer at the thermogram of fig. 2 we can see how some energetic peaks appearing in the aerobic phase tend to reach a rhythm of constant period at the culmination of the anaerobic phase. In the last death phase (the system tends to a closed system) the temporal structure degenerates. The undamped-relaxation oscillations detected by the microcalorimeter were identified as being of the glycolytic-oscillation type by means of fluorimetric recording of the NADH concentration, a metabolic indicator parameter. There are probably oscillations occurring perpetually within individual cells, but these can only be observed in the thermogram (75 $\mu$W) when the culture is brought into synchrony. It is, however, difficult to conceive a mechanism that would give rise to such a maintained synchrony between such a large number of cells. The characteristic situation of Goldbeter's allosteric-enzyme model for glycolitic systems[2] is suggested in order to describe this cooperative effect. The final equations of this model are

$$\frac{\partial\alpha}{\partial t} = \sigma_1 - \Phi + D_\alpha \frac{\partial^2\alpha}{\partial r^2}$$

$$\frac{\partial\gamma}{\partial t} = \Phi - \sigma_2\gamma + D_\gamma \frac{\partial^2\gamma}{\partial r^2} \qquad (1)$$

where $\alpha$ and $\beta$ denote, respectively, the concentrations of the substrate and the product, $\sigma_1$ is the injection rate of substrate, $\sigma_2$ is related to the outflow of product, and $D_\alpha$ and $D_\gamma$ are the diffusion coefficients of the substrate and the product along the single coordinate $r$ . $\Phi$ is a nonlinear function of $\alpha$ and $\gamma$ which depends on the structural hypothesis of the model. A stability analysis of the kinetic equations (1) indicates that, for a large enough value of $L$ (supracellular distances), propagating concentration waves appear in the system. Fig. 3 shows the time evolution of the spatial structure. If the steady-state values of $\alpha$ and $\gamma$ are imposed symmetrically at the boundaries, two sharp wavefronts of the reaction product are formed near the boundaries. These wavefronts move to the centre where they collide, and then the resulting peak decreases until new wavefronts build up near the boundaries. This periodic effect is maintained in time. It is not difficult to see that the central point of the system will undergo relaxation oscillations such as those observed in the microcalorimeter.

We suggest that propagating concentration waves are established between the cells of Flavobacterium. In this regime the system acquires the ability to exceed, in a very short time interval, some threshold of chemical substances, which are then released into the extracellular medium as a pulse. The heat of dilution of this mixture would then be translated into the peaks recorded by the thermogram. In this picture a cell is not a unit that preserves the characteristics of the supra-cellular level. Each cell has a different but cooperative behaviour in the whole system, and this can be considered as a new hierarchy of biological order. A very important experimental feature is that only the cultures of Flavobacterium that displayed rhythmic behaviour (eleven out of sixteen) exhibited a visually observable strong aggregation of cells. It seems, therefore, that nonequilibrium organizations can actually be a mechanism to increase the interaction between cells that were initially independent, and provides a nondeterministic source of new structures[3)] whose survival is determined by subsequent Darwinian selection. We would take the view that the cooperative behaviour of Flavobacterium represents an intermediate step in the biological evolution from a population of free cells to a state of strong and harmonious interaction : the tissue.

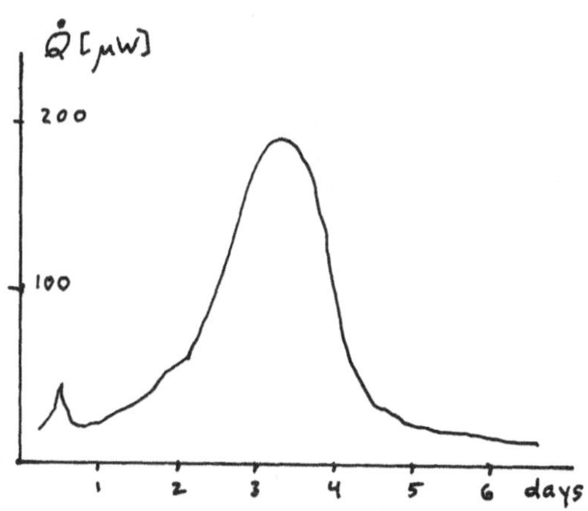


Fig. 1

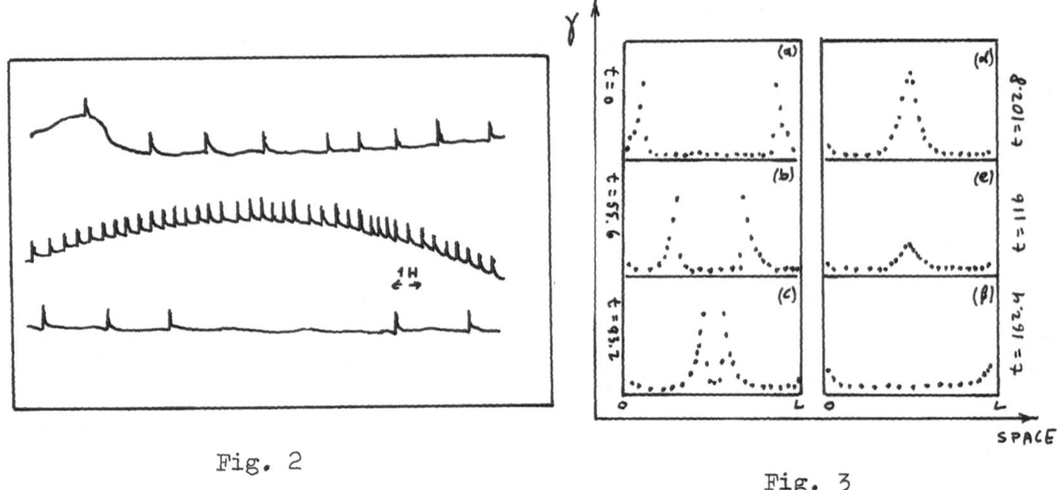


Fig. 2

Fig. 3

REFERENCES

1) J. WAGENSBERG et.al., Inv. Pesq. 42, 162 (1978).

2) A. GOLDBETER, Proc. Nat. Acad. USA 70, 3255 (1973).

3) D. LURIE and J. WAGENSBERG, J. Nonequilib. Thermodyn., in press.

Communications in

# Mathematical Physics



The journal is devoted to the following topics: General relativity, equilibrium and non-equilibrium statistical mechanics, foundations of quantum mechanics, classical and quantum mechanics of finitely many degrees of freedom, Lagrangian quantum field theory and constructive quantum field theory. Mathematical papers are accepted only if they are of direct relevance to physics.

Springer-Verlag
Berlin
Heidelberg
New York

Selected Issues from

# Lecture Notes in Mathematics

Vol. 516: M. L. Silverstein, Boundary Theory for Symmetric Markov Processes. XVI, 314 pages. 1976.

Vol. 518: Séminaire de Théorie du Potentiel, Proceedings Paris 1972–1974. Edité par F. Hirsch et G. Mokobodzki. VI, 275 pages. 1976.

Vol. 522: C. O. Bloom and N. D. Kazarinoff, Short Wave Radiation Problems in Inhomogeneous Media: Asymptotic Solutions. V. 104 pages. 1976.

Vol. 523: S. A. Albeverio and R. J. Høegh-Krohn, Mathematical Theory of Feynman Path Integrals. IV, 139 pages. 1976.

Vol. 524: Séminaire Pierre Lelong (Analyse) Année 1974/75. Edité par P. Lelong. V, 222 pages. 1976.

Vol. 525: Structural Stability, the Theory of Catastrophes, and Applications in the Sciences. Proceedings 1975. Edited by P. Hilton. VI, 408 pages. 1976.

Vol. 526: Probability in Banach Spaces. Proceedings 1975. Edited by A. Beck. VI, 290 pages. 1976.

Vol. 527: M. Denker, Ch. Grillenberger, and K. Sigmund, Ergodic Theory on Compact Spaces. IV, 360 pages. 1976.

Vol. 532: Théorie Ergodique. Proceedings 1973/1974. Edité par J.-P. Conze and M. S. Keane. VIII, 227 pages. 1976.

Vol. 538: G. Fischer, Complex Analytic Geometry. VII, 201 pages. 1976.

Vol. 543: Nonlinear Operators and the Calculus of Variations, Bruxelles 1975. Edited by J. P. Gossez, E. J. Lami Dozo, J. Mawhin, and L. Waelbroeck, VII, 237 pages. 1976.

Vol. 552: C. G. Gibson, K. Wirthmüller, A. A. du Plessis and E. J. N. Looijenga. Topological Stability of Smooth Mappings. V, 155 pages. 1976.

Vol. 556: Approximation Theory. Bonn 1976. Proceedings. Edited by R. Schaback and K. Scherer. VII, 466 pages. 1976.

Vol. 559: J.-P. Caubet, Le Mouvement Brownien Relativiste. IX, 212 pages. 1976.

Vol. 561: Function Theoretic Methods for Partial Differential Equations. Darmstadt 1976. Proceedings. Edited by V. E. Meister, N. Weck and W. L. Wendland. XVIII, 520 pages. 1976.

Vol. 564: Ordinary and Partial Differential Equations, Dundee 1976. Proceedings. Edited by W. N. Everitt and B. D. Sleeman. XVIII, 551 pages. 1976.

Vol. 565: Turbulence and Navier Stokes Equations. Proceedings 1975. Edited by R. Temam. IX, 194 pages. 1976.

Vol. 566: Empirical Distributions and Processes. Oberwolfach 1976. Proceedings. Edited by P. Gaenssler and P. Révész. VII, 146 pages. 1976.

Vol. 570: Differential Geometrical Methods in Mathematical Physics, Bonn 1975. Proceedings. Edited by K. Bleuler and A. Reetz. VIII, 576 pages. 1977.

Vol. 572: Sparse Matrix Techniques, Copenhagen 1976. Edited by V. A. Barker. V, 184 pages. 1977.

Vol. 579: Combinatoire et Représentation du Groupe Symétrique, Strasbourg 1976. Proceedings 1976. Edité par D. Foata. IV, 339 pages. 1977.

Vol. 587: Non-Commutative Harmonic Analysis. Proceedings 1976. Edited by J. Carmona and M. Vergne. IV, 240 pages. 1977.

Vol. 592: D. Voigt, Induzierte Darstellungen in der Theorie der endlichen, algebraischen Gruppen. V. 413 Seiten. 1977.

Vol. 594: Singular Perturbations and Boundary Layer Theory, Lyon 1976. Edited by C. M. Brauner, B. Gay, and J. Mathieu. VIII, 539 pages. 1977.

Vol. 596: K. Deimling, Ordinary Differential Equations in Banach Spaces. VI, 137 pages. 1977.

Vol. 605: Sario et al., Classification Theory of Riemannian Manifolds. XX, 498 pages. 1977.

Vol. 606: Mathematical Aspects of Finite Element Methods. Proceedings 1975. Edited by I. Galligani and E. Magenes. VI, 362 pages. 1977.

Vol. 607: M. Métivier, Reelle und Vektorwertige Quasimartingale und die Theorie der Stochastischen Integration. X, 310 Seiten. 1977.

Vol. 615: Turbulence Seminar, Proceedings 1976/77. Edited by P. Bernard and T. Ratiu. VI, 155 pages. 1977.

Vol. 618: I. I. Hirschman, Jr. and D. E. Hughes, Extreme Eigen Values of Toeplitz Operators. VI, 145 pages. 1977.

Vol. 623: I. Erdelyi and R. Lange, Spectral Decompositions on Banach Spaces. VIII, 122 pages. 1977.

Vol. 628: H. J. Baues, Obstruction Theory on the Homotopy Classification of Maps. XII, 387 pages. 1977.

Vol. 629: W. A. Coppel, Dichotomies in Stability Theory. VI, 98 pages. 1978.

Vol. 630: Numerical Analysis, Proceedings, Biennial Conference, Dundee 1977. Edited by G. A. Watson. XII, 199 pages. 1978.

Vol. 636: Journées de Statistique des Processus Stochastiques, Grenoble 1977, Proceedings. Edité par Didier Dacunha-Castelle et Bernard Van Cutsem. VII, 202 pages. 1978.

Vol. 638: P. Shanahan, The Atiyah-Singer Index Theorem, An Introduction. V, 224 pages. 1978.

Vol. 648: Nonlinear Partial Differential Equations and Applications, Proceedings, Indiana 1976–1977. Edited by J. M. Chadam. VI, 206 pages. 1978.

Vol. 650: C*-Algebras and Applications to Physics. Proceedings 1977. Edited by R. V. Kadison. V, 192 pages. 1978.

Vol. 656: Probability Theory on Vector Spaces. Proceedings, 1977. Edited by A. Weron. VIII, 274 pages. 1978.

Vol. 662: Akin, The Metric Theory of Banach Manifolds. XIX, 306 pages. 1978.

Vol. 665: Journées d'Analyse Non Linéaire. Proceedings, 1977. Edité par P. Bénilan et J. Robert. VIII, 256 pages. 1978.

Vol. 667: J. Gilewicz, Approximants de Padé. XIV, 511 pages. 1978.

Vol. 668: The Structure of Attractors in Dynamical Systems. Proceedings, 1977. Edited by J. C. Martin, N. G. Markley and W. Perrizo. VI, 264 pages. 1978.

# Lecture Notes in Physics

Vol. 44: R. A. Breuer, Gravitational Perturbation Theory and Synchrotron Radiation. VI, 196 pages. 1975.

Vol. 45: Dynamical Concepts on Scaling Violation and the New Resonances in $e^+e^-$ Annihilation. Edited by B. Humpert. VII, 248 pages. 1976.

Vol. 46: E. J. Flaherty, Hermitian and Kählerian Geometry in Relativity. VIII, 365 pages. 1976.

Vol. 47: Padé Approximants Method and Its Applications to Mechanics. Edited by H. Cabannes. XV, 267 pages. 1976.

Vol. 48: Interplanetary Dust and Zodiacal Light. Proceedings 1975. Edited by H. Elsässer and H. Fechtig. XII, 496 pages. 1976.

Vol. 49: W. G. Harter and C. W. Patterson, A Unitary Calculus for Electronic Orbitals. XII, 144 pages. 1976.

Vol. 50: Group Theoretical Methods in Physics. 4th International Colloquium. Nijmegen 1975. Edited by A. Janner, T. Janssen, and M. Boon. XIII, 629 pages. 1976.

Vol. 51: W. Nörenberg und H. A. Weidenmüller. Introduction to the Theory of Heavy-Ion Collisions. IX, 273 pages. 1976.

Vol. 52: M. Mladjenović, Development of Magnetic β-Ray Spectroscopy. X, 282 pages. 1976.

Vol. 53: D. J. Simms and N. M. J. Woodhouse, Lectures on Geometric Quantization. V, 166 pages. 1976.

Vol. 54: Critical Phenomena. Sitges International School on Statistical Mechanics, June 1976. Edited by J. Brey and R. B. Jones. XI, 383 pages. 1976.

Vol. 55: Nuclear Optical Model Potential. Proceedings 1976. Edited by S. Boffi and G. Passatore. VI, 221 pages. 1976.

Vol. 56: Current Induced Reactions. International Summer Institute, Hamburg 1975. Edited by J. G. Körner, G. Kramer, and D. Schildknecht. V, 553 pages. 1976.

Vol. 57: Physics of Highly Excited States in Solids. Proceedings 1975. Edited by M. Ueta and Y. Nishina. IX, 391 pages. 1976.

Vol. 58: Computing Methods in Applied Sciences. Proceedings 1975. Edited by R. Glowinski and J. L. Lions. VIII, 593 pages. 1976.

Vol. 59: Proceedings of the Fifth International Conference on Numerical Methods in Fluid Dynamics. 1976. Edited by A. I. van de Vooren and P. J. Zandbergen. VII, 459 pages. 1976.

Vol. 60: C. Gruber, A. Hintermann, and D. Merlini, Group Analysis of Classical Lattice Systems. XIV, 326 pages. 1977.

Vol. 61: International School on Electro and Photonuclear Reactions I. Edited by C. Schaerf. VIII, 650 pages. 1977.

Vol. 62: International School on Electro and Photonuclear Reactions II. Edited by C. Schaerf. VIII, 301 pages. 1977.

Vol. 63: V. K. Dobrev et al., Harmonic Analysis on the n-Dimensional Lorentz Group and Its Application to Conformal Quantum Field Theory. X, 280 pages. 1977.

Vol. 64: Waves on Water of Variable Depth. Edited by D. G. Provis and R. Radok. 231 pages. 1977.

Vol. 65: Organic Conductors and Semiconductors. Proceedings 1976. Edited by L. Pál, G. Grüner, A. Jánossy and J. Sólyom. 654 pages. 1977.

Vol. 66: A. H. Völkel, Fields, Particles and Currents. VI, 354 pages. 1977.

Vol. 67: W. Drechsler and M. E. Mayer, Fiber Bundle Techniques in Gauge Theories. X, 248 pages. 1977.

Vol. 68: Y. V. Venkatesh, Energy Methods in Time-Varying System Stability and Instability Analyses. XII, 256 pages. 1977.

Vol. 69: K. Rohlfs, Lectures on Density Wave Theory. VI, 184 pages. 1977.

Vol. 70: Wave Propagation and Underwater Acoustics. Edited by J. Keller and J. Papadakis. VIII. 287 pages. 1977.

Vol. 71: Problems of Stellar Convection. Proceedings 1976. Edited by E. A. Spiegel and J. P. Zahn. VIII, 363 pages. 1977.

Vol. 72: Les instabilités hydrodynamiques en convection libre forcée et mixte. Edité par J. C. Legros et J. K. Platten. X, 202 pages. 1978.

Vol. 73: Invariant Wave Equations. Proceedings 1977. Edited by G. Velo and A. S. Wightman. VI, 416 pages. 1978.

Vol. 74: P. Collet and J.-P. Eckmann, A Renormalization Group Analysis of the Hierarchical Model in Statistical Mechanics. IV, 199 pages. 1978.

Vol. 75: Structure and Mechanisms of Turbulence I. Proceedings 1977. Edited by H. Fiedler. XX, 295 pages. 1978.

Vol. 76: Structure and Mechanisms of Turbulence II. Proceedings 1977. Edited by H. Fiedler. XX, 406 pages. 1978.

Vol. 77: Topics in Quantum Field Theory and Gauge Theories. Proceedings, Salamanca 1977. Edited by J. A. de Azcárraga. X, 378 pages 1978.

Vol. 78: Böhm, The Rigged Hilbert Space and Quantum Mechanics. IX, 70 pages. 1978.

Vol. 79: Group Theoretical Methods in Physics. Proceedings, 1977. Edited by P. Kramer and A. Rieckers. XVIII, 546 pages. 1978.

Vol. 80: Mathematical Problems in Theoretical Physics. Proceedings, 1977. Edited by G. Dell'Antonio, S. Doplicher and G. Jona-Lasinio. VI, 438 pages. 1978.

Vol. 81: MacGregor, The Nature of the Elementary Particle. XXII, 482 pages. 1978.

Vol. 82: Few Body Systems and Nuclear Forces I. Proceedings, 1978. Edited by H. Zingl, M. Haftel and H. Zankel. XIX, 442 pages. 1978.

Vol. 83: Experimental Methods in Heavy Ion Physics. Edited by K. Bethge. V, 251 pages. 1978.

Vol. 84: Stochastic Processes in Nonequilibrium Systems, Proceedings, 1978. Edited by L. Garrido, P. Seglar and P. J. Shepherd. XI, 355 pages. 1978